D0944709

SCHIZOPHRENIA AND GENETICS

A TWIN STUDY VANTAGE POINT

PERSONALITY AND PSYCHOPATHOLOGY

A Series of Monographs, Texts, and Treatises

SCHIZOPHRENIA AND GENETICS
A TWIN STUDY VANTAGE POINT

Irving I. Gottesman

Department of Psychology
University of Minnesota
Minneapolis, Minnesota

and

James Shields

Institute of Psychiatry
De Crespigny Park
London, England

With a contribution by
Paul E. Meehl
Department of Psychology
University of Minnesota
Minneapolis, Minnesota

 1972

ACADEMIC PRESS   New York and London

COPYRIGHT © 1972, BY ACADEMIC PRESS, INC.
ALL RIGHTS RESERVED.
NO PART OF THIS PUBLICATION MAY BE REPRODUCED OR
TRANSMITTED IN ANY FORM OR BY ANY MEANS, ELECTRONIC
OR MECHANICAL, INCLUDING PHOTOCOPY, RECORDING, OR ANY
INFORMATION STORAGE AND RETRIEVAL SYSTEM, WITHOUT
PERMISSION IN WRITING FROM THE PUBLISHER.

ACADEMIC PRESS, INC.
111 Fifth Avenue, New York, New York 10003

United Kingdom Edition published by
ACADEMIC PRESS, INC. (LONDON) LTD.
24/28 Oval Road, London NW1

LIBRARY OF CONGRESS CATALOG CARD NUMBER: 72-7684

PRINTED IN THE UNITED STATES OF AMERICA

RC
514
G67

NMU LIBRARY

To V.W.G. and E.J. S.
for patience and faith.

Chapter 4   Case Histories

Chapter 5   Multiple Blind Diagnoses of the Twins and Consensus Concordances

Chapter 6   Clinical Genetic Analysis

Chapter 7   Psychometric Contributions to Genetic Analysis

Chapter 8   Environmental Factors

CONTENTS

FOREWORD

When shortly after the end of World War II Jerry Shields came to work with me in what was to become the Genetics Unit of the London University Institute of Psychiatry, and later the Psychiatric Genetics Research Unit of the Medical Research Council, neither of us had any idea of how genetic problems in psychiatry were to be transformed in the next two decades. In the continent of Europe, including Britain, all psychiatric research started from the foundations, securely based, we all assumed, in the solid and painfully acquired knowledge of phenomenological, or descriptive, psychiatry. There was no question but that schizophrenia, whether one called it a reaction type or a syndrome or even a disease, was a biological as well as a psychological reality. If doctors disagreed whether a particular patient was schizophrenic or not, all still believed that an answer would eventually be given by his later life history, if he could be adequately followed up. Hardly more disputable was the assumption, founded both on the published results of family investigations and on the clinician's personal experience of innumerable family histories, that genetic factors provided a specific contribution to the predisposition to schizophrenia, which might or might not be released by environmental stress. No need was felt to prove this; all efforts were devoted to identifying a mendelian hypothesis, to distinguishing between possibly genetically heterogeneous forms of the illness, and so on.

As we in Europe realized, across the Atlantic there was another great world in which all we held for sure was doubted, and much that we doubted was held for sure. In this world, schizophrenia might be something or it might be, in effect, nothing though dignified with a name. Even the exercise of "diagnosing" it was discredited as an almost meaningless formality. If one were to speak of a patient as schizophrenic, this stigmatized him to the prejudice of his future, and psychiatrically signified little more than a rather pointless description of a state of mind which was itself in a state of flux and could change to quite a different picture at any time. These sick states of mind were known to be the results of traumatic effects, malign life experiences, distorted upbringing, impaired communication, and faults of will and judgment. The biological basis was human nature, the same for us all. If family likenesses showed up in the form these reactions took, this was because family life constitutes an environment much the same for one sib as for another, and one that tends to persist as a cultural entity from one generation to the next.

This is, of course, a grossly oversimplified picture of highly complex ways of thinking and understanding. But it helps to explain why communication between two psychiatric cultures was impeded by misunderstandings on both sides, by impatience with unfamiliar ideas, and all too much satisfaction with one's own indoctrinated preconceptions. On the one side, a rather unthinking assumption and, on the other, an equally unthinking rejection of genetic concepts divided the two psychiatries of the English-speaking world.

Fortunately the scene is now entirely changed. Collaborative research between the psychiatries of east and west has become possible; and, having once begun, it has done much to bring our two psychiatries together and to unite us all in a free communication system. In the collaboration between workers at Bethesda and Copenhagen and in the researches reported here by Irving Gottesman and James Shields, we can see how fruitful such collaboration can be. It seems that at last we have got one basic problem out of the way. The radical researches on foster children, which we owe to Kety, Rosenthal, Schulsinger, Wender, Welner, and others, have both disproved the null hypothesis (which denies any genetic contribution to schizophrenia) and have shown how the nebulous phenotypic effects of that contribution may be clarified and made more precise.

There is one painful lesson which the European worker has had to learn from American discipline: the need for blindfolding the judge. One can try to put out of one's mind considerations that should be temporarily disregarded, but it is safer and better not to be aware of them until the code can be broken. This is now standard practice in therapeutic trials, but is necessary also in etiological investigations. It is not only by blindfolding techniques that one aims at results which will be accepted as valid. One needs also to gather as much information as possible independent of clinical judges, such as the hard data of hospital admissions and discharges, the results of objective tests, and tape-recorded interviews. When Irving Gottesman joined us in the Psychiatric Genetics Research Unit, he brought with him expertise in all the techniques of the clinical psychologist and a determination to use them to the full, even though they tended to make the investigation, and especially the field study part of it, a very tough assignment.

In this book, the reader is put in full possession of all the information available to the researchers; and he is therefore in a position to check their results at every point. Writing as a lover of case histories and the vivid light they throw on the personalities and the lives of the human beings who are reduced to nameless units in a statistical analysis, I should like to emphasize the information value of Chapter Four in which the life stories of the twins are told. This is required reading for anyone who wants to appreciate how solid is the basic material which, in other chapters, is analyzed and assayed with such subtlety and care. The authors have made a unique departure in providing case histories of the dizygotic as well as the monozygotic pairs. This is one of the important ways in which the twin method is developing as workers become aware of the wider problems to which it can be applied.

Some sceptical reader may ask whether we have not by now had enough in the way of twin investigations and whether the population-based Scandinavian studies have not said the last word. The last word will never be said. The twin method will continue to evolve as knowledge advances. It is still our one and only fully controlled experiment in which the genetic predisposition is held constant while the environment varies. With greater understanding of the pathogenesis, we shall be able to put more subtle questions to be answered by these experiments of nature; and the more we know, the more informative they are likely to prove.

All twin investigations are unique in one way or another, and often in many ways. We have learned much, for instance, from the different locations of twin and genetic family studies and from being able to compare their varying results in different countries and cultures in Europe and America. The way in which concordance rates vary from study to study, for instance, has not been adequately explained, but has been a starting point for much new thinking. At the present time, probably no genetic investigation would produce as valuable results as a good twin study in, say, India or China.

As soon as a new technique appears, twin material suggests itself as just the place where it could be applied with best advantage for increasing our knowledge. In this book the Object Sorting Test and the Minnesota Multiphasic Personality Inventory have been applied with results which tell us much both about what is psychometrically accessible in schizophrenia and about the tests themselves, their significance, and their applicability. The technique of blindfold diagnosis has been developed in the Gottesman–Shields investigation with remarkable flexibility and comprehensiveness. It has added security to the validity of a diagnosis of a schizophrenic illness, even across psychiatric cultures; and it has illuminated the way in which the clinician's "personal factor" enters into the diagnostic process and influences it, without reducing it to idiosyncratic subjectivity.

As will be gathered from this book, twin research is relevant to almost all aspects of etiology: to comparative epidemiology related to countries and cultures and social hierarchies; the intrafamilial and extrafamilial sources of variance, psychodynamic, social, and organic; the problems of heterogeneity and the existence of phenocopies; the possible modes of manifestation of schizophrenic equivalents; the basic but still unsolved problems of the "schizoid"; and the relationship of overt personality traits and liability to illness.

Have the authors been able to come to conclusions about the nature of the genetic predisposition? This is too basic a problem to be solved out of hand; but they have certainly brought us much nearer to a theoretical model which adequately accounts for all the available information gathered to date, while remaining one that is refutable and one that suggests lines of enquiry for the future. Without saying that other possible models are out of date, the authors propose a polygenic theory, with a large part of the genetic variance controlled by probably a small number of genes of major effect. This is thought of as the basis of those schizophrenic illnesses which appear without gross and obvious cause, such as poison, infection, or brain lesion. The genetic predisposition is regarded as the main source of the specificities of the schizophrenic illness. The environmental contribution, whose importance is repeatedly emphasized, is thought of as less specific. A stress which in one constitution precipitates a schizophrenia will in another bring about a neurotic illness.

The genetic model of the schizophrenic predisposition in no way and nowhere excludes the environmental model. Indeed, both are needed; and they are complementary to one another. The geneticist and the environmentalist are working along different dimensions in a field of interaction whose complexity of interaction cannot possibly be covered in unidimensional terms. One of the fascinating aspects of the case analyses in this book is the way in which particular environmental factors can be incriminated in different cases. Those who are concerned with environmental research in schizophrenia should find much encouragement in these pages and fresh ideas worth trying out on other case material.

Schizophrenia and Genetics is a major contribution to etiological investigation in schizophrenia, and I am proud that the work was organized and carried out in the Medical Research Council's Psychiatric Genetics Unit at the Maudsley Hospital while I was its Director. The Unit and the Hospital together supplied the source material, but it was the imagination and enterprise of the authors that turned potentialities into accomplishment. The Gottesman–Shields collaboration was from the start of that happy kind which is one of the rare rewards of scientific work in which the marriage of true minds admits no impediment. I have been fortunate in having been able to help them. My own reward has been abundant. I have seen their work prosper; arising from their fellowship I have experienced the unique honor of the publication, under their editorship, of my own papers in *Man, Mind and Heredity*; best of all it has been to count them as my friends.

Institute of Psychiatry
London, England

Eliot Slater

PREFACE

Little did we imagine when we initiated our trans-Atlantic correspondence a decade ago that it would lead to accelerated personal growth, awareness of the complexities of schizophrenia, deep personal friendships, and a series of scientific communications culminating in this book. Our Anglo-American effort gives a new instance, we hope, of Louis Pasteur's aphorism—"Science has no country." Unlike other monograph-length treatments of schizophrenic twin studies, this one is the product of an American clinical psychologist and a British psychiatric social worker, both now better characterized professionally by the neologism "behavioral geneticists."

This book is intended to integrate the definitive findings of our twin study. From that vantage point we range into a critical review and discussion of the current state of thinking about the roles of genetic factors and of environmental factors in the etiology of schizophrenia. Regrettably, the roles assigned by vocal partisans on each side foster polarization more appropriate to politics or religion than to science. Our book is directed toward the as yet uncommitted students of psychiatry, psychology, social work, and human genetics, as well as to their teachers who may wonder why the Nature–Nurture Controversy still rages in respect of the origins of human behavior, normal and abnormal. We have tried to write the book in a gently persuasive way, answering questions that would logically be raised by a listener in one of the above disciplines but avoiding the harsh cross examination atmosphere associated with adversary proceedings. Perhaps we undermine our own purpose with such a dialectic approach; strident voices always carry so much farther than reasoning ones.

A virtual torrent of research findings from twin, family, and adoptee studies of schizophrenia has inundated the field of psychopathology since 1963; few professionals, let alone laymen, are aware of these labors, and it is still possible to find fellow professionals who maintain that there has been no progress into the causes of schizophrenia or mental disorders generally since World War II, or who believe that Kallmann's 1938 book on the genetics of schizophrenia was the most recent data from the genetic quarter. We undertook our study of schizophrenia in twins in order to find out what the effects would be on the results when care was taken to avoid, or to make provision for, the alleged sources of error and bias in the earlier "classical" studies conducted before 1953. We added many refinements to the classical twin method of Francis Galton such as tape-recording interviews, multinational blind diagnoses of histories, and the use of personality tests. In essence we reaffirmed "the big facts" of the earlier studies but differed in varying ways from them on details and interpretation.

We have gone to great lengths to cast our findings in the setting of modern quantitative human genetics and to show our appreciation for the contributions of psychodynamic concepts to our understanding of the phenomenology of schizophrenia. For reasons which will become clear to the reader, we advocate a renewal of a biological emphasis by those concerned with the origins and treatment of severe mental disorders but not an emphasis that would lead to biology becoming a dog in the ideological manger. Our intellectual debt to many scientific experts in many fields will be obvious to all who read this book.

Newcomers to this particular mixture of psychiatry, psychology, and human genetics may wish to forego some of the technical material too foreign to their background in a few of the chapters and content themselves with the chapter summaries so as to maintain continuity. Chapter 3 details the sampling and zygosity procedures, Chapter 5 reports our cross-national diagnostic experiment, Chapter 6 gives the results in depth of our clinical-genetic analyses, and Chapter 7 does the same for the psychological tests administered to the twins and their relatives.

More than anything else this book reports where the field of psychiatric genetics has been and where it is at with respect to schizophrenia. It would be presumptuous to forecast where the field is going. If our perspectives increase the probabilities of "true positive" research efforts and decrease the probabilities of "true negative" enterprises, we will consider ourselves successful.

Irving I. Gottesman
James Shields

ACKNOWLEDGMENTS

We owe our greatest debt to Dr. Eliot Slater. Without his support we would not have had access to the Maudsley Twin Register or to the extensive hospital records of the twins and their relatives. It was Slater who, in 1948, suggested that a question about twinship should be incorporated on the front sheet of Hospital notes for each patient; the Medical Committee of the Maudsley and Bethlem Royal Joint Hospital agreed. The information thus obtained on more than 45,000 patients (1948–1964) was forwarded to us in the genetics section of the Department of Psychiatry in the Institute of Psychiatry; from 1959 to 1969, the section was the Medical Research Council's Psychiatric Genetics Research Unit directed by Slater; the Department of Psychiatry was headed by Sir Aubrey Lewis during most of those years and latterly by Sir Denis Hill. We are grateful to all of them as well as to the many psychiatrists and social workers who garnered the information about twinship, and to Mrs. Marjorie Perkins who made sure that we received the information. To our other mentor, Regents' Professor Paul E. Meehl, we owe a debt for setting us an example of critical thinking, for inspiration, and for listening. We humbly acknowledge the selfless participation and cooperation of the twins and their families; we hope this book vindicates their trust in us. A genetic-family study inevitably reflects the quality and competence of the history takers and recorders; in piecing together such information we are indebted to a corps of professional workers at the Maudsley and other places who sent us reports or to whose case notes we were given access.

We would also especially thank D. S. Falconer, I. M. Lerner, B. A. Maher, D. Rosenthal, G. Allen, C. Smith, V. E. Anderson, S. Wright, J. M. Thoday,

T. Dobzhansky, J. H. Edwards, Ø. Ødegaard, and M. Bleuler for the stimulation of their views. Further, we thank the National Institute of Mental Health, University of North Carolina School of Medicine, and the University of Minnesota for helping to make our research possible.

In addition to some of the former reading parts of our manuscript we benefited from the comments of the following: L. Erlenmeyer-Kimling, E. Essen-Möller, E. H. Hare, L. L. Heston, and L. R. Mosher. Brendan A. Maher read the entire manuscript in its very roughest form and did his best to make us write clearly. While quite willing to acknowledge the assistance we have received over the years, we must ourselves shoulder the blame for any remaining shortcomings. It would be impossible to mention by name everyone who has helped us since the start of the Psychiatric Twin Register in 1948, and we apologize to any whose contributions have been inadvertently omitted in these acknowledgments.

For helping us prepare the manuscript for publication we especially thank Claire Daly, Marjorie Johnson, R. F. Johnson, Mary Anne Larson, G. Miner, Vivian Phillips, Vera G. Seal, Elizabeth Shields, and Bertha Storts. Blood for zygosity diagnosis was collected for us by cooperating physicians and technicians, and the actual blood-grouping was performed in the "Mecca" of that world through the courtesy of R. R. Race, FRS, and Ruth Sanger, FRS, at the MRC Blood Group Research Unit, The Lister Institute. For many other tasks which guaranteed the successful completion of our research we also thank the following: Beatrice Rouse, Sue Nicol, Kristin Arnold, J. Kahn, J. W. Patten, G. K. B. Nias, Barbara Baxter, Berry Harrison, Mary Crotty, M. E. L. Malherbe, D. B. Connell, the late Marjorie A. Brown, M. B. St. Clair, D. W. K. Kay, M. B. H. Whyte, J. L. T. Birley, E. Kringlen, K. Abe, J. S. Price, Margit Fischer, C. Flach, and A. Tellegen.

1
PROLOGUE

Schizophrenia research embodies a microcosm of the vexing problems that confront the behavioral sciences, but particularly those disciplines concerned with psychopathology. We are ignorant of the ways in which to prevent schizophrenia because we continue to be ignorant about its etiology. Despite recognizable descriptions of the syndrome in ancient Hindu treatises ca. 1400 B.C., and 76 years after its more modern description as dementia praecox by Kraepelin, we are still grappling with such basic issues as the classification and diagnosis of Eugen Bleuler's (1911) "group of schizophrenias" (Stengel, 1967; Katz *et al.*, 1968). Despite brilliant advances in molecular biology, cytogenetics, neurochemistry, and brain-behavior phenomena generally, we cannot pinpoint any necessary biological defect in all or most schizophrenics. Despite selfless expenditures of time and energy by gifted psychotherapists and methodological sophistication among social scientists, we cannot specify any necessary life experience that is common to all or most schizophrenics, whether at the level of the family or of a culture. Casting the problem of etiology in an interactionist framework defies an easy solution because we still must specify which element(s) in the genotype interact with which element(s) in the internal and/or external environment (as well as when and how) to produce the phenotype we recognize as a schizophrenic one.

Furthermore, the observation that schizophrenia may well be etiologically heterogeneous, with many biological and environmental causes, could lead to research nihilism and obscurantism. Should the observation be taken to mean that 100 causes each account for 1% of the cases, or that one cause accounts for 99% of the cases and the remaining 1% has 99 causes, or what? One concrete example in this regard may be a comfort. Mental retardation is clinically and etiologically heterogeneous. Between 2 and 3% of the population have IQ's

1

under 70, but no more than 10% of the retarded become hospitalized. Some 200 single gene substitutions have so far been identified (McKusick, 1971) with discrete retarded phenotypes, but they only account for some 2% of hospitalized cases. Down's syndrome (mongolism) has an incidence at birth of 1 in 600; other relevant chromosomal abnormalities are much rarer. We must also allow for various exogenous causes such as maternal rubella and birth injury, and for a small percentage of cases just under the psychometric threshold of IQ 70, which are phenocopies associated with family neglect or deprivation of stimulation. This still leaves the vast majority of cases in the total population of retarded—some 70%—that represent the tail end of the normal distribution for intellectual ability, and are most probably polygenic in origin (Penrose, 1963). So far no Mendelian forms of schizophrenia have been identified and no major departures from a normal chromosomal picture have been reported. We shall reserve our own views about the weight to be accorded different causes until after an attempt has been made to integrate our findings with those in the literature. If one type of schizophrenia is not predominant, the data collected so far on heterogeneous samples of schizophrenics contain so much noise that the signal permitting the discernment of mode of transmission of the disorder will not be detectable.

The wish for a simple etiology with a then obvious rational treatment and eventual prevention is ever present. The "stuff" of which such dreams are made emanates from our coveting the successes with pellagra, paresis, tuberculosis, poliomyelitis, and phenylketonuria. It may turn out that our ambitions are doomed to failure at this stage of our technology and conceptualizations; if that is the case the important tasks become the elimination of unprofitable lines of inquiry, the reordering of priorities for new research as a function of our current understanding of the facts, and the generation of refined, sharply focused, testable hypotheses which, in toto, will add enough strands to the network of information about schizophrenia to allow its ultimate comprehension. Complexity, ambiguity, and uncertainty on the one hand, and the chance to contribute to the solution of an important enigma on the other, are the lot of researchers into the causes of schizophrenia. Diabetes mellitus is a "familial" disorder with an obscure etiology and would be a more apt condition to look to for possible analogs. Incidences in the general population and among relatives of index cases are similar to those found for schizophrenia; it is not clear whether diabetes is a qualitative departure from normal glucose metabolism or simply the tail of a normal (Gaussian) distribution; the frequency with which the disorder is diagnosed is markedly influenced by trait-relevant environmental factors; finally, all possible modes of genetic transmission have been suggested and supported by reputable experts. Although the manifestations of the diabetic genotype are more palpable and tractable in a laboratory than those of schizophrenia, diabetes has been branded "a geneticist's nightmare" (Neel *et al.*, 1965). At least in this respect schizophrenia has an analog with diabetes!

INTENTIONS AND BELIEFS

Minimally, it is our intention to present a careful, in-depth account of our own research into the etiology of schizophrenia. Our subjects were twins who became schizophrenic, their co-twins, and other immediate relatives. Our method was a contemporary version of the classical twin method proposed by Sir Francis Galton in 1875; we use the method as an experimental technique and not as mere observation. The independent variable, manipulated by the experimenter, is the genotype; the dependent variable is the behavioral phenotype—schizophrenia or the psychological status of the co-twins and other relatives. (More of the twin method later.) Suffice it to say that our own research is not a series of case studies, although we shall present fairly detailed case histories for each of our 57 pairs of twins; Rosenthal (1963) presented an exhaustive case study for the Genain quadruplets, and our histories will appear to be superficial compared to that heroic account by 25 investigators. Neither is our study epidemiological in the sense of Kringlen's (1966, 1967), since we did not depend on any kind of national registration system for twinning or psychiatric illness. However, our sampling was systematic and, like all samples, selected in certain respects that will become clear. Our observations were clinical but not casual. For the first time in this area of research we have used and relied on psychometric devices to aid in assessment of subjects, and have tried to avoid "contaminated" diagnoses by using blind judgments of 6 experts for each twin; further, we have added to the reliability of information from interviews by tape-recording them from the majority of our twin subjects.

ZEITGEIST IN THE 1960's

Our study of schizophrenic twins was designed in 1962 and executed over a 7-year period, although the study was prospective in the sense that our index cases were registered systematically along with all twins among consecutive admissions to the Maudsley Hospital from 1948 to 1964. Through the foresight of Eliot Slater, efforts were made to obtain current and later information on the family and co-twin of each twin admitted to the inpatient, outpatient, or children's services in anticipation of building up a sample large enough to permit some kind of improved replication of his 1953 investigation. It is important for a more complete understanding of our work to depict the zeitgeist that characterized the early 1960's with respect to the probable etiology of schizophrenia. In an influential article summarizing the relevance of biochemistry to schizophrenia, Kety (1959) found that no biochemical lesion had as yet been demonstrated, despite numerous claims by enthusiasts. Although some psychochemical theories seemed to hold promise, the frequency of methodological errors and technical limitations prevented finding the needle in the haystack. His review was readily regarded as heralding the demise of a "genetic point of view"; the null hypothesis had been *proved*, schizophrenics were biochemically like non-schizophrenics; therefore their genes were the same. Kety himself, however, was not so persuaded: "Although the evidence for genetic and therefore biological factors as necessary and important components in the etiology of many or all schizophrenias is quite compelling, the signposts pointing the way to their discovery are at present quite blurred and . . . illegible [p. 1595]."

Next in time came the book edited by Jackson (1960), *The Etiology of Schizophrenia*, containing a critique by him of the "literature on the genetics of schizophrenia." Even today, a decade later, the chapter is considered in many quarters to be the definitive rebuttal to genetic thinking and an excuse to ignore any developments generated by genetic hypotheses; the critique dealt especially harshly with Kallmann's (1938, 1946) findings and conclusions. Jackson said his critique was written particularly "to question the assumption that seems to be very widely made that there is overwhelming factual evidence for a strong genetic component in the etiology of schizophrenia [p. 80]." The 45 pages of criticism were balanced by an 11-page review of genetic evidence by Böök, a well-known Swedish psychiatric geneticist who had made important contributions to the area. Although a few of Jackson's points were well taken, most have since been shown to have no force (Gottesman & Shields, 1966b; Shields *et al.*, 1967; Clausen, 1968; Eisenberg, 1968; Rosenthal, 1968). [As partial support for rejecting twin studies, Jackson noted the *false* implication of important genetic factors for the etiology of mongolism (Down's syndrome) from exceedingly high concordance rates in identical compared to fraternal twins when the critical factor was known [sic] to be intrauterine development—since 1959 it has been shown that all cases are genetic and caused by an extra chromosome or surplus chromosomal material.]

Confidence in the weight to be accorded genetic factors was next called into question by the penetrating, careful, and balanced reanalyses of twin and family studies by Rosenthal (1959 *et seq.*). His review of the 5 major twin studies conducted between 1928 and 1953 led him to say that the MZ concordance rates were "misleadingly high" and that "heredity does not account for as much of the variance with respect to what is called schizophrenia as some have alleged [1962a, p. 132]." His chief criticisms were that the samples were biased in various ways and that the diagnoses of probands and co-twins could have been influenced by knowledge of zygosity (see Chapter 2). Rosenthal concluded, however, that the twin studies provided the best evidence we have for the importance of genetic factors and should not be rejected because they contained weaknesses.

When Tienari, in 1963, reported a 0% concordance rate for schizophrenia in his Finnish sample of 16 pairs of male identical twins, followed by Kringlen's 1964 report of a 25% MZ rate for 8 Norwegian pairs, psychodynamicists were comforted, and skepticism and doubt about earlier twin and family studies conducted by genetically minded researchers were understandably raised to new levels. Such was the zeitgeist in which we conducted our field work and contemplated the meanings of our findings.

LEVELS OF EXPLANATION

In our opinion, no one or two investigators or disciplines will be able to solve the precise problem of schizophrenia's etiology; our aim is to attempt a partial synthesis of the facts at a level best known to us and to present a partial theory of the etiology of schizophrenia based on these facts as of the end of 1971. This will require going beyond our own findings and using information from other studies of twins and both biological and foster families of schizophrenics, especially those in which "genetic thinking" was emphasized, whether concurring or dissenting. We are not reluctant to go beyond our own research findings for heuristic purposes. It will remain for others, both above and below behavioral genetics in the pyramid of sciences, to integrate our views into the overall picture. Our judgments are channeled by our beliefs. Our beliefs are eclectic—not an opportunistic eclecticism, but rather the necessary, flexible strategy demanded by the diversity of phenomena encompassed by the etiology, course, and outcome of schizophrenia. Thus we value explanations for behavior that range from sociological to physicochemical, and recognize that the gene-to-behavior pathway is exceedingly complex (Gottesman, 1968a).

Is it not anachronistic, in this age of molecular biology, to look at *whole* human beings and their *behavior* when genetic aspects of variation in a trait are emphasized? Simpson (1964) and Dobzhansky (1968, p. 1), among others, have observed that "There are two approaches to the study of the structures, functions, and interrelations of living beings—the Cartesian or reductionist and the Darwinian or compositionist." Dobzhansky did not mean that biological sciences could be dichotomized in this way; he meant only that biological phenomena have Cartesian and Darwinian aspects, and further, that the two aspects were both necessary and complementary. Mayr (1964), another eminent evolutionist, elaborated the analytical and systems approaches to biology in another way. In his view the biologist studies systems of increasing complexity ranging from molecules through organ systems and individuals to populations, species, and species aggregates. It becomes obvious, on contemplation, that a system at any level is composed of elementary units that are themselves the systems of the level lower down. Mayr goes on to conclude:

> On each level it is equally legitimate to study either the system as a whole or the elementary units of the system, but we will not get the whole truth unless we study both. It is fortunate, both for physics and biology, that systems at higher levels can be studied with profit long before the elementary units at the lower levels are fully understood. The past history of biology has shown that progress is equally inhibited by an anti-analytical holism or by a purely atomistic reductionism [p. 1235].

We shall approach schizophrenia at the level of the schizophrenic individual and shall alternate between compositionist and reductionist views as seems most fitting to the particular aspect of schizophrenia focused on at any particular time. Our reductionism will not take us "into the skin" of our subjects; our holism will encourage us to look at the implications of our data for populations, and eventually for the species.

WHAT IS MEANT BY A
SPECIFIC GENETIC ETIOLOGY?

It is difficult to convince even the informed behavioral scientist that genetic factors are both relevant and important to the etiology of schizophrenia if his knowledge does not include the differences between the classical Mendelian genetics of discrete entities and the genetics of continuously distributed traits. The two kinds of variation observed in man are associated with genes differing in the directness of their effects on the observable phenotypic characteristic. A large number of genes in the realm of medical genetics (McKusick, 1971) have been identified, each of which regularly produces easily recognized pathological phenotypes that fall into a class or classes that do not overlap with the normal phenotype, and so are qualitatively different. The effects of these so-called major genes are not easily deflected from their course, and the number of links in the causal chain connecting them to their specific traits are few. These Mendelian pathologies are caused by the familiar dominant and recessive segregating genes and have in common the fact that they are very rare in the general population, of the order of 1 in 10,000 or rarer.

The establishment of a genetic etiology for such rare disorders is fairly straightforward, but a few examples can clarify frequent misunderstandings of what are necessary conditions for such claims. Huntington's chorea is caused by a dominant gene; one parent of each index case (proband) is also affected only if he or she lives long enough; in a population of siblings and children of probands, 50% are expected to develop the disorder at some point in their lives. Although genetic, the disease is not congenital; no biochemical error has been discovered;

unaffected gene carriers cannot be identified; we are unable to trace the pathway from the gene to the abnormal behavior; and no rational or empirical therapy has been devised. Phenylketonuria (PKU) is a type of mental deficiency caused by the presence of two recessive alleles (genes at a specific locus): 25% of the siblings are expected to be similarly affected. A specific inborn error of metabolism exists detectable at birth, and carriers (heterozygotes) can be identified biochemically but not behaviorally; an elevated rate of cousin marriages exists among proband parents. Although the disease is genetic, neither parent is affected, and there is not likely to be a positive family history; the fact that severe retardation in PKU can now be prevented by an environmental factor, early diet manipulation, makes it no less a genetic disorder of phenylalanine metabolism. For both dominant and recessive conditions the expectation in co-twins of identical (MZ) twins is 100% and in co-twins of fraternals (DZ) is 50% and 25%, respectively.

The bulk of gene-determined human variability arises from the simultaneous occurrence of many small discontinuities, polygenic effects, which, when put together and smoothed by environmental effects, are not individually detectable (cf. Mather, 1964; Morton, 1967) and result in (bell-shaped) continuous variation in a trait. The multiple genes or polygenes of quantitative genetics (Falconer, 1960; Roberts, 1964) are not different in kind from Mendelian (major) genes; each has a small, multiply mediated effect on trait variation when compared to the total variation observed for that trait. The expression of the trait, e.g., height, blood pressure, or disease resistance, depends much less on which unspecific genes in the specific trait system an individual possesses than on the total number pulling him toward the tail of the distribution of values. Polygenic systems contain certain features that contrast with single-gene substitution. The traits are especially sensitive to internal and external environmental influences, thus reducing the correlation between genotype and phenotype. "... as the branches of a tree sway more in the wind than the trunk, so the more remote gene effects become progressively more sensitive to all kinds of influences of the environment [Grüneberg, 1952, p. 110]." Another feature is the capacity to store or conceal genetic variation and thus protect the genes from the effects of natural selection when deviants at the tail of a distribution are removed from the breeding population (Lerner, 1958); given a 2-locus, 4-gene system, two parents with genotypes Aa Bb would be in the middle of the distribution for a trait determined by the *number* of capital letter genes they had, but they could produce children very different from themselves in both directions, AA BB and aa bb.

It sometimes happens that a continuous distribution of genotypes gives rise to discrete phenotypes for some common disorders and congenital malformations. This can best be explained by the assumption of a continuously distributed liability and a threshold. The discontinuity of phenotypes is *not* genetic, but results from the value of a variable exceeding a threshold that may be physiological, biochemical, developmental, or mechanical (Grüneberg, 1952; Edwards, 1960, 1963, 1969; Crittenden, 1961; Falconer, 1965; Carter, 1965, 1969b). Disorders fitting these stipulations are called quasicontinuous or threshold characters and are exemplified by cleft palate, arterial rupture, overt diabetes, ulcer, and, in our opinion, schizophrenic psychosis. It is important to note that some of these disorders have a tendency to escalate in severity over their initial dysfunction because of a kind of vicious feedback mechanism; the overt bimodality formed by this "break-away" at the end of a continuum may then be mistakenly interpreted as evidence for Mendelian heredity (Edwards, 1960). We shall return to this point in Chapter 10.

According to the definition of a common genetic disorder, schizophrenia is common, with a lifetime incidence or risk of about 1% or more, markedly different from the .004% of PKU or the .005% of Huntington's, but not so different from the .1% of harelip or clubfoot. As a working hypothesis, we shall adopt the idea that the genetic basis of common disorders is to be found in the ubiquitous variability of our species and not necessarily in any specific major pathogenes. We shall defer discussion of the concept of incomplete penetrance until later (see Chapter 10). The connection between polygenic variation and disease incidence in its simplest form consists, for example, of continuous variation in blood pressure associated with a continuously increasing risk for (essential) hypertension. The properties of the relevant dimension underlying the continuous liability to schizophrenia are hypothetical at the present time.

It is obvious from the previous discussion that the classic methods of genetics for discrete entities cannot be applied to the study of schizophrenia in order to prove a genetic etiology; the approach must be biometrical and data on frequencies must be converted to measures of similarity between relatives (e.g., Falconer, 1965). The biometrical approach to genetic problems traces back to Galton's (1886) attempts to assess the genetic contribution to differences in stature between fathers and sons. Data on the genetic background of schizophrenia become meaningful as a function of the incidence of the disorder in different genotypes. Genotypes are identified indirectly but accurately (for groups) by our knowledge of the proportion of genes shared in common by the different blood relatives of representative schizophrenic index cases (probands), and this gives us the basis for the primary method of studying common disorders —the twin-family method. The closest relatives are monozygotic (MZ) twins who have all their genes in common; dizygotic twins (DZ), full siblings, parents, and children—first-degree relatives—share half their genes, on the average, with probands; whereas second-degree relatives of probands—grandparents, grandchildren, uncles, aunts, nieces, nephews, and half-siblings—share one-fourth their genes. Familial aggregation is the *sine qua non* for a genetic explanation for schizophrenia: the closer the relationship, the higher the expected risk for the disorder. However, the genes held in common are paralleled to an unknown extent by environments held in common, thus preventing any premature ascription to heredity for the clustering in families. How far could the raised incidence in relatives be accounted for by a common family milieu? It is here that twin and foster child studies enter. We shall examine this evidence in detail in Chapter 2; suffice it to say at this point that MZ and DZ co-twins of schizophrenic probands, despite sharing virtually the same ecology with their twin siblings, are schizophrenic themselves at markedly different rates—MZ co-twins are from 2 to 5 times as often affected as DZ co-twins. This, in turn, is not due to subtle within-family environmental factors as shown by studies of MZ twins reared apart and adoption studies in which the increased rate prevails even when the relative is not brought up in the same home as the schizophrenic. The fact that concordance in MZ pairs is far from complete shows that genetic factors are not sufficient, or alternatively or additionally, that only some schizophrenias are genetic.

Among our expectations for a disorder of polygenic etiology are the following: a sharp drop in the proportion affected as one passes from first-degree to second-degree relatives rather than one-half the first-degree rate predicted by dominant gene theory; the more severe the case, the greater the risk to relatives; the risks to relatives will increase as the number of other family members affected increases (Carter, 1969a, b).

Even if genetic factors are implicated in the etiology of schizophrenia, how important are they? We can raise but not definitively answer this question at this point. Some perspective can be gained though about some misunderstandings surrounding the question from a look at Table 1.1. It is misleading to take as a guide to genetic importance the risk for siblings of probands affected with Mendelian pathologies compared with the risk to a member of the general population. The comparative risk of 10,000 for Huntington's chorea, for example, makes the figure of 25 for clubfoot, or 10 for diabetes mellitus or schizophrenia *seem* trivial. Penrose (1953) and Crittenden (1961) have cautioned against such a conclusion for common chronic diseases, and the latter has shown

TABLE 1.1

Comparative Risks in Siblings for Mendelian Disorders,
Congenital Malformations, and Common Diseases

Condition	Population risk (a)	Proband's sibling risk (b)	Comparative risk (b/a)
Huntington's chorea	.00005	.5	10,000
Phenylketonuria	.00004	.25	6,250
Clubfoot	.001	.025	25
Diabetes mellitus	.003	.028	9
Schizophrenia	.0085	.087	10

that a comparative risk as low as twice the population figure for breast cancer is compatible with "a fairly high degree of genetic control," and we (Gottesman & Shields, 1967) reached the same conclusion for a comparative risk of five for schizophrenia.

Meehl (1962, 1972) summarized what is *not* meant by positing a specific genetic etiology and it bears repeating.

1. The etiological factor always, or even usually, produces clinical illness.
2. If illness occurs, the particular form and content of symptoms is derivable by reference to the specific etiology alone.
3. The course of the illness can be materially influenced only by procedures directed against the specific etiology.
4. All persons who share the specific etiology will have closely similar histories, symptoms, and course.
5. The largest single contributor to symptom variance is the specific etiology. [Meehl, 1962, p. 828]

Not one of the above statements need be true for schizophrenia to have a specific genetic etiology and none can be validly invoked in opposition. The working hypothesis that we will hold, then, maintains that a genetic contribution is necessary but not sufficient for schizophrenia to occur. Furthermore, the genes are conceived of as specific to schizophrenia, i.e., whatever else they may do, they increase the liability to schizophrenia rather than to high blood pressure, affective disorder, etc. The genetic contribution may or may not be the largest contributor to the variance of liability; in analysis of variance terms, it means an interaction effect preventing other variables from causing schizophrenia when the specific genetic etiology is missing. Since our model imposes the requirement of necessity on the genetic component of schizophrenia, it follows that the component is *important* regardless of any quantification that may subsequently emerge.

All characters are acquired, and heredity and environment interact in the development of all of them. When we speak of genetic transmission, we are addressing the real problem of how far differences in the genotype and differences in the environment contribute to the development of such a character, even if it is schizophrenia. If genetic differences make a contribution it is important to try to find out what we can about their nature. This holds however much or little we know about environmental factors. And of course the converse holds. The better one factor is studied, the more the ground is delimited in which the effects of other factors can be sought.

SCHIZOPHRENIA *QUA* SCHIZOPHRENIA

It is so easy to become dazzled by the extraordinary phenomenology of schizophrenia that testable questions about the etiology of the class (as opposed to cases) of the disorder become lost among fascinating speculations (e.g., Sullivan, 1962; Searles, 1965). Even the existence of schizophrenia has been called into doubt. Karl Menninger (1963) would have all mental disorders unitary with variation in symptom picture merely reflecting levels of decompensation; Szasz (1960) would have us exorcise mental illness altogether from our vocabulary, relegating it to the category of "myth." Brown (1967), reviewing findings in social psychiatry, noted that too often unacceptable conclusions drawn from inadequate evidence become quoted unquestioningly, thus leading to a confusion between speculation and knowledge; some biologically oriented research on schizophrenia is equally at fault. Presently there is no substitute for the longitudinal, intensive observation of good-sized samples of schizophrenics in the same setting for gaining an appreciation of what schizophrenia is all about. We have, for example, the writings of the elder (1911) and younger Bleuler (1972) on the Burghölzli Swiss population, Shakow's (1969, 1971) on the Worcester population, and Ødegaard's (1972) experience at Norway's Gaustad Hospital. The latter has seen ". . . a picture of the functional psychoses as an ocean of individual cases without natural boundaries, but not at all without an orderly pattern. For the recognition of this pattern the conventional diagnoses have proved most helpful, which was to be expected, as they have been worked out by eminent clinicians with an immense experience. But they are helpful in the same way as the degrees of latitude and longitude for ocean navigation and not as unsurpassable walls erected between two parts of a population [p. 266]."

Nosology or classification in psychopathology is admittedly difficult; the issues involved have been grappled with by Gruenberg (1968), Shakow (1968), and Strömgren (1965a). A diagnosis of schizophrenia cannot be made unambiguously because no laboratory or psychometric findings and no one or two clinical symptoms are pathognomic, thereby making its diagnosis by inexperienced clinicians unreliable. Bleuler made the useful distinction between primary and accessory symptoms, the latter being the results of efforts by the patient to compensate for the primary effects of the schizophrenic process. Kraepelin's supposed criterion of deterioration has been excluded for some time now. Schizophrenia does have a variable age of onset, a variable course, variable symptoms, and variable outcomes. The appearance of borderline, schizoaffective, pseudoneurotic, pseudopsychopathic, or pseudoretarded cases does not invalidate the existence of schizophrenia any more than twilight invalidates the existence of night and day. It is probably these kinds of phenomenal flux that have led to exaggerated maligning of classification enterprises in psychiatry; in our opinion, no adequate experiments have been completed yet to justify such pessimism. We believe our faith in conscientiously carried out diagnoses will be vindicated by our own work (Chapter 5) and that of others (e.g., Shepherd *et al.*, 1968; Cooper *et al.*, 1969). It may not be sufficiently appreciated that schizophrenics are individuals who can have neurotic symptoms, be pacifists for nondelusional reasons, suffer the effects of colds and broken legs, and, after years of hospital confinement also suffer the effects of arteriosclerotic brain damage and social isolation.

The syndrome of schizophrenia enjoys the status of an "open concept" (Cronbach & Meehl, 1955; Meehl, personal communication; Pap, 1958) and need not be strictly defined operationally in order to retain its legitimacy as a concept. Gruenberg (1968) warned that some operational definitions are circular, imprison our thinking, and obscure issues. Our own definition of schizophrenia will be made explicit by our twins' histories in Chapter 4 together with our comments on the consensus diagnoses of our 6 judges.

A failure to distinguish at all times between genesis and pathogenesis can only delay efforts to identify the etiology of schizophrenia. A case in point, hopefully not a misleading one, is general paresis (Bruetsch, 1959); it took almost 100 years for the firm connection to be made between the heterogeneous symptom picture in affected patients and a syphilitic infection of the central nervous system. Meanwhile, most experts, including Virchow and Maudsley, believed that such diversity could not possibly have a unitary etiology, and such causal agents were invoked as heredity, alcohol, smoking, menopause, and overwork, *each supported by research data.* Our preferred model for construing the syndrome of schizophrenia permits the clearer separation of etiological and pathogenetic-phenomenological considerations. It comprises a network of events connected by sequential causal arrows. A chain of consequences, or a *pedigree of causes* (Grüneberg, 1947), is set into motion by a genetically caused predisposition and culminates in the set of symptoms recognized as the schizophrenic syndrome. There is little merit in calling such a model a "medical model"; to do so may even dissuade some people from seeing the utility of it. The idea has much in common with the gene-to-behavior pathway illustrated by Fuller and Thompson (1960). Ideally the pedigree of causes would also incorporate the epigenetic or developmental ideas of Waddington (1957).

Such a model has been worked out for diabetes mellitus (Neel and Schull, 1954, p. 170) and clearly shows that there are many points in the chain and its branches for therapeutic intervention without ameliorating the root cause, insulin deficiency; it also shows that many features of diabetes are due to efforts at compensating for the loss of energy from sugar catabolism by increased catabolism of proteins and fats. Our proposed analogous network for schizophrenia would allow room for feedback loops or vicious circles and elevate the concept of "chance" to a more important role; Edwards (1963) feared that the confusion of controllable environmental variation with chance variation would lead to undue optimism about the effects of environmental modification: "Chance, . . . is no less scientific in biology than in the physics of elementary particles. In both, the configuration of any system will affect only the probability of any future configuration [p. 631]." Our construction makes it clear that psychotherapy, phenothiazines, or a good mother may each contribute to symptom amelioration without necessarily casting light on etiological questions. It is our hope that progress in refining broad genetic theories about the etiology of schizophrenia will benefit from advances in knowledge about the pathophysiology and developmental psychopathology of schizophrenia.

2

EVIDENCE RELEVANT TO
A GENETIC ETIOLOGY FOR
SCHIZOPHRENIA: AN OVERVIEW

At the turn of the century Gregor Mendel's (1822–1884) revolutionary ideas about the genetic basis for the transmission of physical resemblances between parents and offspring were rediscovered and widely circulated some 35 years after the monk's garden experiments, coinciding with the publication of the equally revolutionary ideas of Sigmund Freud (1856–1939), *The Interpretation of Dreams*. Independently of these major developments, Charles Darwin (1809–1882) had revolutionized thinking about man's status in the universe with his theory of evolution enunciated in *The Origin of Species* (1859). These three movements were destined to overlap, intertwine, and greatly influence thinking about normal and abnormal human behavior as the twentieth century unfolded. A neuropsychiatry dominated by the natural sciences eagerly embraced, applied, and misapplied the neat principles of Mendelian genetics to their burdensome, unsolved "diseases": insanity, feeblemindedness, and alcoholism. Regrettably, unwarranted eugenic fervor, Social Darwinism, and racism were stimulated by the political distortion of some of these biological ideas; we categorically disassociate our thinking from such evils (cf. Dunn, 1962; Haller, 1963). Eugen Bleuler (1857–1939) was influenced by both Freud and the great organic psychiatrist Emil Kraepelin (1856–1926), who gave us the modern classifications of dementia praecox and manic-depressive psychoses, when he reconceptualized dementia praecox as "the group of schizophrenias" (1911) and rejected Kraepelin's criterion of deterioration.

15

The concerns and questions of the first workers on the genetic aspects of schizophrenia are very much the same as those of today: the mode of inheritance, the role of environmental factors, etiological heterogeneity, determiners of resistance, and the identification of "neuropathic equivalents." Bleuler, like Freud and Kraepelin, believed the origins of schizophrenia to be basically organic in nature, but he was not content with imprecise generalizations. Granting that the schizophrenias were commonly familial, he said

> . . . if an adherent of an "infectious theory" of this disease should choose to say that there is no hereditary factor in schizophrenia but merely an infection from some common source, or if someone else cares to assume that modifications of the psychic or physical factors produced by communal living produce such accumulations of disease in a given family group, we would be unable to produce any proof to the contrary. Such skeptics could observe that in many cases, even after the most thorough study, no evidence of any hereditary *anlage* and no individual predisposition (such as a seclusive, withdrawn character structure) has ever been proven. And yet heredity does play its role in the etiology of schizophrenia, but the extent and kind of its influence cannot as yet be stated. In order to be able to accomplish something more than what has already been done on this question of heredity, we first of all would need a workable concept of heredity. [1950, p. 337]

In that same year, 1911, Rosanoff and Orr published the first family pedigree study designed to explain the etiology of insanity in Mendelian terms. Although they eliminated toxic and gross organic causes of insanity from their proband material, they pooled all types of functional neuropathy including epilepsy and feeblemindedness; the (retrospectively) primitive methods of these pioneers revealed what they were looking for: *". . . the fact of the hereditary transmission of the neuropathic constitution as a recessive trait, in accordance with the Mendelian theory, may be regarded as definitely established."* And so the modern era of psychiatric genetics was ushered in.

A burgeoning literature on the relatives of schizophrenics, epidemiology, biochemistry, experimental psychology, and neurophysiology precludes other than a selective review of the literature that we perceive as being antecedents of our own study. We hope to highlight facts that will maximize the opportunity for a choice among alternative explanations for the basic observations forming the initial input for the various schools of thought. Areas we omit completely may be followed elsewhere (Maher, 1966; Venables, 1968; Mandell & Mandell, 1969; Smythies, 1968; Heath & Krupp, 1967), whereas topics we cover have been dealt with in more depth by Mishler and Scotch (1963), Hare (1967), Kohn (1968), Rosenthal and Kety (1968), Brown (1967), Shields (1967, 1968), Strömgren (1965b), Zerbin-Rüdin (1967), Rosenthal (1970), Slater and Cowie (1971), and Gottesman and Shields (1966b). Our review will focus on epidemiological research, genetic family studies, fostering studies, twin studies, and psychodynamically oriented family studies.

EPIDEMIOLOGICAL CONTRIBUTIONS

Between-culture and within-culture rates for the development of schizophrenia give us some very useful benchmarks for asking questions about the etiology of schizophrenia. If cross-cultural incidences are more or less the same despite the gross differences in social organization and child-rearing practices, it suggests a species-specific characteristic, a genetical predisposition fairly evenly distributed. Alternatively, it could suggest an as yet unspecified universal similarity in cultural practices by a subset of each population. Apparent cross-cultural stability in the rates of alcoholism and general paresis led to the false implication of genetic etiologies. As seen in Table 2.1 the rates tend to be the same. In 19 earlier European studies based on field work with the relatives of normal persons, there was fairly good agreement that the lifetime expectancy for schizophrenia, whether hospitalized or not but narrowly diagnosed, was about .85%. There have been exceptions (cf. Murphy, 1968) such as Böök's (1953) estimate of 2.85% for an isolated region of Sweden north of the Arctic Circle and Garrone's (1962) estimate of 2.4% for Geneva. The former may be explained by a mixture of unique attributes of the genetics and environments of the isolate and the latter by the assumption of a uniquely wide risk period, 15-70, leading to too large an age correction.

In England and Wales hospital statistics for first admissions (1952-1960) show that the accumulated lifetime risk for admission of a man to a health service hospital with a diagnosis of schizophrenia was 1.12% by the age of 65, the expectation for a female being 1.10% (Slater & Cowie, 1971). A wider concept of schizophrenia led Yolles and Kramer (1969) to estimate that the lifetime risk

TABLE 2.1

Expectation of Schizophrenia for the General Population[a]

Date	Country	Age-corrected total (N)	Expectation (%)	S.E.
1931	Switzerland	899	1.23	.368
1936	Germany	7,955.5	.51	.088
1942	Denmark	23,251	.69	.054
1942	Finland	194,000	.91	.021
1946	Sweden	10,705	.81	.087
1959	Japan	10,545	.82	.088
1964	Iceland	4,913	.73	.121

[a]From Slater, 1968.

of schizophrenia for a person born in the United States in 1960 was between 2% and 6%; these figures include both inpatients and outpatients.

Within industrial cultures there are wide social class differences in the incidence of schizophrenia today; the incidence in the lowest class (5-point scale) is 3 to 4 times greater than in the 2 highest classes, whereas the prevalence is some 8 times greater (Hollingshead & Redlich, 1958). Are the obvious social stresses of life in urban ghettos in New Haven, Detroit, or London sufficient to cause this excess of schizophrenia? If stress were causal, the patients would be expected to have grown up in such surroundings and would tend to be of the same lower social class as their fathers. If prodromal symptoms of schizophrenia prevented the patients from attaining the social class expected from their educational background (social selection) or led to a fall in occupational level (social drift), the social class of their fathers could well be the same as that in the general population. Goldberg and Morrison (1963) found that schizophrenics' fathers were distributed occupationally like all fathers in England and that both social selection and drift were operative for the patients themselves. Dunham (1965) reports similar results for Detroit. Careful work by Turner and Wagenfeld (1967) with the psychiatric register for private and public patients near Rochester, New York, showed that downward mobility does indeed account for the excess of schizophrenics at the bottom of the social ladder; 36.4% of the total male sample were downwardly mobile relative to their own fathers, only 6.8% showed downward movement within their own work histories; 11.6% of the schizophrenics moved downward, relative to their fathers, to the lowest occupational level compared to 4.9% of the control group of normals. Hare (1967) concludes from his review of the evidence that downward mobility, rather than socioeconomic conditions, causes the high schizophrenia rates observed in the lower classes. Kohn (1968) refuses to accede to a social drift hypothesis for fear it would remove stress as a cause of mental disorder; no one has interpreted the above data to mean that stress is unrelated to schizophrenia. After agreeing that the evidence favors two propositions—lower class origins are conducive to schizophrenia and most lower class schizophrenics come from higher socioeconomic origins—Kohn proposes what he calls a sophisticated drift hypothesis, with which we cannot but agree. Some people genetically or otherwise predisposed to schizophrenia show the effects early and never achieve expected occupational levels, thus suggesting an interaction between genetic predisposition and early social circumstances. The fact is that social migration is inseparable from gene migration, and the model for social class differences in the gene pool for talent outlined by Gottesman (1968c) may be applicable. Genetic vulnerability accumulates in the lower class gene pool where interaction with rampant social stresses may precipitate cases that could stay compensated in better environments.

GENETIC FAMILY STUDIES

Attribution of genetic factors as causal for the familial occurrence of a disorder is quite hazardous in the absence of a supporting network of confirmatory evidence. Kuru, an acute degenerative disease of the CNS, occurs in a familial pattern among the Fore tribe of Western New Guinea in the absence of any obvious infectious or toxic factors. An *ad hoc* genetic theory could be made to fit and was proposed (cf. the review by Kirk, 1965); all females with the dominant gene K are at risk but only homozygous males, KK, will be affected, and the gene frequency is .45 or higher. Some voices were raised against accepting a genetic hypothesis too readily (Williams *et al.*, 1964). Recent studies suggest that Kuru may be due to a form of slow acting virus, since chimpanzees have developed a Kurulike disease a year or so after being inoculated with cerebral tissue from Kuru victims (Gajdusek *et al.*, 1967). It is now known (Gajdusek, 1970) that the religious practice of eating dead relatives helped to mimic dominant gene transmission.

Each well-done family study of schizophrenia strengthens the network of basic facts from which hypotheses can be generated. We are indebted to Rüdin (1916) for the first attempt to study schizophrenia, as differentiated from other psychoses, according to Mendelian principles. Since most patients had nonschizophrenic parents, recessivity was the most likely possibility. In 701 sibships with unaffected parents Rüdin found an age-corrected risk for schizophrenia of 4.48% in the brothers and sisters of schizophrenics and psychoses of other kinds occurring almost as frequently, 4.12%. Two independent recessive gene pairs gave the best-fit, one-fourth of one-fourth or 6.25%, to the observed risk of 4.48% and was the first genetic theory to be put forward. His other findings are given in Table 2.2. One of the chief conclusions of his Munich school was to establish the

TABLE 2.2

Findings of Rüdin (1916)[a]

Parental morbidity	Number of sibships	Expectation (%) in sibs of schizophrenics of	
		Schizophrenia	Other psychoses
Neither schizophrenic	701	4.48	4.12
One schizophrenic	34	6.18	10.30
One psychotic (nonschizophrenic)	133	8.21	8.21
One alcoholic (neither psychotic)	109	7.80	5.20
Both psychotic or alcoholic	10	22.72	—

[a]From Shields, 1967.

essential genetic independence of *typical* schizophrenia and *typical* manic-depressive psychosis. By restricting themselves to diagnostically clear starting cases in an attempt to secure genetic as well as clinical homogeneity, they may have underestimated the incidence of other disorders in relatives and overestimated the genetic discreteness of clinical syndromes. Ideas in psychiatry have changed and it is easy to point out that the Munich school paid little attention to internal psychological processes or to interpersonal relationships which may or may not be relevant to *etiology* (cf. Shields, 1968).

Before pooling the family risks in the literature so as to permit some generalizations, let us touch on the problem of etiological heterogeneity. Schulz (1932) reinvestigated most of Rüdin's original cases, omitting misdiagnoses, and found that 6.7% of the sibs had certain schizophrenia, 8.2% if doubtful cases were included. Like later investigators he found a tendency for Kraepelinian subtypes to be associated within families; but the majority of affected relative pairs were not isomorphic, and it seemed that the risk of schizophrenia of any type was increased for the sibs of probands of each type. When he subdivided the cases according to reported precipitating factors, he found the sib risk of 6.7% reduced to 2.9% in the 55 families in which head injury had preceded the proband's schizophrenia. When no precipitating factors of any kind were noted, the risk to sibs went to 10.0% including doubtful cases; later (Wittermans and Schulz, 1950) he showed that the sibs of recovered schizophrenics had only a risk of 3.3%, and the children a risk of 7.4%. Other workers have found the expectation of schizophrenia in relatives to be reduced as they moved from index cases close to the core of dementia praecox to those on the periphery. The risks found for the sibs and children according to subtype are given in Table 2.3, from Kallmann's (1938) Berlin study and Hallgren and Sjögren's (1959) rural Swedish survey; relatives of paranoid probands are least at risk, but paranoid schizophrenics are most likely of all subtypes to become parents (Reisby, 1967).

TABLE 2.3

Expectation (%) of Schizophrenia for Sibs or Children of Schizophrenics according to Kraepelinian Type of Schizophrenia in Proband[a]

Study		Hebe-phrenic	Catatonic	Paranoid	Other[b]	All types
Kallmann (1938)	Sibs	12.6	13.4	8.9	9.5	11.5
(1087 probands)	Children	20.7	21.6	10.4	11.6	16.4
Hallgren and Sjögren (1959) (247 probands)	Sibs	7.0	8.5	4.7	2.3	7.0

[a]From Shields, 1967.
[b]Simple (Kallmann) or Atypical (Hallgren & Sjögren).

A comprehensive review of the literature by Zerbin-Rüdin (1967), daughter of the pioneer, draws together a great deal of pertinent information in useful form. Some pooled data on risks to first-, second-, and third-degree relatives are given in Table 2.4. The sample sizes are such as to give fairly small standard errors.

There is the expected large difference in risks between first- and second-degree relatives. Unexpected is the difference between parents, 5.5%, and sibs and children. The difference can be explained largely by the fact that parents are selected for a degree of mental health through having married and having children. When correction is made (Essen-Möller, 1955) for the actual number of "risk lives" involved, the risk in parents of schizophrenics becomes 11%, close to that of sibs and children. The somewhat higher risk to children than sibs may be due to sampling error. The similarity among the three estimates of risk across three generations is a fact significant for genetic theories and difficult for environmental theories to embrace.

It has regularly been found that when a parent is affected it is more often the mother than the father, and this has been cited in support of psychogenic

TABLE 2.4

Expectation of Schizophrenia for Relatives of Schizophrenics[a]

Relationship	Total relatives[b]	Schizophrenic			
		No.		%	
		1[c]	2[d]	1[c]	2[d]
Parents	7675	336	423	4.4	5.5
Sibs					
All	8504.5	724	865	8.5	10.2
Neither parent schizophrenic	7535	621	731	8.2	9.7
One parent schizophrenic	674.5	93	116	13.8	17.2
Children	1226.5	151	170	12.3	13.9
Children of mating Schiz. × Schiz.	134	49	62	36.6	46.3
Half-sibs	311	10	11	3.2	3.5
Uncles and aunts	3376	68	123	2.0	3.6
Nephews and nieces	2315	52	61	2.2	2.6
Grandchildren	713	20	25	2.8	3.5
First cousins	2438.5	71	85	2.9	3.5

[a] After Slater and Cowie, 1971.
[b] Age-corrected sample sizes.
[c] Diagnostically certain cases only.
[d] Also including probable schizophrenics.

causation or transmission. However, the observation is in keeping with the fact that women marry earlier than men and become schizophrenic later than men, thus having more years of possible fertility. Slater (1968) using Kallmann's 1938 data has calculated that female and male nuclear probands have 17.4% and 18.4% definitely schizophrenic children, although the number of children involved were 241.5 and 130.5 (age-corrected totals) for the sexes, respectively. The risks in offspring of paranoid plus simple parents were 8.7% for mothers and 9.8% for fathers. Bleuler (personal communication) reports similar findings, but Reed *et al.* (1972, in press) found a higher risk for psychosis in the offspring of psychotic mothers than fathers. Mednick *et al.* (1971) report potentially useful data on the infants of 57 schizophrenic mothers and 26 schizophrenic fathers identified from 9006 consecutive births on a maternity service (about two-thirds of the parents did not appear to be mentally disordered until some time after the delivery); various indicators of CNS immaturity distinguished the schizophrenics' offspring from controls—but the proportion of abnormal infants was independent of the sex of the psychotic parent.

The risk in second-degree relatives shows a considerable drop from that in first-degree relatives; a dominant gene theory would predict that the former would be one-half the latter. The pooled average risk in first cousins of 3.5% is unexpectedly high for third-degree relatives and is perhaps unreliable.

TABLE 2.5

Summary of 5 Studies of Offspring of Two Schizophrenic Parents[a]

Investigator	Total families[b]	Children			
		Ever-born	Surviving to age 15 and over	Questionable schizophrenia	Schizo-phrenia
Kahn, 1923	8	26	17	2	7
Kallmann, 1938	12	55	35	3	13
Schulz, 1940	23	92[c]	59[c]	5	13
Elsässer, 1952	15	72	56	3	12
Lewis, 1957	7	27[d]	27[d]	0	4
Total:	65	272	194	13	49

[a]After Erlenmeyer-Kimling, 1968.

[b]Excludes infertile marriages and those discarded by investigators because children were too young or because children could not be traced.

[c]One child of doubtful paternity (and also doubtful schizophrenia) excluded by Schulz in his calculations is also excluded here.

[d]Lewis does not give number ever-born and includes only those children who were age 20 or over at investigation.

When neither parent is schizophrenic the best estimate of the risk of schizophrenia in the sib of a schizophrenic is 9.7%. When, in addition to the proband, one parent is schizophrenic the risk goes to 17.2% for the sibs of the proband. When both parents are schizophrenic, the risk to their children is between 36.6% and 46.3% (Slater, 1968; see also Rosenthal, 1966 and Erlenmeyer-Kimling, 1968). Although the extremely chaotic environment provided by the dual mating of schizophrenics might account for the raised risk from 17.2% to 46.3%, it can be shown that in what must be an almost equally disturbed setting—one parent schizophrenic and the other psychopathic—the sibling risk for schizophrenia is only 15% (Kallmann, 1938). Environmental theories about the etiology of schizophrenia would be compelled to predict that virtually all of the children from dual matings would be psychiatrically disturbed if not psychotic. Elsässer found that 70% of the nonpsychotic children were apparently quite normal. A recessive gene theory would predict 100% affected, whereas a simple dominant gene theory would anticipate 75% of the children to be schizophrenic. The data about the children from these dual matings studies are so rare and important that they deserve tabular presentation in Table 2.5.

Matings of other psychotic combinations of parents, other than Sc × Sc, provide useful information about etiological heterogeneity and specificity (Elsässer, 1952). 20 dual manic-depressive matings yielded 14 psychotic offspring, but only one was schizophrenic; 19 Sc × M-D matings, however, yielded equal numbers of schizophrenic and manic-depressive children, 8 of each.

A need exists for more data on the half-sibs of schizophrenics where there is a control for contemporaneity of generation but where there is only a 25% gene overlap, and the data may be further divided according to whether the shared parent was schizophrenic or not. Rüdin (1916) found only a .6% half-sib risk. Kallmann (1938) found 7.6%; however, when they shared a schizophrenic parent, 24% of 21 half-sibs of a schizophrenic were affected, while only 1.7% of 57.5 half-sibs were affected when the common parent was healthy.

From the very beginning of genetic family studies on schizophrenia investigators have been plagued by the problem of how to characterize the disorders other than schizophrenia observed among proband relatives. It led to the use of the term *schizoid*. Currently a concern is expressed about what disorders are in or out of the schizophrenic *spectrum*, a term replacing the older, but no more explicit, *Kreis*. Assigning the label schizoid to a known relative maximizes the possibility of contaminated diagnoses and experimenter bias. The problem of overlap between schizophrenia and other abnormalities has been accentuated since the field was expanded to include other than *typical* cases as probands (cf. Slater, 1953; Ødegaard, 1963). We shall return time and again throughout the remainder of the book to the issues raised by the concepts of schizoid and spectrum.

TWIN STUDIES

To what extent could the raised incidences in relatives of schizophrenics be accounted for by a common family milieu or by psychological contagion? The strongest evidence implicating genetic factors in the etiology of schizophrenia comes from twin studies. If the family environment, the time of birth into such a setting and the sex of the child are significant contributors to the raised incidence in families, there should be little or no difference between the incidence in MZ co-twins and DZ co-twins of the same sex as the proband. The reported differences in concordance between genetically identical twins and genetically dissimilar twins of the same sex require a detailed consideration at this point. In the past ten years the ten systematically conducted twin studies in the literature on adult schizophrenic twins besides our own have been criticized, defended, and reevaluated at length by many authors. Neither space nor inclination will permit unnecessary redundancy. We shall concern ourselves with a brief account of the salient points in each study, devoting more attention to the research appearing since our detailed review (Gottesman & Shields, 1966b). First we must orient the newcomer to the methodology and criticisms of twin studies.

Methodology of Twin Studies

Since MZ twins have identical genotypes, any dissimilarity between pair members must be a result of the action of the environment, either prenatally, including extranuclear cytoplasmic differences (Storrs & Williams, 1968), or postnatally. Anything less than 100% concordance among MZ pairs living through a risk period permits us to exclude genetic (nuclear) factors as sufficient determinants of schizophrenia. DZ twins, either same-sex (SS) or opposite-sex (OS), on average will have half their genes from a common source but will also have certain environmental factors in common such as birth rank and age of mother, thus providing a means for controlling such factors not otherwise possible. When MZ and DZ SS twins from sizable and representative samples are contrasted on measures of a disorder, a means is available for the evaluation of the effect of different environments on the same genotype or for the expression of different genotypes under the same environment. The twin method depends on the accurate separation of SS DZ pairs from MZ, and this can be routinely accomplished nowadays by the quantification of blood grouping (Race & Sanger, 1968) and fingerprint patterns. Earlier researchers did not have the advantages of serological objectivity, but their clinical judgments, when based on the simultaneous examination of both twins were not likely to be much in error (Essen-Möller, 1941a). The representativeness of a sample of twins, unlike most samples in psychiatric research, can readily be established by an internal comparison of the proportion of the two kinds of twins with the expectation from vital statistics for the country.

Basically the comparison of MZ and DZ intrapair resemblances provides evidence for possible genetic factors, whereas the histories of MZ pairs that differ elucidate environmental factors. The possibilities inherent in twin research are extensive and go far beyond mere nose counting and the calculation of concordance rates. No human material other than identical twins provides an analog to the clones and inbred lines of lower organisms and animals. Just some of the strategic implications of twin results will be listed (cf. Allen, 1965; Jinks & Fulker, 1970).

1. If genetic differences are of no importance to the etiology or manifestation of schizophrenia, the rates in MZ and DZ co-twins of probands should be the same. Such is the case with measles (both high) and conversion hysteria (both near zero). Ideally a study of Kuru in twins might have prevented a false genetic theory.

2. If genes are important in the manifestation of schizophrenia, MZ co-twins should be affected more often than DZ co-twins. Such a result proves the importance of genotype unless it can be shown either that the MZ twins as such are especially predisposed to schizophrenia or that the environments of MZ twins are systematically more alike than those of DZ twins in respects which can be shown to be of etiological significance for schizophrenia.

3. The comparison of twin intrapair resemblance could lead to the discovery of subtypes or components of schizophrenia that may be either more under genetic control than others or under environmental control.

4. The variability of abnormal personality in the MZ co-twins of typical schizophrenics should lead to the identification of schizophrenic "equivalents" or schizotypes (Meehl, 1962). Such developments could lead to the psychometric or physiological detection of compensated or premorbid cases.

5. The comparison of MZ twins who differ totally or quantitatively in outcome permits inferences about the role of previous life experiences in affecting outcomes. If experiences *i, j, k*, etc., apply equally to both twins, *none of them* relate to the difference in outcome. Although the goal of many classical human genetic studies has been to establish the mode of inheritance of a condition, the twin method alone is not equipped to do this.

Some Criticisms of the Twin Method

How important is the sharing of a more similar environment by MZ pairs than DZ pairs to the higher MZ concordance rate? Three ways in which environmental factors might contribute have been advanced: monozygosity per se, the effects of identification, and the sharing of a more similar ecology.

1. It is said (Jackson, 1960) that MZ twins as such may be at a special risk for schizophrenia on account of such problems as confusion of identity and weak ego formation. If true, MZ twins should be over represented in samples of

NMU LIBRARY

schizophrenics compared to their frequency in the general population. Rosenthal (1960) reviewed Luxenburger's and Essen-Möller's conclusions on this issue and agreed that there was no excess of MZ twins in schizophrenic samples; other twin researchers have since obtained similar results.

2. A second way in which psychological factors might contribute to higher MZ concordance rates is through the identification of one twin with another. Theoretically, this could work in either direction; a potential patient could be stopped from becoming psychotic by strong ties to his healthy twin, or a well twin might be driven crazy by his schizophrenic twin. The identification theory need not imply an excess of MZ twins, but there are too few cases of *folie à deux,* in the sense of shared delusions, in unselected series of concordant pairs to support the theory. Female pairs are said to be specially prone to identification. We will take up this point later (pp. 31 and 307).

The high correlation for age of onset in concordant pairs may be interpreted to mean that one twin has caused the other to break down or that both responded to the same stress (Abe, 1969). Yet some pairs only become concordant after a lapse of two or three decades (Essen-Möller, 1970).

3. A third way in which the environment can contribute to higher MZ concordance rates is from the possibility that identical twins have a greater chance of both being exposed to a triggering situation than DZ pairs. For example, MZ twins may have been treated more alike within the family with a morbid parent picking on both. Kallmann (1946) reported higher rates in pairs living together for 5 years or more before the onset than in pairs separated during that period.

The above discussion is not intended to imply that psychological or ecological causes for schizophrenia are nonexistent or minimal. It does serve to strengthen the view that genetic factors account in the main for the higher MZ concordance rates to be discussed in the next two sections (cf. Rosenthal, 1962a).

Earlier Twin Studies (1928-1961)

Luxenburger

Luxenburger was the first to draw attention to the importance of collecting an uninterrupted series or total sample of psychiatric conditions in twins, and to collect and investigate one himself. The most detailed account of his own work was contained in his provisional report of 1928. He obtained his cases from the Munich Psychiatric Clinic and from the resident population of Bavarian mental hospitals, checking the twinship of all patients in the sample retrospectively by means of birth records. There was no excess of twins as such among the 16,000 patients screened. Being an old material, several pairs were lost through uncertainties of zygosity or clinical diagnosis. MZ pairs did not show as high a concordance as was found in pooled single case reports from the literature at that time (Luxenburger, 1930), nor, when concordant, did they show the photographic similarity that one might have been led to expect from previous reports. Nonetheless MZ pairs were clearly more often both affected with schizophrenia than DZ pairs. Best estimates of the proportion of concordant pairs actually found are 58% of 19 MZ pairs and none of the 13 SS DZ pairs. Luxenburger later increased and revised his sample in various ways, excluding many doubtful cases, but without reporting his findings in any detail. In one report (1934) MZ concordance appeared to have sunk to 33%, rising in a 1936 report to 51%. Like most other studies, his was based on a small number of pairs.

Rosanoff

As early as 1911, Rosanoff was interested in the heredity of mental illness. In 1934 he and his colleagues reported on a schizophrenic twin study based on twins from all over the United States and Canada. It is virtually impossible to evaluate the sample, except to say that it was obviously incomplete. Diagnostic standards appear to have differed from those used in most of the other earlier studies, and there was no personal investigation. Surprisingly, findings are comparable with those in the other earlier studies; 61% of 41 MZ pairs and 13% of 53 SS DZ pairs were concordant for schizophrenia. Concordance was higher in female pairs than in males, and the samples contained an excess of females. Rosanoff believed that discordance was often to be accounted for by brain injury suffered by the future schizophrenic twin.

Essen-Möller

Essen-Möller's study (1941b) was based on consecutive admissions to Swedish mental hospitals, and twinship was checked from the birth registers. There were only 7 certain and 4 doubtful schizophrenics from MZ pairs. At the time of the original report none of these were concordant for strictly diagnosed schizophrenia, although two have since become so (1970), and 7 of the 11 had (or now have) illnesses far from trivial with schizophrenic features. If such illnesses are counted as concordant, Essen-Möller's rates are 64% for MZ pairs and 15% for 27 SS DZ pairs. Essen-Möller considers character traits to be more under genetical control than psychosis itself.

Kallmann

The largest of the schizophrenic twin studies is that of Kallmann (1946). He sampled the resident population of New York State Hospitals in 1937 and subsequent admissions to those hospitals until 1945. Opposite-sexed pairs were certainly missed and probably many other pairs as well; but from the proportion of MZ to SS DZ pairs there is no evidence of the excess of MZ pairs which might have been expected had there been a grossly biased ascertainment in favor of concordant MZ pairs as some critics have supposed. Kallmann's sample included more females than males. He did not report concordance for the sexes separately. However, Kallmann stated that there was no difference between the rates for male and female pairs. From the way in which he presented his findings he left himself open to many criticisms as to his method of diagnosing schizophrenia in the twins. It was suggested that he may have used looser standards for schizophrenia when it came to diagnosing an MZ co-twin than one known to be DZ. From later information provided by Kallmann (Shields, Gottesman, & Slater, 1967) it can be shown that this was not so. At the start of his study 50% of 174 MZ pairs and 6% of 517 DZ pairs were concordant for a hospital diagnosis of schizophrenia. By the end of the study, and with some diagnostic revision by the author—actually more often in DZ than in MZ pairs—59% of MZ pairs and 9% of DZ pairs were concordant for what Kallmann called "definite schizophrenia." All but three of the MZ co-twins called "definite" had been hospitalized. When cases of "doubtful schizophrenia" were included the rates were increased to 69% and 10%, respectively. When the rates for "definite schizophrenia" were corrected for age they rose to the much cited figures of 86% and 15%. There was almost certainly an overcorrection here in the case of the MZ pairs.

We have seen that the uncorrected rates for "definite" schizophrenia were already substantial. Despite what most people previously believed, Kallmann did not count any pairs twice to reach these figures because both twins might have been probands. Kallmann noted that the probands in the discordant pairs tended not to have deteriorated, unlike those from concordant pairs. One of the more likely reasons for Kallmann's high rates is that his sample from the state hospitals was overweighted with severe, chronic, deteriorated cases. Among reasons for discordance put forward by Kallmann himself was the greater physical debility of the eventually schizophrenic twin. Environmental effects were also suggested by the fact that when MZ twins had been living apart for 5 years or more before the onset of schizophrenia in the first twin, concordance (in the sense of age-corrected morbid risk) was 78%, as compared with 92% in pairs where the twin had not been living apart in this way.

Slater

Contemporaneously with Kallmann, Slater (1953) in London was carrying out a twin study in the years immediately before the war, based on the resident population and subsequent admission to the London County Council mental hospitals. Enquiries as to twinship were made of relatives, retrospectively. Investigation of cases already ascertained was completed in 1947-1949. Unlike Kallmann, Slater in his 1953 report presented case histories and the information, including fingerprint data, from which the zygosity of the twins had been decided. The actual concordance rates were not very different from Kallmann's. In terms of uncorrected pairs, they were 65% of 37 MZ pairs and 14% of 58 SS DZ pairs. Concordance was higher in female than male pairs, especially among the residential part of the sample in which there was an excess of females (Gottesman & Shields, 1966b). Outcome in concordant pairs was not so similar as in Kallmann's study; indeed considerable differences were observed. Rosenthal (1959), analyzing Slater's case histories, noted that the discordant male pairs, often paranoid, were more often mildly affected, had better premorbid personalities, and less often had a family history of psychosis than concordant male pairs, which suggested etiological heterogeneity within cases diagnosed as schizophrenic. Slater himself analyzed his material in some detail to try to discover which aspects of the psychosis were most under genetic control, and which aspects of the environment or the previous personality were associated with poorer outcome. Slater did not consider all the unaffected MZ co-twins to always be schizoid. However, the characteristic of being schizoid is notoriously difficult to define and assess. Slater approached the matter by trying to identify which traits could best be called schizoid by examining in what sort of ways abnormal relatives of schizophrenics distinguished themselves from the abnormal relatives of his nonschizophrenic probands.

Inouye

At the International Congress of Psychiatry held in Montreal in 1961 Inouye (1963) reported the results of a twin study from Japan. Unlike Luxenburger, Essen-Möller, Kallmann, and Slater, he did not aim at collecting an uninterrupted series of twins; such a procedure inevitably involves the inclusion of some pairs in which both twins cannot be examined clinically by the investigator. Though he restricted himself to those pairs he could interview personally, Inouye nevertheless thought his sample was representative. Concordance rates were similar to those reported from Western countries: 60% in 55 MZ pairs and 18% in 11 SS DZ pairs. MZ concordance was higher when there was a severe chronic or recurrent schizophrenia (74% and 86%, respectively) than when the proband had a mild chronic or transient schizophrenia. In this last group concordance was only 39%. Inouye considered all his unaffected MZ co-twins to be schizoid.

Table 2.6 summarizes the main results in the earlier twin studies of schizophrenia.

Recent Twin Studies

We now come to the recent twin studies reported since 1963 which have been based on population registers, or on consecutive admissions to hospitals catering for other than severe, textbook varieties of schizophrenia, or on a population of armed forces veterans. Here the investigators have generally found it necessary to give a range of concordance rates to do justice to their data. Table 2.7 shows the ranges which they themselves report.

TABLE 2.6

Concordance Rates in the Earlier Twin Series[a]

Investigator	Date	MZ pairs[b]			SS DZ pairs[b]		
		N	C	%	N	C	%
Luxenburger	1928	19	11	58	13	0	0
Rosanoff et al.	1934	41	25	61	53	7	13
Essen-Möller	1941	11	7	64	27	4	15
Kallmann	1946	174	120	69	296	34	11
Slater	1953	37	24	65	58	8	14
Inouye	1961	55	33	60	11	2	18

[a]After Gottesman and Shields, 1966b.

[b]N represents the total numbers of pairs; C represents concordant pairs, i.e., those pairs in which both members were schizophrenic.

Population-based psychiatric twin studies were done in Finland, Norway, and Denmark. In all three countries there are registers of all twins born in certain years, and in Norway and Denmark though not in Finland there are also nationwide registers of hospitalized psychotics.

Tienari

Tienari's (1963, 1968) study in Finland was based on all male twins born between 1920 and 1929. All surviving twins were screened, and nearly all those MZ pairs living in an area comprising half the population of Finland were personally investigated by Tienari, whether they were abnormal or not. Those considered to be or to have been psychiatrically ill were included in his 1963 report. Since the study was restricted essentially to pairs in which both twins could be interviewed, there was a large drop-out of cases owing to mortality at various ages (not just infancy), emigration, and failure to trace the twins. In view of the large differences between Tienari's results and those of other workers, it is difficult to be sure that there was not some selective missing of concordant pairs. Whereas one might have expected a low concordance rate from a small sample of male twins not overweighted with chronic hospitalized cases, one would not on any theory, genetic or environmental, have expected zero concordance for schizophrenia in 16 MZ pairs and about 10% (much as in other studies) in 20 DZ pairs. Although no MZ co-twin was diagnosed as schizophrenic in 1963, 3 were called borderline and a further 9 schizoid. It might be, for genetic or environmental reasons not at present known, that genetic factors are less penetrant in this population than in others. The findings become a little less unexpected when it is seen that Tienari included as schizophrenic, syndromes occurring in probably organic psychoses. By 1968 Tienari had followed up his cases. One pair was then concordant for schizoaffective psychosis, giving a

TABLE 2.7

Concordance Rates in Recent Schizophrenic Twin Studies[a]

Investigator	MZ pairs		DZ pairs	
	N[c]	% concordant	N[c]	% concordant
Tienari (Finland)[b]	16	6-36	20	5-14
Kringlen (Norway)	55	25-38	90	4-10
Fischer *et al.* (Denmark)	21	24-48	41	10-19
Pollin *et al.* (U.S.)[b]	80	14-35	146	4-10

[a]Range of uncorrected pairwise concordance rates reported by investigators (various criteria, including borderline features, when diagnosed).
[b]Males only.
[c]N represents number of pairs.

minimum concordance rate of 6%, and 4 co-twins were possibly borderline. The maximum concordance reported by Tienari, 5/14 (36%), was obtained when 2 probands suspected of being organic were omitted and the functional borderline schizophrenics were included in the count. A further follow-up (1971) yielded results even more in line with the other recent twin studies.

Kringlen

In Norway, Kringlen (1967, 1968) matched the Central Register for Psychosis managed since 1936 by Ø. Ødegaard with the register of all twins born between 1901 and 1930. His study merits close attention for a number of reasons: The sampling was excellent and includes the largest number of psychotic twins personally investigated among the modern studies; all functional psychoses were included in addition to schizophrenia; and his two volume work includes case history, pedigree, clinical analysis, and rating scale data. Kringlen identified 55 pairs of MZ twins where one or both had been registered as schizophrenic (45) or as having a schizophreniform psychosis (10). In 14 of these pairs (or 25%) both twins were registered. After personal investigation, 17 (or 31%) of the second twins were found to be schizophrenic or schizophreniform and a further 4 to have borderline states, bringing MZ concordance, more broadly considered, up to 38%. In 90 same-sex DZ pairs concordance ranged from 4% to 10%. It was just as high in opposite-sex pairs, which is evidence against the sex identification theory of schizophrenia. Kringlen also found that concordance was as high in male pairs as in female pairs, unlike what was found in some of the earlier less well controlled studies. It is interesting that in studies based on consecutive admissions to a hospital or on a register where there is also good ascertainment of twinship, MZ concordance in the two sexes is practically equal (Shields, 1968). As had other investigators, he found that concordance rates were somewhat higher when schizophrenia in an MZ proband was rated as severe; it ranged from 60% for the most severe group to 25% for the most benign.

There is reasonably good agreement among Tienari, Kringlen, Slater, Inouye, Fischer, and ourselves in that the premorbidly more submissive, more dependent twin is the one that tends to fall ill with schizophrenia in discordant MZ pairs. Low birth weight was not found in these studies to be so important as it appears to be in the rather specially selected but intensively investigated sample of discordant MZ twins studied by Pollin *et al.* (1966) at the NIH (see p. 45). According to Kringlen's study, the future schizophrenic seemed to have been predisposed to mental illness compared with his twin in a rather nonspecific way. The same sort of factors differentiate the sick twin from the well twin in discordant neurotic pairs in other studies. There is no general tendency for submissive, mother-dependent twins to become schizophrenic.

The range of Kringlen's MZ rates, 25-38%, is about half that of Kallmann's uncorrected rates of 50-69% described above. It is this contrast that probably led Kringlen to conclude that the genetic factor in schizophrenia which he recognizes does not play as great a role as had been assumed before 1960. He goes on to claim that the genetic factor is relatively weak and nonspecific. An alternative explanation is possible. Table 2.8 indicates the degree of specificity found in pairs where both twins were psychiatrically abnormal.

For schizophrenia, not including schizophreniform psychoses, 14 co-twins (31%) had the same *type* of psychosis; remarkable, but undisclosed by this table, is the degree of *subtype* specificity. In 13 of the 14 concordant pairs both members were diagnosed by Kringlen as belonging to the same subtype: hebephrenic, 2 pairs; catatonic, 1; paranoid, 2, catatonic-paranoid, 7, and "mixed," 1. The four partly concordant co-twins of schizophrenics consisted of 1 borderline, ? schizophrenic, 2 borderline psychotics (schizoid), and 1 paranoid-depressive reactive psychosis (the partner of the latter was also depressed and

TABLE 2.8

Graded Concordance in Kringlen's (1967) Norwegian Twin Study

Status of second twin	Psychosis of first twin							
	Schizophrenia		Schizophreniform		Reactive		Manic-depressive	
	N^a	%	N^a	%	N^a	%	N^a	%
Same psychosis	14	31	3	30	2	14	2	33
Similar disorder	4	9	0	0	2	14	2	33
Other disorder	12	27	4	40	3	22	0	0
Normal	15	33	3	30	7	50	2	33
Total:	45	100	10	100	14	100	6	100

[a]N represents number of pairs.

paranoid). For the schizophreniform psychoses, each of the three concordant pairs consisted of twins who were also schizophreniform.

For the reactive and manic-depressive groups, none of the co-twins had an illness with schizophrenic features. The pattern of specificity, however, continued; two concordant reactive pairs had the same subtype, and two concordant manic-depressive pairs each had manic-depressive circular illnesses. In each of the four psychotic groups the "other disorders" suffered by the second twin consisted of a variety of neurotic and character-disordered conditions without discernible specificity, and with a prevalence not far removed from the population base rate.

Treating the 21 concordant pairs in the top row of Table 2.8 as a group, the degree of association for the form taken by the psychosis in the co-twin (20/21) and the absence of any expected crossing-over to a psychosis from a different group of the four, are very difficult to reconcile with Kringlen's claim that genetic factors are nonspecific.

Fischer et al.

Fischer, Harvald, and Hauge (1969) reported the first completed Danish study of schizophrenic twins. It was based on a national twin register started in 1954 by Harvald and Hauge of all same-sex pairs born between 1870 and 1920 and a national psychiatric inpatient register started around 1940 by T. Kemp. By cross-matching the two registers, 78 twin schizophrenic probands were ascertained from a starting sample of 6723 traced pairs of same-sex twins where both had survived up to their sixth birthday. After omitting eight probands of uncertain zygosity, they were left with 70 probands belonging to 62 pairs: 21 MZ pairs and 41 SS DZ pairs. The sample is very representative in respect of twin types and sexes of the probands, 32 males and 38 females. Only probands who met strict criteria for chronic or process schizophrenia were included although schizophreniform, paranoid and atypical psychoses in co-twins led to data analysis for quantitative concordance for other than process schizophrenia. Some personal interviewing was conducted for cases in which the histories provided ambiguous information. The age of the sample together with the fact that all discordant pairs have been followed for an average of more than 24 years eliminates the requirement for age correcting the risks in co-twins. One disadvantage of the age of the sample was that 43/124 (35%) of the twins were dead at the time the study was reported.

When the concordance analysis is restricted to a Kraepelinian-like diagnosis of schizophrenia the MZ rate was 24% (4/21) compared to one in SS DZ pairs of 10% (4/41). The latter are simple pairwise rates reported to permit comparison with other studies so reported. The authors themselves appropriately calculated the rates according to the proband method (cf. Chapter 3), which were 36% (9/25) for MZ and 18% (8/45) for DZ pairs. Fischer *et al.* then broadened the concept of concordance in a reasonable way to include co-twins who "suffered from schizophreniform, paranoid or atypical psychoses." It is somewhat surprising that five of the MZ probands had ages of onset of 40 or more and were still considered to be process schizophrenics. Grade II probandwise concordance rates were 56% and 26% [pairwise rates were 48% (10/21) MZ and 19% (8/41) DZ]; the Grade II DZ rate is the highest reported in the literature. If, finally, concordance is extended to include cases where the co-twin had a psychiatric abnormality without having been psychotic (e.g., neurotic, "nervous," or "odd"), the probandwise rates were 64% MZ and 41% DZ. Sex differences in concordance were not conspicuous with 4 of 10 MZ male pairs and 6 of 11 MZ female pairs being concordant at the Grade II level.

They found no association in their sample between severity and concordance; severity was inferred from age at first admission and outcome status (7-point scale). They suggest that the reason why they failed to detect such an association, found in other twin studies, is that heterogeneity with respect to environmental influences in their national sample may obscure genetic relationships which appear more clearly in a material with a more uniform environmental background. Fischer (1972) went on to suggest that severity judged by "final outcome" in her study may have been influenced by the "institutionalization syndrome" associated with custodial care in the days before active methods of physical treatments for schizophrenia; she cautioned against naive comparisons across twin studies with respect to severity.

Pollin et al.

The preliminary results of a twin study by Pollin *et al.* (1969) are based on the Twin Panel of the National Academy of Sciences-National Research Council (Jablon *et al.*, 1967). The Panel was obtained by the Follow-up Agency of the NAS-NRC by matching all white U.S. males who had served in the armed forces during a period spanning World War II and the Korean War (1941-1955) with a register of all twin births, 54,000, 1917-1927, in 42 states. In a white male cohort born in 1920, about 86% survived to 1942 and of these about 80% were inducted, according to Jablon *et al.* The panel consists of 15,909 pairs where *both* twins passed the military screening process. Jablon *et al.* acknowledge that the panel is neither representative nor cross-sectional, and from our point of view it is selected for the psychiatric health of both twins at the time of induction. However, it is a well-defined epidemiological sample consisting of all twins and not merely of pairs where one was schizophrenic. Zygosity was validly diagnosed by the NAS-NRC using a combination of the twins' own opinions and anthropometric and fingerprint data, but in 31% of pairs there was insufficient data for categorization. The study is so far based entirely on chart diagnoses available in the VA records or obtained in response to NAS (mailed) questionnaires when the twins were aged 38-48. A total of 364 schizophrenics were found among the 31,818 individuals, a prevalence of 1.14%. For a diagnosis of schizophrenia 11/80 MZ pairs (13.8%) were concordant, compared with 6/146 (4.1%) of DZ pairs; 9/112 (8.0%) of pairs of unknown zygosity were also concordant. From an MZ/DZ concordance ratio of 3.3, similar to that in other studies of schizophrenia and higher than the ratio for psychoneurosis (1.3), the authors concluded that there was a genetic loading in schizophrenia higher than in any of the 11 other psychiatric or medical diagnoses analyzed in their study.

A unique aspect of the VA Panel is that the prevalence of schizophrenia in this population of twins can be calculated. It is 91/9426 = 0.965% in MZ twins, and 152/1248 = 1.22% in DZ twins. Thus, MZ twins are not more prone to schizophrenia than DZ twins. These rates may be compared with a prevalence of 2.18%, calculated from the proportion of all living males aged 35-44 in New York State in 1960 who have ever had a first admission for schizophrenia (Deming, 1968). The latter population was unselected for health. From the VA Twin Panel data we can calculate that schizophrenia is 25 times as frequent in the MZ co-twin of a schizophrenic (22/91 or 24.1%) as in the population of VA MZ twins. Prevalence in the DZ co-twins of schizophrenics (12/152 or 8.0%) is 6.5 times its population prevalence. This epidemiological method of analysis adds perspective to the weight to be accorded to genetic factors in accounting for differences in the risk for schizophrenia between relatives of schizophrenics and an unrelated population of their peers. The diagnosis of schizophrenia in wartime by nonpsychiatrist physicians was probably made more loosely in this study than in the others discussed here. This would have the effect of increasing the heterogeneity of the schizophrenic sample and lowering the concordance rate for schizophrenia. Hoffer and Pollin (1970) reported that a further 21 MZ and 10 DZ co-twins received psychiatric diagnoses. These figures provide an upper range of concordance for psychiatric conditions of 40% and 11%. Table 2.7 counts only the 17 MZ and 8 DZ co-twins who had been hospitalized. Work in progress on the sample shows that a review of claims folders forces a marked upward revision in the MZ rate for schizophrenia; new pairwise rates are 26/95 MZ, 27%, and 6/125 DZ, 5% (Allen *et al.,* 1972).

IDENTICAL TWINS REARED APART AND FOSTERING STUDIES

It is quite simple to perform cross-fostering experiments in mammals other than man so as to separate the effects of cultural inheritance from parents to offspring from biological heredity for behavioral traits. Some strains of mice that differ with respect to fighting with litter mates over food retain their belligerence or submissiveness when cross-fostered and reared by mothers and with sibs of the different mouse strain. Scott and Fuller's (1965) extensive experiments with the social behavior of dog breeds leads them to conclude that dogs retain the behaviors of their breed despite cross-fostering, isolation-rearing, and adoption into a genetically different sibship later in life. Analogous data from schizophrenic individuals have been scarce or nonexistent until the last few years.

Despite the mass of evidence from twin and family studies implicating genetic factors in the etiology of schizophrenia, it is still necessary to evaluate the extent of environmental transmission. The presence of psychosis or of noxious, irrational modes of behavior in the family might account for the familial aggregation of schizophrenia. Studies in this section do not, by themselves, provide genetic evidence. They speak to the issue of the genotype × environment interaction by examining effect or lack of effect of specific environments in modifying the expression of a genotype.

A case in point is alcoholism, which twin studies suggest to be partly under genetic control (Partanen *et al.*, 1966). In an informative study conducted by Roe, Burks, and Mittelmann (1945) 36 children who had one biological alcoholic parent had been fostered out before the age of 10. When followed up at a mean age of 31, none of the children were alcoholic and only three used alcohol "regularly." In commenting on the quality of foster home care, Roe *et al.* observed a marked relationship between the present personality adjustment of children and the degree of love and affection they had received from foster parents; they also noted that some foster children had ratings of "well adjusted" despite having foster parents who were antagonistic and indifferent.

MZ Twins Reared Apart

Identical twins reared apart where one or both later become schizophrenic constitute a strategic population, but information from such pairs should not be taken as the crucial experiment proving or disproving a genetic etiology for schizophrenia. At the time of Jackson's (1960) review he could find only two bona fide pairs of MZ twins reared apart and they were both concordant. At the present time 17 pairs have been reported who were separated from each other at ages ranging up to 7 years. All but one pair (Craike & Slater, 1945) were discovered in the course of ordinary twin studies and not biased by a special search for such pairs. Eleven of the 17 pairs are concordant or 10 if we omit the special case study pair, yielding a pairwise concordance rate of 62%. The cases are set out with their sources in Table 2.9. Mitsuda's research (1967, p. 8) furnishes half of the pairs; his series of schizophrenic MZ pairs reared together showed a concordance rate of 8/11 compared to a similar rate of 5/8 for those shown in the table who were reared apart from each other.[1]

[1] Inouye (1971) has reported on 10 pairs of Japanese MZ twins with schizophrenia separated before 5 years of age; 6 pairs were reunited at various ages after 2 or more years' separation. Four co-twins were schizophrenic and three had disorders resembling schizophrenia.

TABLE 2.9

Schizophrenia in MZ Twins Reared Apart

Investigation	Age at separation	Number of pairs		Source
		Concordant	Discordant	
Kallmann, Germany, 1938	Soon after birth	1	–	Daughters of schizophrenic proband
Essen-Möller, Sweden, 1941	7 years	1	–	In consecutive series of schizophrenic twins, followed up by Kaij (1960)
Craike and Slater, U.K., 1945	9 months	1	–	Single case report
Kallmann and Roth, U.S., 1956	Not stated	1	–	Index pair in series of of preadolescent schizophrenics
Shields, U.K., 1962	Birth	1	–	From consecutive series of Maudsley Hospital twins
Tienari, Finland, 1963	3 years and 8 years	–	2	All twin births
Kringlen, Norway, 1967	3 months & 1 year, 10 months	1	1	Index case in twin series
Mitsuda, Japan, 1967	Infancy	5	3	All investigated twins with psychiatric illness

The argument that both identification and common family environment were major causes of the MZ rates being higher than the DZ rates as reported in the previous section are seriously undermined by the data in Table 2.9. The separated MZ twins did not have the same opportunity for identifying with each other and they were reared by mothers of different personalities in families with differing role structures. That their homes were not ideal or that they were both reared in lower class niches is irrelevant to the present argument. So far studies of persons reared apart from their schizophrenic biological relatives speak in the same direction as those of twins brought up apart.

Fostering Studies

Karlsson

In the course of his pedigree study of schizophrenia in Iceland, Karlsson (1966) discovered that 17 biological sibs of schizophrenics, one of whose parents was also a psychotic, had been reared apart from their parents either by relatives or nonrelatives. Of the 17, 5 became schizophrenic, a surprisingly high rate. In 8 cases in which an individual who eventually became schizophrenic had been brought up in a foster family it was possible to compare the incidence of schizophrenia in his biological and his foster sibs. Of 29 biological sibs, 6 became schizophrenic; of 28 foster sibs raised with a schizophrenic peer, none became similarly affected.

Heston

A better controlled study that included extensive personal fieldwork of a contemporary sample of children who had been born between 1915 and 1945 in Oregon State Psychiatric Hospitals to chronic schizophrenic mothers was conducted by Heston (1966; Heston & Denney, 1968). It is all the more remarkable because it was done in a country without a national registration system, without a staff and generous funding, on a highly mobile sample, and virtually single-handed. The children were separated from their mothers within the first 3 days of life and reared by nonmaternal relatives or in various institutions and adoptive homes. Of the 74 experimental subjects located in this fashion, 15 were found to have died in infancy or childhood and others had had contact later on with maternal relatives, leaving a final sample of 47 children, 30 males and 17 females, who were followed up to a mean age of 36. A control group of children taken away from their mothers who had no record of any psychiatric illness was matched individually with the experimentals for sex, type of placement, and length of time in child care institutions. The 50 control children were also followed up to a mean age of 36. Of the 47 foster-reared children of schizophrenic mothers, 5 were found to be schizophrenic themselves, an age-corrected rate of 16.6%, while none of the matched control group were psychotic. Heston's principal findings are given in Table 2.10. It can also be seen that there were considerably more abnormalities of kinds other than schizophrenia in the children of both groups. Heston believed that eight of the nine male children of schizophrenics who were diagnosed as sociopathic personalities would be more

TABLE 2.10

Psychiatric Disorders in Foster-Home Reared Children[a,b]

	Mother schizophrenic (N = 47)	Controls (N = 50)
Mean age	35.8	36.3
Schizophrenia	5	–
Mental deficiency, IQ < 70	4	–
Sociopathic personality	9	2
Neurotic personality disorder	13	7

[a]After Heston, 1966.
[b]N represents number of children.

accurately described by the "old-fashioned" term, *schizoid psychopath.* Evaluations were accomplished blindly by two psychiatrists plus the author with a fourth called in to resolve differences. Psychosocial disability was objectively assessed with the Menninger Mental Health Sickness Rating Scale (MHSRS). More than two-thirds of the 97 subjects were interviewed and tested with the Minnesota Multiphasic Personality Inventory (MMPI).

For comparison, the results of a follow-up study of children of schizophrenic mothers born before the onset of the mothers' illnesses and not removed from them systematically are of interest. Reisby (1967) did a census study of all female schizophrenics in three large rural mental hospitals in Sjaelland, Denmark on a given day in 1963, finding 428, of whom 136 were mothers. Some 73% of the mothers were of the paranoid subtype; 51% had been hospitalized for more than 20 years, 76% more than 5 years. Definite information was obtained on 322 children born to 132 women; 33 died as children and 11 as adults leaving 278 alive at the time of follow-up. The national psychiatric register, police records, and death registers were searched for the names of the schizophrenics' children. Since the search method did not permit determination of neurosis or character deviations, the figures for these categories must be considered to be underestimates, but 13 were neurotic and 9 were sociopathic. Based on 201.8 risk lives, the risk for schizophrenia (3.5%) plus schizophreniform psychoses (6.9%) came to 10.4%.

In addition to the national Danish register for ascertaining physical and mental disorders, two others, a register of adoptees and the *Folkeregister* (current addresses of the populace) have made possible joint U.S. NIMH-Danish fostering studies similar to Heston's.

Rosenthal et al.

Rosenthal *et al.* (1968) searched the mental disorder register for the names of the biological parents of all 5500 children put up for nonfamilial adoption in the greater Copenhagen area from 1924 to 1947. 69 parents, 23 fathers and 46 mothers, were found to be psychotic. At the time of their initial report the adopted children of 34 schizophrenics and 5 manic-depressives had been examined and assessed over a 2-day period filled with interviews, psychological and psychophysiological procedures.

At a mean age of 31, 8 (21%) of the 39 adopted children of psychotics were blindly diagnosed as schizophrenic or borderline schizophrenic, compared with only 2% of 47 carefully matched control adoptees whose biological parents were normal. Our detailed analysis of the findings of Rosenthal *et al.* are given in Table 2.11.

"It is worth pointing out that if such findings were reported in the usual study, where the diagnostician knows which subjects are Index Cases and which are Controls, they would be mighty suspect indeed. Here there is no question of the

TABLE 2.11

Diagnosis of Adoptees by Diagnosis of Parent[a]

| | 39 Index cases | | | | 47 Controls | |
| | One parent chronic schizophrenia or ?chronic schizophrenia | | One parent other diagnosis[b] | | Both parents normal | |
Diagnosis of offspring	No. diagnosed	Cumulative total	No. diagnosed	Cumulative total	No. diagnosed	Cumulative total
Schizophrenia	3	3 (11%)	–	– (0%)	–	– (0%)
Borderline schizophrenia	3	6 (22%)	2	2 (17%)	1	1 (2%)
?Borderline schizophrenia	2	8 (29%)	–	2 (17%)	2	3 (6%)
Schizoid or paranoid tendencies	1	9 (33%)	2	4 (33%)	4	7 (15%)
None of above	18	27	8	12	40	47
Total offspring so far diagnosed	27	(100%)	12	(100%)	47	(100%)

[a]Data of Rosenthal et al., 1968.

[b]Acute schizophrenia (5), borderline schizophrenia (2) and manic-depressive or ? manic-depressive (5).

diagnostician being influenced by such information [Rosenthal et al., p. 387] ."
If they had had to rely solely on psychiatric hospitalization to define their experimental adoptees as affected, they would only have found three of whom one was schizophrenic.

Table 2.11 shows that diagnoses of schizophrenia, without reservations, are found only in the adopted children of chronic or ? chronic schizophrenics. The addition of ? borderline and schizoid or paranoid diagnoses in this study seems to dilute the strength of the discrimination between the index and control adoptees. The observed 11% rate of schizophrenia in the index adoptees is the same as Heston's uncorrected rate. Differences between the studies are the apparent absence of mental deficiency, sociopathy, and neurosis in the Danish offspring and the less malignant nature of their schizophrenic illnesses. All the Oregon schizophrenic offspring were hospitalized. The reasons could be associated with the psychotic parents in the two studies. These parents differed in diagnosis, sex, and their clinical state when the children studied were born. Four possible reasons may be mentioned. (1) Unlike the Danish psychotic parents, those from Oregon were by selection all chronic schizophrenics. Their eventually schizophrenic children may therefore have perhaps been more predisposed genetically to have a typical process schizophrenia. (2) Although a third of the Danish psychotic parents were men, all the Oregon schizophrenic parents were women. The children of the Oregon mothers were all born in mental hospitals when the mothers were already chronically ill, whereas in the Danish sample the children were born on average 11 years *before* the first hospital admission for psychosis in a parent. Harmful prenatal and perinatal factors operating in the Oregon mothers might therefore account for some of the differences between the Oregon and Danish children. (3) Some of the sociopathy and mental deficiency in the Oregon children might also be associated with the gene contribution of their fathers, i.e., the kind of men who had impregnated chronic schizophrenic women. (4) A uniformly lower social class was probable for the Oregon parents (L.L. Heston, personal communication).

The Danish registry system has also lent itself to another important and elegant study by the NIMH-Copenhagen group, differing in strategy from the one described above but starting from the same list of adoptees.

Kety et al.

Kety *et al.* (1968) identified *adoptees who became schizophrenic* and then using only documents, determined the status of their biological and adoptive parents, siblings, and half-siblings. They wanted to find out whether schizophrenia in the adoptees could be accounted for by schizophrenia in the adoptive families to whom they were unrelated biologically, or by schizophrenia in their biological families with whom they had had no contact.

From a starting sample of 5483 children adopted by non-relatives, 507 were identified from the national psychiatric register as having been hospitalized; a three-judge consensus diagnosis of schizophrenia based on hospital abstracts was made for 33 who then became the index cases. Controls were carefully matched and selected from among the remaining adoptees whose names were not on the psychiatric register. Through the *Folkeregister* 463 biological (306) and adoptive (157) relatives of the 66 adoptees were found, and their names were compared with those in psychiatric registers and hospital records as well as police records. The information obtained was summarized blindly for each relative and submitted to a four-judge panel. Then a consensus was reached regarding membership in the schizophrenic spectrum, defined as chronic, acute, or borderline (pseudoneurotic) schizophrenia, definite or uncertain, together with inadequate personality (a diluted form of borderline schizophrenia); other diagnoses were also made and recorded. Our analysis of the findings in their preliminary report are given in Table 2.12 and include the authors' "extended spectrum," defined in the table.

Of the 150 biological relatives of adopted schizophrenics, 13 had a spectrum disorder (8.7%), of whom 7 were definite schizophrenics, 4 were uncertain, and 2 were inadequate personalities. Only 2 adoptive relatives of the index group had

TABLE 2.12

Psychiatric Abnormalities in Biological and Adoptive Relatives[a]

	Biological relatives				Adoptive relatives			
	Parents	Sibs	Half-sibs	Total	Parents	Sibs	Half-sibs	Total
Index group: Relatives of 33 schizophrenic adoptees								
Schizophrenia spectrum disorders[b]	3	1	9	**13**	2	0	0	**2**
Extended spectrum disorders[c]	9	0	3	12	2	1	0	3
Other psychiatric diagnoses	3	0	2	5	6	1	0	7
No psychiatric diagnosis	48	1	71	120	53	6	3	62
Total relatives identified	63	2	85	**150**	63	8	3	**74**
Control group: Relatives of 33 nonschizophrenic adoptees								
Schizophrenia spectrum disorders[b]	2	1	0	**3**	0	3	0	**3**
Extended spectrum disorders[c]	2	0	2	4	3	0	0	3
Other psychiatric diagnoses	6	1	3	10	2	0	0	2
No psychiatric diagnosis	53	3	83	139	58	14	0	75
Total relatives identified	63	5	88	**156**	63	17	0	**83**

[a]Data of Kety et al., 1968.

[b]Schizophrenia (chronic, acute, or borderline), certain or uncertain; inadequate personality.

[c]Character disorders, psychopathy (including imprisonment and delinquency); suicide; failure to reach consensus diagnosis, 2 out of 4 judges favoring certain schizophrenia.

spectrum disorder: 1 uncertain schizophrenia and 1 inadequate personality. The 13/150 was significantly greater (p = .0072) than the amount of disorder found in the biological relatives of controls, 3/156 (1.9%), or in the adoptive relatives of index cases, 2/74 (2.7%). The 8.7% rate is unexpectedly high for a group consisting of parents selected for fitness, and half-sibs who are only second-degree relatives, many of them still young.

The variety of psychiatric conditions among the relatives of the index group is reminiscent of some of Heston's findings, again suggesting that the genetic potential for schizophrenia may manifest itself in ways other than in overt schizophrenia. If the 19 index and 20 control cases separated from their biological families in the first month of life are looked at separately, a new finding, embarrassing to environmental theories, emerges. Of the biological relatives of the index cases 10% had spectrum disorder compared to none in the biological relatives of controls; furthermore, more than half of the schizophrenia spectrum disorder was observed in their *paternally* related half-sibs with whom they shared only genes and not an *in utero* environment. The authors concluded that the prevalence of schizophrenia in *naturally reared* relatives of probands is a manifestation of genetically transmitted factors, probably polygenic in nature.

Three other studies of related interest deserve mention. Wender, Rosenthal, and Kety (1968) examined ten pairs of adoptive parents of schizophrenics in the United States. They were found to have less psychopathology than biological parents who reared schizophrenics, though possibly more than adoptive parents of normals. Higgins (1966) studied 50 children of schizophrenic parents, comparing fostered and nonfostered for maladjustment; both groups had the same amount. The children were too young to be at risk for schizophrenia. The study was part of Mednick and Schulsinger's (1968) continuing prospective study of schizophrenics' offspring. Fischer (1971) studied the offspring of 7 normal, reproducing MZ co-twins of schizophrenics, i.e., the children of individuals with a schizophrenic genotype (see p. 309 for exceptions) but not a schizophrenic phenotype. In their 25 children were 3 cases of schizophrenia, an age-corrected morbid risk of 12.3%. This risk did not differ from the one of 9.4% found for the children of MZ schizophrenic twins in the same study (Fischer *et al.,* 1969) and was close to the value reported for the children of schizophrenics generally (e.g., 10.4% in Reisby's 1967 Danish survey).

PSYCHODYNAMIC FAMILY STUDIES

Discordant MZ Twins

In the section on twin studies (p. 25) we stated that the histories of discordant MZ twins could show which environmental factors could be *excluded* as causal. Their histories can also suggest hypotheses as to which life experiences *may* be causal. Findings here require replication, and any hypotheses generated need to be tested both on nontwin schizophrenics and on singletons unselected for schizophrenia but who were subjected to the supposedly causal life experience. Only if confirmed, can the environmental factor in question be accepted as etiologically related to schizophrenia, independently of the genetic predisposition. The use of identical twins permits the control of genetic factors, not their exclusion.

In the classical twin studies, most of the genetic investigators made a point of looking at discordant MZ pairs. However, their conclusions may have been limited by their interests, samples and methods. For the most part they were dealing with large, often old, epidemiological material, where accurate information about neonatal factors, interpersonal relationships, IQ scores and so on, now considered highly relevant, would have been difficult if not impossible to elicit.

Since 1963 Pollin and his colleagues (Pollin *et al.*, 1966; Pollin & Stabenau, 1968) at the Twin and Sibling Unit of the NIMH have been carrying out intensive studies on MZ twins discordant for schizophrenia. As the investigators themselves say, their studies do not provide answers about the relevance of genetic factors. But if genetic factors are relevant, studies of discordant pairs are still extremely valuable if they can tell us why only one twin falls ill in pairs where both have the genetic predisposition for the disorder; they tell us about genetic-environmental interaction, and such data are important for rational prophylaxis.

The Twin and Sibling Unit's sample was obtained by means of repeated nationwide mailings describing the study to American psychiatrists. To be accepted, both twins and both parents had to be willing and able to come to Bethesda for extensive, multidisciplinary inpatient psychiatric evaluation over a 2-3 week period, including over 100 hours of interviews per family. On completion of clinical evaluation a social worker visited the home community, living for a brief time with the family. By 1968, 15 pairs had been studied where the Index case was schizophrenic and the certainly MZ control twin had been free from psychosis. The time during which the twins had been discordant for psychotic symptoms ranged from 2½-21 years, mean 8.8 years. Ten of the Index twins were "hardcore" schizophrenics, as judged by a panel of five psychiatrists, and five are described as probable or borderline schizophrenics. Investigation has been completed and follow-up information obtained for 14 pairs; surprisingly 3 of them were from sets of triplets.[2] Analysis has focused so far on birth weight, the "overall picture of the family and specifically the life course of the twins," neurological status, protein-bound iodine (PBI) and the lactate/pyruvate ratio. In

---

[2] Further sampling problems are intimated by the average Verbal IQ of 126.5 in their male co-twins (Mosher *et al.*, 1971).

at least 12 of the 15 pairs the Index twin was the lighter at birth by 10 gm or more, with a median difference of 255 gm (less than 10 ounces).

From the consistency of his observations Pollin infers the following sequence of events in the development of schizophrenia in the Index twins:

i. Intrauterine differences lead to birth weight differences and an associated relative physiological incompetence (e.g., lower PBI).

ii. This induces in the parents attitudes and relationship patterns which accentuate dependency and lead to incompatible role expectations in the less-favored twin.

iii. Via slowly increasing ego weakness these developments lead to excessive anxiety and lowered competence, reflected for instance in the relatively poorer school performance of the twin who had the lower birth weight, thereby increasing his predisposition to schizophrenia and possibly to other psycho-pathologies as well.

iv. In the Index schizophrenics some of the current physical and bio-chemical observations may be secondary to the schizophrenia or to medication.

Pollin (Pollin & Stabenau, 1968) notes that, by the very fact of both parents and both twins being available, his sample may have "provided ... a skewed and somewhat atypical population, so that the group we are seeing represents only one of several ... pathways to schizophrenia [p. 328]." He believes that his findings and his hypothesized sequence of events can be explained on models which lay most stress either on the supposed neurological deficiency or on the contradictory and incompatible role expectations induced by the parents. But whichever model is preferred, and whether genetic factors are involved or not, he believes that their work clarifies "why one of a pair of identical twins would encounter increasingly greater amounts of stress than his co-twin when con-fronted by customary life situations [p. 331]."

In trying to assess how far Pollin's conclusions about birth weight hold for his own sample we must remember that the intrapair differences in birth weight were often very small. In some pairs they represent only a trivial proportion of the total weight of the Index twin. Of the 14 pairs for which data are recorded only 4 differed by 480 gm (17 ounces) or more and 4 differed by less than 6 ounces. In only 2 pairs, of which one was from triplets, the control was 20% or more heavier than the Index twin, in 6 pairs 15% or more, and in 9 pairs 10% or more. It seems doubtful whether neurological deficiencies sufficient to account for much of the later psychopathology can be directly attributed to these differences in birth weight which are not unusual for normal MZ twins. If this is so, more emphasis should be placed on psychodynamic rather than neurological inferences.

Furthermore, the reported birth weight findings may be accounted for by bias in sampling. Neither the association between lower birth weight and schizophrenia in discordant MZ pairs, nor the consistency in the association between birth weight differences and other variables such as family relationships and intellectual competence are as marked in other more representative series of twins. However, Pollin and Stabenau (1968) pooled data from single case reports and from other sources including concordant pairs differing in severity. They calculated that in these pooled data the schizophrenic (or, as they should more correctly have added, the more severe schizophrenic in the concordant pairs) was the lighter at birth in 29 cases and the heavier in 18. They saw further confirmation of their findings with regard to birth weight and school perform- ance in the work of Lane and Albee (1965, 1966) on schizophrenics and their siblings. The latter have since (1968) retracted their conclusions that preschizo- phrenics could be distinguished from their same-sex siblings by early intellectual deficits. The findings in the Pollin study which are most consistent with those of other studies are that the future schizophrenic was the more neurotic child, the more submissive, the more sensitive and the less outgoing of the pair.

Family Studies

It is impossible to do more than enumerate in a very condensed manner some of the more promising models proposed by investigators focusing on inter- personal factors in the etiology and pathogenesis of schizophrenia. Our main reasons for trying to broach these topics, however inadequately in the space allowed, is to ask: To what extent do the ideas from psychodynamic family studies successfully compete with genetic ideas in explaining similar raw obser- vations? To what extent may specific interpersonal factors be implicated as either etiological in their own right or as importantly involved in the (psycho)- pathogenesis of schizophrenia in individuals with a particular genetic predis- position? How far can the insights of psychotherapists help to explain discord- ance in MZ twins?

Clarifying summaries or criticisms of the facts and principles that support an important role for interpersonal factors in the understanding of schizophrenia may be found in Alanen (1966), Brown (1967), Clausen (1968), Kreitman and Smythies (1968), Lidz, Fleck, and Cornelison (1965) and Rosenthal (1963). A sociologist expressed his perspective as follows:

> One cannot *prove* the importance of interpersonal factors without very intensive longitudinal data; all that one can hope to do is to demonstrate a high degree of plausibility for a particular formulation ... (which) requires very intensive, well conceptualized study of the patient in his social matrix; the delineation of linkages between life experiences and the thoughts and behaviors that manifest schizo- phrenia, *and* the systematic elimination of alternative explanations that might account for the coexistence of particular life experiences and schizophrenic sympto- matology [Clausen, 1968, p. 252].

Two explanations of the connection between abnormal behavior in a parent and schizophrenia in his offspring need to be considered as alternatives to psychogenic causation of the latter by the former. The abnormal parental behavior may be a manifestation of genes shared with the child, or it may have been the result of the stress of coping with a schizophrenic (cf. Bell, 1968).

It is obvious that somehow interpersonal, dynamic and motivational variables play a part in at least the phenomenology of schizophrenia; however, the validity and reliability of the constructs proffered by proponents in their explanatory attempts leave much to be desired and frustrate those who wish to put the ideas to test. Furthermore, the dynamic investigators report surprisingly high rates of abnormality in the relatives of schizophrenics; are the terms of reference and the patients the same as those of phenomenological "descriptive" psychiatrists who had made genetically oriented studies of schizophrenic families? At Yale, Lidz *et al.* (1965) found that 9 of 15 intensively studied therapy inpatients had at least one parent who was "more or less schizophrenic"; of 15 siblings of the same sex as the patient, 3 were schizophrenic, 5 borderline, and only one adequately adjusted. In Finland, Alanen (1966) in a detailed study of the families of 30 schizophrenics and 30 neurotics found that 4 of the parents of schizophrenics had overt schizophrenia, 12 were borderline schizophrenics or had other functional psychoses, and a further 26 had "disturbances more severe than psychoneurosis," mostly grave schizoid or other character neuroses; only one parent in 60 was normal. 70% might thus fall into the schizophrenic spectrum described on page 42. Of 49 siblings, 4 were schizophrenic, 6 borderline and 10 schizoid or paranoid, while only 9 of 28 same-sexed sibs were normal. Waring and Ricks (1965) studied, besides other groups, the families of 20 child guidance patients who on follow-up were found to be chronic schizophrenics. 55% of the mothers and 40% of the fathers were assessed as psychotic or borderline, with 20% of all parents showing "clear evidence of psychosis." These figures are at variance with the earlier work, according to which only about 5% of the parents had schizophrenia or probable schizophrenia. Manfred Bleuler (1930), a sensitive clinician, studying the families of schizophrenics in a private American mental hospital some 30 years before Lidz, found only four certain and three doubtful schizophrenics in 200 parents of 100 cases. Kallmann (1946), who has been criticized for the looseness of his diagnosis, found among 1191 parents of schizophrenics only 108 who were schizophrenic and 413 who were schizoid, i.e., 44% at most in a schizophrenic spectrum.

The reason for the difference may partly be a wider concept of schizophrenia used by the psychodynamic workers than by the European genetical workers, and partly selection in favor of abnormal families. There remains the possibility that it may also reflect the intensity and mode of investigation. One does not know how much abnormality would be elicited if "normal" families could be sufficiently motivated to submit themselves to lengthy psychoanalytically oriented investigation. There is the possibility that the interpersonal investigative process itself, with its transference relationships and opportunities for experimenter influence (R. Rosenthal, 1966), may elicit primary processes in subjects particularly susceptible in an ambiguous situation (cf. Sargant, 1957). To paraphrase Jackson (1960, p. 66), there may be no better nidus for the transmission of irrationality than the analyst's couch. It could be argued further that the dynamic theorist has the parent of the schizophrenic in a double bind: The mother, if married to a man who might be characterized as inadequate, is to blame whatever she does; if she criticizes her husband she denigrates him, bringing about a skew marital interaction, whereas if she covers up she is "masking" and thereby transmitting irrationality.

Some of the differences in rates of abnormality may be resolved by experimental methods. As supposedly pathological types of family structure or categories of defective communication become quantifiable the testing of hypotheses using suitable controls becomes possible. The double bind construct was found to be unreliable (Schuham, 1967). Wynne (1972) and Singer and Wynne (1966) and Mishler and Waxler (1968) have devoted considerable attention to devising methods of measuring types of thought disorder and family interaction.

One way to abstract a common theme in the current "life experience" and dynamic views is to look at them as implicating a failure of learning which then is causal (Rosenthal, 1963). This would include failure of socialization, failure of cognitive integration, and failure to contain anxiety (Mednick, 1958). In early cognitive development the child, exposed to irrational, thought-disordered parents, has his own ability to think straight compromised; such deficits are supposed to directly engender thought processes characteristic of schizophrenia later in life. A related formulation addresses the problems of ego strength and individuation among children reared by "schizophrenogenic" mothers, the latter tying them symbiotically or catching them in binds between conflicting demands or conflicting parents. From the earlier attention devoted to the schizophrenogenic mother, psychodynamic theorists, finding fathers of schizophrenics to be disturbed as frequently as mothers, later implicated the faulty structure of the nuclear family (e.g., Lidz' "marital schism" and "marital skew").

As already noted, the establishment of a connection between some kind of family interaction or faulty learning process and schizophrenia in *ex post facto* studies does not help to evaluate the role of genetics. The earlier psychodynamic theories (e.g., Jackson, 1960) strongly denied any significant contribution to schizophrenia from genetic variation. They regarded their models as competing with, rather than complementing genetic hypotheses. In 1965 Lidz, Fleck, and Cornelison believed that "a genetic . . . predisposition to schizophrenia increasingly rests upon preconception and tradition, while evidence points to environmental and social factors [p. 430]." Since then, Lidz has somewhat modified his position (Rosenthal 1968, p. 415). Alanen (1966) accepts the possibility of genetic factors in schizophrenia, but like most others of a psychodynamic persuasion believes them to be non-specific. For instance, "a child who is innately more passive than his siblings . . . owing to this very fact, is selected by the mother as the child onto whom she most definitely directs her own dependent needs" (Alanen, 1968, p. 209).

Like Lidz, Alanen regards schizophrenia as the end of a continuum of severity of psychopathology and not as qualitatively different from other psychiatric disorders. But whereas Laing (1964) looks upon schizophrenia as a way of life, Lidz has described it as a "deficiency disease." Insofar as there may be specific genetic factors for schizophrenia, these tend to be conceptualized as related to tendencies to abnormal perception or thinking. Singer and Wynne (1965) see possible interaction between a genetic predisposition to amorphous, fragmented, or other types of thought disorder and learning phenomena which exacerbate these predispositions:

> We assume that the individual's biologic capacities for focusing attention and for perceiving, thinking, and communicating gradually are shaped and modified by interchange with the environment during development. This viewpoint is *epigenetic*: The interchanges or transactions at each developmental phase build upon the outcome of earlier transactions. This means that constitutional and experiential influences recombine in each developmental phase to create new biologic and behavioral potentialities which then help determine the next phase. If the transactions at any given developmental phase are distorted or omitted, all the subsequent developmental phases will be altered because they build upon a different substrate [p. 208].

It is now almost trite to say that some form of interaction between genes and environment is the proper frame of reference in which to discuss the familial distribution of schizophrenia. A plausible and widely held hypothesis of the kind of interaction in question is similar to that suggested by Singer and Wynne above, namely augmentation of vulnerability through the combination of an unusual genotype in the future schizophrenic with unusual patterns of behavior on the part of the very parents who supplied him with his genes. However, the recent adoption and fostering studies (pp. 37-43) throw a new light on this interpretation of the family data. No evidence was found for the operation of the one particular environmental influence that had generally been assumed to be highly important—the presence of schizophrenia or related disorders in the rearing family. Focus must thus be shifted to individual vulnerability reacting with nonspecific and culturally common if not ubiquitous factors. What for most people are valuable learning experiences—"an inoculation" to stimulate mechanisms for coping adequately with later life situations—may be idiosyncratically apperceived by the schizophrenia-prone individual as disruptive. Chance events, influencing the inner state of an individual or the external stresses he encounters at a particular time or in a particular family setting can, without any great stretch of the imagination, be seen as capable of carrying only one of a pair of MZ twins over a threshold of liability. It is not difficult to conceive of this as one kind of reason for discordance in MZ twins and one which interpersonal studies may help us to understand. All four Genain quadruplets (Rosenthal, 1963) had the same genes, all were grossly exposed to the same abnormal family environment, and all became schizophrenic; but a combination of circumstances, among which the different patterns of interpersonal relationships within the family were prominent, was associated with schizophrenias ranging from mild with recovery in Myra to severe and deteriorating in Hester. Other interpersonal constellations might have resulted in some of the Genain children being schizophrenic and others not, or in still different combinations of severity.

Yet, specific schizophrenia-provoking stresses cannot be ruled out. Some kinds of environment might produce a typical schizophrenic syndrome in persons with no more of a genetic predisposition than the average member of the population. Hypotheses as to the nature of such stresses require testing by selection of such posited high-risk groups from the general population without reference to a schizophrenic relative.

Following Kety *et al.* (1968), we have stated that schizophrenia in the rearing family has not been proved to be an environmental stress which augments the genetic predisposition to schizophrenia. But other specific kinds of stress may possibly do so. It is arguable that adoption, by provoking problems of identity, might be such a stress, given the diathesis. Since the majority of schizophrenics do not have a similarly affected close relative, such a hypothesis would demand a slight increase in the rate of schizophrenia among adopted children not known to have a schizophrenic biological relative. To the best of our knowledge such an increase has not been reported (Heston, Denney, & Pauly, 1966; Bratfos, Eitinger, & Tau, 1968; Rosenthal, 1971).

ONWARD

Our selective review of the literature in this chapter, together with our section in Chapter 1, zeitgeist in the 1960's, sets the stage for the detailed reporting of our own twin study of schizophrenia, first published in preliminary form in 1966 (Gottesman & Shields, 1966a). Its distinctive features, once again, are its sampling method—a series consisting of all same-sexed twins from 16 years' consecutive admissions to outpatient and short-stay inpatient psychiatric facilities—its use of a consensus blind assessment by a panel of diagnosticians, and the employment of psychological testing and tape-recorded interviews. This combination of methods has not been used before in a study of psychiatric abnormality in MZ and DZ twins.

3
THE MAUDSLEY
SCHIZOPHRENIC TWIN SERIES

SAMPLING

Our sample of schizophrenics was selected from the Maudsley Twin Register maintained by the Psychiatric Genetics Research Unit (directed by Dr. Eliot Slater), situated at the Maudsley Hospital. The register sets out to be a complete, uninterrupted series of twins in Luxenburger's (1928) sense. In other words, it aims at being, so far as possible, a total sample of patients consecutively admitted to any of the facilities of the hospital, (1) who were born one of a multiple birth; and (2) whose co-twin was of the same sex as the patient and survived until the age of 5 in the case of Children's Department patients, or age 15 in the case of adults. Insofar as the sample is complete and the twins referred to the Maudsley are representative of Maudsley patients as a whole, the cases on the register can be regarded as a representative sample of a psychiatric population with defined characteristics, i.e., those of the Maudsley Hospital during the 16 year period of ascertainment, 1948-1964. These characteristics are described more fully in the triennial reports of the hospital, compiled first by C. P. Blacker and then by E. H. Hare.

The Maudsley Psychiatric Population

The Maudsley and Bethlem Royal Joint Hospital is a postgraduate teaching hospital, exclusively for psychiatry, consisting of a large outpatient department that is situated in the south London district of Camberwell, and a short-stay inpatient department (about 400 beds, half of them at Bethlem; median stay about 2 months). Toward the end of the period there were some 4000 discharges per year, about 370 of them from the Children's Department. A little over a third of adult patients are admitted as inpatients, some of them directly, i.e., without first attending as an outpatient. The inpatient beds are intended primarily for cases with good prognoses. Admission is restricted to voluntary or informal patients. All treatment is free as part of the National Health Service. Compared with the United States there is little private practice of psychiatry in the United Kingdom, and there is no separate treatment for war veterans.

As a teaching hospital of repute, the hospital may receive patients from any part of the country or even from abroad. There is no local catchment area from which all psychiatric patients must be admitted to the Maudsley. However, the hospital sets out largely to provide a local service. Almost two-thirds of the referrals of outpatients come from general practitioners and of these 70% are from general practitioners in south London, the majority from Camberwell and neighboring districts. The second-largest source of outpatients is self-referrals (18%), also mostly local. Inpatients, when not admitted via the OPD, most frequently come from St. Francis Hospital nearby, an emergency psychiatric unit, previously known as an Observation Ward (OW). Though not part of the Joint Hospital the OW is staffed by psychiatrists from the Maudsley. During the greater part of the period, virtually all the acute psychiatric cases from south London would be admitted to St. Francis Hospital. It should be noted that members of the Genetics Unit have had no influence on which patients are admitted to the Maudsley Hospital, and the Unit has no inpatient beds.

There are more female than male inpatients, but slightly more male than female outpatients. About 40% of patients are under age 35 on admission. There is a slight excess of patients from the highest, and (among males) the lowest social classes. However, the social class distribution of patients is not grossly different from that of London as a whole. Most patients are diagnosed as falling within the broad categories of neurosis or "character disorder, etc." About one-half of inpatients and one-quarter of outpatients are diagnosed as psychotic, and they are about twice as often called depressive as schizophrenic.

Though this characterization does not allow for changing trends during or after the period, it probably gives a sufficient description of the psychiatric population sampled by the Maudsley Twin Register at the time of this inquiry. No claim can be made that the Maudsley Hospital clientele is representative of all psychiatric cases everywhere. Unlike psychiatric populations sampled previously in twin studies, the Maudsley population from which our twins were obtained consisted largely of outpatients. It is different again from the private clinic cases on which most psychodynamic family studies have been based.

Ascertainment of Twinship in Maudsley Patients

Before World War II, Slater arranged for every new patient attending the Maudsley Hospital to be asked whether he was born a twin. After the war, Professor (now Sir Aubrey) Lewis and Dr. C. P. Blacker agreed to the reinclusion of the question about twinship in the case front data to be completed for every patient. This has usually taken the following form: Was patient born a twin? (Yes, No, NK). If Yes, was twin of same sex, of opposite sex, or sex NK? In the case of outpatients the questions about twinship were usually asked by a psychiatric social worker who would normally see every patient or an accompanying relative at intake. Direct inpatient admissions were asked about twinship in the course of history-taking. Starting with adult outpatients in April 1948 and extending to include all patients in January 1949, the answers to the above questions were recorded on punched cards when the patient was discharged. Throughout the period the transcription officer responsible for punching the cards on which the triennial reports were based was Mrs. Marjorie Perkins who passed on to the Genetics Unit the names of all twins. Many cases were of course also notified to the Unit before the patient was discharged, and checks were also made of the punched cards. For the purposes of the schizophrenic twin series reported here we looked at all cases on the Register up to the end of March 1964. During the first part of the period the proportion of patients of whom it was not known whether they were a twin was 2%. From 1958 onward the proportion was 7%. One of the main reasons for the increase is probably the expansion of the Emergency Clinic at which a social worker did not routinely seek information about twinship. Co-twins of patients missed through being recorded as "twinship not known" are likely to have been few. Probably most of them died in infancy, since composition of family was nearly always inquired about, and had there been a surviving twin the fact is likely to have been noted. We believe we can have missed few patients with twins who survived. Cases are most likely to have been missed in the earliest months before notification got under way, and in the later months when we will have missed any twins currently attending for the first time and not yet notified to the Unit. Any bias introduced through lack of completeness is unlikely to have been serious. Between November 1950 and June 1953 registration of same-sex twins, similar to that at the Maudsley, was also made from consecutive admissions to Belmont Hospital, a Neurosis Center on the outskirts of London. Twenty cases with a neurotic or "character disorder, etc.," diagnosis were incorporated into the Maudsley Twin Register.

It may be asked whether more accurate ascertainment of the twinship of patients might not have been obtained by searching for their birth certificates.

Birth certificates in England and Wales, however, do not indicate twinship. Even if they did, the method, though it might have provided a useful check on whether many twins had been missed, would not under existing conditions have been a preferred substitute for direct routine personal inquiry made of patient or relative at the time of admission. As with all methods the use of birth records has its own difficulties. The parish of birth is not regularly recorded for all patients. Some patients are not born in the country or part of the country for which birth records may be accessible. In the absence of an existing and immediately available national register of twins of all ages or of an efficient record-linkage system, search for twinship would have to be made retrospectively, i.e., after discharge of the patient from hospital. The problem would then arise, in the case of patients identified as twins, of following the co-twin up to his death—many are likely to have died young—or to his present place of residence by means of official records. The latter are not available except in Scandinavia. In trying to ascertain this information, retrospective postal inquiries of patients or relatives are likely to be less efficient and reliable than inquiries made when the patient is first seen at the hospital.

Number of Twins on the Register

At the closing of the intake for the schizophrenic twin study, there were 479 cases with various psychiatric diagnoses on the register. It is difficult to say exactly how many individual patients at the Maudsley, Bethlem Royal, and Belmont Hospitals were screened in order to obtain this number of twins on the psychiatric register, since hospital statistics are mostly given in the form of spells of treatment, and a patient may have many such spells. We estimate that it must have been about 45,000 over the period in question. Many times this number of persons from the *general* population would have to be screened in order to obtain 479 psychiatrically disordered twins.

Before describing how the schizophrenics were selected from the register we must first try to assess whether the twins were representative of the psychiatric sample from which they were drawn.

Representativeness of Twins on the Register

The frequency of twins on the register was analyzed by sex and zygosity for the first time by Shields in 1954. This analysis related to 10,296 Maudsley children and adult outpatients, 1949-1951, and Belmont inpatients, 1951-1953. The percentage of all patients who were twins was 2.18%. The expected percentage of MZ pairs, calculated from the proportion of opposite-sex pairs according to Weinberg's method, was 26.5%. These figures were in good agreement with what would be expected in the United Kingdom population—2.39%

of persons come from multiple births; 28.6% of all twins can be expected to be MZ and 35.7% same-sex DZ (Waterhouse, 1950).

A more detailed analysis was undertaken for Maudsley patients for the years 1952-1957 (Shields, unpublished). The proportion of twins of any kind among all 26,948 discharges was 2.2%, of whom 26.3% would be expected to be MZ. The proportion of females was insignificantly higher among the twins (55.2%) than among Maudsley patients as a whole (51.3%). The diagnostic distribution of the twins was similar to that of all Maudsley patients when grouped under the broad categories of psychosis, neurosis and "character disorder, etc." These findings still held when patients whose co-twins had died in infancy (31% of all twin patients) were omitted.

In the three triennial reports now available for the years 1958-1966, the proportion of patients reported as being twins of any kind has remained constant at close to 2.1% and no significant deviation from the expected proportion of MZ and DZ pairs has been found. This last finding has also been confirmed in clinical reports from the Unit on twins belonging to different diagnostic groups of same-sex twins. The series of twins diagnosed as neurosis or personality disorder up to 1958 (Shields & Slater, 1966) comprised 80 MZ and 112 same-sex DZ pairs. Heston and Shields (1968) reported on 5 MZ and 7 SS DZ pairs in which the proband was homosexual.

Selection of Schizophrenics and Final Sample Size

The 479 cases on the register at March 31, 1964, included 67 from the Maudsley Children's Department, where schizophrenia is not used as a diagnosis, and 20 patients from Belmont Hospital. Of the remaining 392 Maudsley adult patients, 47 (12%) had at some time been diagnosed at the hospital as suffering from schizophrenia and coded *300* under the then current International Classification of Diseases. Over the same period it was calculated that about 11% of all Maudsley patients were so diagnosed. Hence there is no significant excess of schizophrenic twins such as might be expected, either if there were preferential referral of such twins to the hospital as being cases of special research or teaching interest, or if twins per se were especially susceptible to schizophrenia. These 47 schizophrenic twins consisted of 22 males and 25 females, which is close to the equal proportion of males and females diagnosed as schizophrenic at the hospital. Anticipating our later findings, 17 came from MZ pairs and 26 from same-sex DZ pairs, zygosity being uncertain in 4. Again this is in good agreement with expectations. At the level of Maudsley diagnosis there is no excess of MZ pairs, which again supports the view that our material was not selected because of an alleged special interest in MZ schizophrenics.

The group of schizophrenic probands finally included by us in the present study differs from the above 47 Maudsley diagnosed cases by reason of both exclusions and additions.

Three female schizophrenics were excluded because they were black immigrants from Ghana, Jamaica, or Barbados, and their inclusion might have introduced extraneous heterogeneity. Two were apparently DZ and discordant. In the third pair there is no information on zygosity or concordance. Three Caucasian schizophrenics were omitted because, despite all efforts, there was insufficient information about their co-twins. In two males, there was no information about zygosity or concordance. In the case of a female proband— who had a late-onset paranoid illness with good recovery—information about zygosity was inconsistent. The normal co-twin had died at 35, 13 years before the first signs of schizophrenia in the proband. The family (studied extensively by us) was of interest because of the occurrence of late onset illnesses with schizophrenic and affective features in the mother and in one if not two of the brothers, as well as in the proband.

The series was increased by the inclusion of 21 cases discovered to have been diagnosed as schizophrenic after their discharge from the Maudsley. By the start of the investigation, so far as possible all the neurotic and personality disordered twins and all children on the register had already been followed up by the Unit, some of them for long periods. So, too, had many of the depressives (da Fonseca, 1963) and twins falling into other diagnostic categories. Thirteen were known to have been diagnosed as schizophrenic: 10 had been so diagnosed in mental hospitals; 3 others[1] had been subsequently diagnosed as schizophrenic only by Slater in the course of the follow-up of the neurotic and personality disordered twins. He made these diagnoses without knowing the zygosity or concordance of the pair. All other cases on the register where schizophrenia had been suspected, where schizoid personality had been diagnosed, where paranoid features had been mentioned in the formulation or where there were other reasons to think that the subject might have become schizophrenic, were followed up by us in 1963-1964. Eight were found to have been diagnosed as schizophrenic in a mental hospital.

Of the 21 cases added, one (MZ 13) was originally from the Maudsley Children's Department and one (MZ 15) a former Belmont case. Their previous Maudsley (or Belmont) diagnoses were as follows: 3, paranoid state; 1, sensitive personality; 1, schizoid psychopathy; 3, depression with paranoid features; 1, schizophrenic-like puerperal psychosis; 1, alcoholic hallucinosis; 1, "diagnosis uncertain," schizophrenia suspected; 2, psychopath or delinquent, ? borderline

---

[1]These are DZ 2 and 17, where chronic paranoid delusions remained in the absence of depression, and DZ 25 where the proband had previously been diagnosed as schizophrenic in a mental hospital, though the Maudsley diagnosis was schizoid psychopath.

psychotic; 3, affective disorders; 3, anxiety states; 1, hysteria; and 1, behavior disorder.

Since every case on the register was not followed up to the start of the present investigation (and sometimes follow-up information was incomplete), it cannot be claimed that our sample includes every Maudsley twin who had by that time been diagnosed as schizophrenic, but it appears to be a reasonably complete and quite unselected representation of such cases.

By increasing our numbers in this way we have substantially raised our sample size. In support of our having included cases not diagnosed as schizophrenic at the Maudsley, it can be pointed out that some of our Maudsley-diagnosed probands (MZ 21*A*; DZ 9, 15, 23), like those later added, were not coded as *300* when they *first* entered the register, but only when they relapsed and were sent back to the Maudsley. Other registered cases who relapsed were sent to different hospitals and there diagnosed as schizophrenics. There is no good reason for including the former and excluding the latter. Our intention was to obtain a broadly diagnosed group of schizophrenics and then to submit their histories to outside judges for final independent evaluation with the expectation that some might be rejected. Although a Maudsley diagnosis of schizophrenia is generally regarded as being made according to stricter and more conservative criteria than the same diagnosis made in many other British hospitals, it is worth noting that not all our Maudsley cases coded *300* were firmly diagnosed as schizophrenic. In DZ 20, for instance, the diagnosis was uncertain. In MZ 20 and DZ 21 the original diagnosis of schizophrenia was changed on later readmission to the Maudsley. Sometimes a nonschizophrenic diagnosis would be offically coded (e.g., MZ 17), although many psychiatrists present at diagnostic case conferences at the time regarded the patient as already schizophrenic. It is known (Clark & Mallett, 1963) that about a third of Maudsley-diagnosed depressives who relapse are later called schizophrenic. Although we shall not normally treat the cases diagnosed at the Maudsley separately from the others, sufficient information will be given to enable the reader to do so if he wishes.[2]

With the exclusion of 6 and the addition of 21 cases, the final number of schizophrenic probands included in the main investigation therefore becomes 62, of whom only 2 were eventually not accepted as schizophrenic by the consensus of our 6 judges. The main series of 62 consists of 31 males and 31 females: 28 probands had MZ co-twins and 34 had DZ co-twins. Since four MZ pairs and one DZ pair were represented by two twins, both meeting the defined criteria and so entering the series in their own right, the 62 probands come from 57 pairs. Most of our analyses will be in terms of raw pairs. Pairwise distribution of the sample by zygosity and sex may be seen in Table 3.1.

[2] The 21 added probands (*A*-twins, except where stated otherwise) were: MZ 3, 9, 10, 12, 13, 14, 15, 17, 18, 19 *B*, 24; DZ 1, 2, 3, 10, 12, 17, 24, 25, 27, 33.

TABLE 3.1

Sex and Zygosity of Final Sample

Pairs	MZ	SS DZ	Total
Female	11	16	27
Male	13	17	30
Total:	24	33	57

A-Twins and B-Twins, Probands and Co-twins

The pairs have been numbered separately by zygosity, MZ 1-24 and DZ 1-33. The numbering is unrelated to the order in which they were registered. In each pair the first twin to become a proband in the investigation has been designated as the A-twin. A-twins must therefore have become Maudsley patients since 1948 and have been diagnosed as schizophrenic by the time of our follow-up. B-twins, if schizophrenic, may also be probands. The five cases in which this was so are MZ 7B, 19B, 21B, 22B, and DZ 7B. The term co-twin normally refers to the twin partners of probands, so that all B-twins and 5 A-twins (MZ 7A, etc.) are co-twins. In a sense the investigation is concerned principally with the co-twins of our defined schizophrenic probands.

Source of Referral

We shall now examine in more detail the 57 investigated pairs with respect to how they first came to our notice. Objection could be raised to the arguments we have used above to support the representativeness of the twins on the register on grounds that MZ schizophrenics form but a small proportion of all such twins. A few unrepresentative pairs might have been directly admitted as inpatients, for instance by psychiatrists at the Observation Ward who might consider them to be of particular interest. If biased in the direction of concordance, their inclusion might artificially inflate the MZ concordance rate for schizophrenia, but their presence might not be detected in the larger numbers of the register as a whole.

The best test is to compare the original sources of the MZ and DZ pairs in our present sample. If there were any bias of the kind just mentioned, one would expect to see it in a relative scarcity of MZ pairs referred by the GP's and an excess of direct admissions to the wards. The sources of A-twin referrals may be seen in Table 3.2.

It can be seen that practically the same proportions of MZ and DZ pairs came via the general practitioner. Somewhat more MZ pairs first came to notice as the

TABLE 3.2

Concordance by Source of Pairs

Source of referral of A-twin	MZ				DZ			
	N	Percent of total	B-twin Schizophrenic N	%	N	Percent of total	B-twin Schizophrenic N	%
OP via GP	9	38%	3	33	13	39%	1 }	11
OP other[a]	2	8%	1	50	6	18%	1 }	
IP direct, OW	7	29%	3	43	5	15%	0 }	7
IP direct, other[b]	6	25%	3	50	9	27%	1 }	
Total:	24	100%	10	42	33	100%	3	9

[a]Example: probation officer, self-referred.
[b]Example: other hospital or OPD.

result of the direct admission of one of the twins from St. Francis Hospital Observation Ward (OW), but the difference is not significant and concordance in these pairs is virtually the same as that in the other MZ pairs.

Any previous hospitalization of a co-twin elsewhere is more likely to have been known to persons arranging a direct admission than to those referring a twin to Outpatients. Such a history was certainly known in the case of three MZ pairs referred from the OW and in two MZ pairs and 1 DZ pair of which the A-twin was admitted directly from another hospital or outpatient clinic. Nevertheless, concordance for schizophrenia is fairly evenly spread throughout the various referral groups. It is only slightly and insignificantly more frequent in pairs ascertained through direct admission of the first twin (7/27) than in the remainder that came via the Outpatient Department (6/30). Within the context, concordance means that both twins have been hospitalized and diagnosed as schizophrenic (Grade I concordance of our previous report, Gottesman & Shields, 1966a).

It is possible that there may have been a tendency to admit a patient to the wards from another department of the hospital because he was a twin, since a rather high proportion of probands have been admitted at some time. It is difficult to find a sample of nontwins comparable with respect to age and diagnosis that would enable one to discover whether this was really the case. However, Maudsley inpatient admission occurs as often in DZ probands (71% of 34) as in MZ probands (70% of 27, i.e., excluding MZ 15 from Belmont).

There was no bias in favor of both twins becoming probands when both were diagnosed as schizophrenic at about the same time (Gottesman, 1968b, p. 40). It is possible that persons with a history of previous psychiatric admission elsewhere might have been more readily admitted to the Maudsley if they were twins than if they were not. Nevertheless, when these probands were omitted from the analysis MZ concordance was very close to what it was in MZ pairs taken as a whole. Our inclusion of pairs where the proband was diagnosed as schizophrenic at another hospital during the follow-up period led to our obtaining a slightly lower concordance than we would have otherwise (see p. 305).

We may therefore conclude that we have been unable to detect any serious source of bias in our sample.

Summary

The Maudsley Twin Register is based on 16 years of consecutive admissions from the year 1948 to the large outpatient departments and short-stay inpatient departments of the Maudsley Hospital, a psychiatric teaching hospital in south London. Though not restricted to patients from a defined catchment area, the hospital largely provides a local service. The main characteristics of Maudsley patients have been briefly outlined.

On admission every patient is routinely asked whether he was born a twin. The twins on the register are representative of Maudsley patients as a whole. There is no excess of twins or of MZ relative to DZ pairs. By March 1964 there were 479 patients with same-sex twins on the register where the co-twin had survived to an age of risk. This was based on a psychiatric population of around 45,000.

Forty-seven twins had been diagnosed as schizophrenic at the Maudsley, close to the proportion of all adult patients so diagnosed. Of these, 3 non-Caucasians and 3 twins for whom there was inadequate information about the co-twin were omitted from our sample. Twenty-one twins who had been diagnosed as schizophrenic since discharge from the Maudsley were added. The resulting 62 probands come from 57 pairs and are evenly divided by sex. So far as we can judge there was no serious bias caused by the direct admission of concordant MZ pairs.

ZYGOSITY DETERMINATION–METHODS AND RESULTS

The twin method assumes that a pair can be classified as monozygotic or dizygotic with little possibility of error. Critics have pointed out that errors in assignment of zygosity could easily arise if assessment were casually made on insufficient investigation and not checked by objective means. The effect of such errors, if random, would be to obscure any real difference between MZ and DZ pairs. But Rosenthal (1962a) has suggested that a bias might be introduced in the direction of wrongly calling concordant pairs MZ and discordant pairs DZ. Nevertheless, it has been pointed out (Slater, 1957; Shields, 1962) that decisions

TABLE 3.3

Extent of Zygosity Investigation

	Number of pairs		
	MZ	DZ	Total
Blood-grouped and fingerprinted	10	6	16
Blood-grouped	10	5	15
Fingerprinted	1	4	5
Twins seen by same observer	3	8	11
History	–	10	10
Total:	24	33	57

about zygosity are likely to be among the most reliable of those that have to be made in clinical psychiatric and psychological twin studies. It is by far easier to agree that a pair is DZ than to agree over ratings of the twins on traits such as aggressiveness and schizoid tendency, or whether one twin was favored by the mother. Recent work (Cederlöf *et al.*, 1961; Jablon *et al.*, 1967) has shown that even the answers to a postal questionnaire as to whether the twins are "as alike as two peas" agree well with the results of serology. Good agreement is also obtained when an experienced observer compares the twins with respect to their resemblance in facial features, hair and eye color, and other characteristics highly determined by heredity (Essen-Möller, 1941a).

Blood groups and fingerprints were the objective characters used by us as aids in determining zygosity, in addition to resemblance in appearance. Table 3.3 shows the extent of investigation.

Blood-grouping was normally carried out for us by Dr. R. R. Race and Dr. Ruth Sanger of the MRC Blood Group Research Unit, London. The twins were generally grouped on the ABO, MNS, P, Rhesus, Lutheran, Kell, Lewis, and Duffy systems, and lately on Kidd and Xg.[3]

In 31 pairs (54%) both twins were blood-grouped; this total includes 1 pair (MZ 3) grouped elsewhere in 1948 on the ABO and Rh systems only. In 11 pairs the twins differed in at least one blood group and were thus proved to be DZ.[4] In the remaining 20 pairs, identical in all groups tested, the probability of their nevertheless being DZ was calculated, using the method of Smith and Penrose

[3] Antisera generally used: Anti-A, -A_1, -B, anti-A+B, anti-M, -N, -S, anti-P_1, anti-C, -C^w, -c, -D, -E, -e, anti-Lu^a, anti-K, anti-Le^a, -Le^b, anti-Fy^a; and sometimes anti-s, anti-Fy^b, anti-Jk^a, -Jk^b, anti-Xg^a.

[4] Our previous statement that 20 DZ pairs had been blood-grouped (Gottesman & Shields, 1966b) was incorrect.

(1955). The method takes into account the relative frequency of MZ and DZ twins at birth and the knowledge that the twins were of the same sex. Without considering resemblance in fingerprints or general appearance, the probability of being DZ ranged in the 19 fully grouped pairs from .106 down to .026 (median .060), depending on the groups with respect to which the twins were alike. The blood-group findings in all cases confirmed those reached independently on the basis of resemblance in appearance.

In 21 pairs (37%) both twins were fingerprinted. From their ridge counts the probability of being DZ was calculated, using the method of Slater (1963). Unlike the case with a blood group difference, a difference in ridge count does not establish dizygosity beyond doubt. In pairs whose zygosity is known on other grounds, the distributions of MZ and DZ ridge-count differences overlap. In our present blood-grouped and fingerprinted material of 16 pairs there was agreement as to predicted zygosity in all cases but one; only in DZ 25, differing in blood groups, did the fingerprints slightly favor monozygosity. In the five pairs fingerprinted but not blood-grouped, fingerprint findings and clinical impression agreed.

It would therefore appear justified to combine the odds in favor of dizygosity derived from the ridge counts (Slater method) with the blood-group odds in pairs alike serologically. With certain reservations (Gottesman, 1963a), this can be done in the manner recommended by Smith and Penrose for the combination of information from serology and quantitative traits. The chance of dizygosity was thereby considerably reduced to a range between .033 and .003 (median .014) in the 9 fully grouped MZ pairs in question. It was .080 in the case of MZ 3, fingerprinted but blood-grouped on only two systems.

These calculations, it must be repeated, make no allowance for the degree of likeness in appearance. This was attested by the same observer in 18 out of the 20 blood-grouped pairs classified as MZ. In the other two the history strongly suggested monozygosity.

It will be seen from Table 3.3 that a higher proportion of MZ pairs, 21/24 (83%), were blood-grouped or fingerprinted than DZ pairs, 15/33 (45%). A greater effort was made to test twins in these respects when they were thought to be MZ than to test those who were obviously DZ by reason of clear differences in other genetically determined aspects of appearance.

The three MZ pairs who were not blood-grouped or fingerprinted were each examined by the same person (IG). None of the three co-twins had been hospitalized for a schizophrenic illness. According to some critics, the investigator, knowing the hypothesis under test, *might* have been biased to call such discordant pairs DZ. However, they were as similar to their twins as "peas in a pod" or "two drops of water" which, as noted above, are good indicators of monozygosity according to Cederlöf *et al.* (1961) and Kringlen (1967), and there was no temptation to consider them to be DZ twins who were more similar in their appearance than most. There were 18 DZ pairs of twins who were not tested in either of these respects. In 10 pairs reliance was placed on a history of difference in appearance without direct personal confirmation, but with substantiating statements of specific differences in build, coloring or features, sufficient to exclude monozygosity.

We may conclude that if there should be misclassification of zygosity in the sample it is very unlikely that it has been of a frequency such as to affect the findings significantly.

METHOD AND EXTENT OF INVESTIGATION AND FOLLOW-UP STATUS

Investigation before 1963

In the course of previous work done in the Genetics Unit of the Institute of Psychiatry, mostly from 1950 onward, 40 of the 57 pairs in the present sample—21 MZ and 19 DZ pairs—had already been investigated to some degree. Most of the blood-grouping and fingerprinting eventually carried out was done prior to 1963 (in 33 out of 36 pairs). In some of these pairs the work of the Unit consisted merely of organizing the zygosity determination, while the staff of the hospital took what history was required. In five pairs the previous work amounted to no more than asking whether the twins were so alike as to be mistaken and whether the co-twin had ever been psychiatrically treated—and of paving the way for possible further inquiries later. However, fairly detailed premorbid and follow-up histories were obtained from twins and relatives in 27 pairs, or about half the total. While most of this previous work was done by JS, some pairs were seen by other workers in the Unit only. One pair (DZ 23) was also in Slater's 1936-1939 series (published 1953) and another (MZ 1), in which both twins were dead by the time of our follow-up, had been studied by M. A. Brown and others on behalf of A. J. Lewis as long ago as 1931-1938.

The information collected over the years on the above pairs proved its usefulness when, on the arrival of one of us (IG) in the Unit in 1963, we came to gather our present schizophrenic twin series and to investigate it more systematically. Inevitably, a few twins, some of them fortunately seen previously, had now died or were no longer available for other reasons. It may be worth noting that in at least six pairs the main information concerning premorbid history was obtained prospectively, in the sense that it was obtained before schizophrenia had even been suspected in either twin. These pairs include DZ 12, originally seen when the twins were 14 as part of a study of normal schoolchildren (Shields, 1954).

Investigation 1963-1965

For the purposes of statistical analysis the investigation of a pair was considered to have been closed at the time of its investigation by us between November

1963 and July 1964. Exceptions were made in the case of 3 pairs, previously not personally investigated by him, whom IG was able to see later in Ireland in July 1965. Any subsequent information, obtained anecdotally or recorded in the Maudsley Hospital notes by July 1968, has, however, been noted in the Case Histories as a matter of interest.

Investigation in 1963-1965 included the completion and elaboration of such background and later information as was already recorded in the extensive Maudsley Hospital charts or Genetics Unit files. Further information was obtained directly from the twins or their relatives and, either then or later, from the charts of many hospitals all over the world where probands, co-twins, or relatives had been treated. Altogether, relatives other than co-twins were seen at some time in 47 pairs, and information by mail was obtained in a further three pairs. In one pair as many as five relatives were seen when a group of the sibs unexpectedly confronted an author to "protect" the stability of a co-twin. We interviewed about 75 relatives ourselves and a further 23 were contacted by others working in or on behalf of the Unit.

In an attempt to gather some of the background data in a standard form, a structured, multiple-choice history schedule (Briggs, 1959) was completed by a relative, or occasionally by more than one relative, in 34 pairs, 14 MZ and 20 DZ. The main informants were parents (17), sibs (8), spouses (3), and, with no relative available, one of the twins (6).

Number of Twins Seen

Whenever possible, interviews with each proband and co-twin were tape-recorded with a portable, battery-powered Uher 4000-S recorder. The twins were usually interviewed for an hour or more, but only some 30 minutes was recorded while following a semistructured interview schedule (see Appendix A). The latter was intended to elicit the subtle and obvious signs of schizophrenia and schizotypy (Rado, 1962; Meehl, 1962, 1964), attitudes toward self, parents, and twin, and personality strengths and weaknesses. As will be seen from Table 3.4 recordings were made with 75 twins of whom 60 were paired; 4 twins were too paranoid to allow recording while (regrettably) handwritten notes had to suffice for one of the most colorful probands (DZ 6A) when the equipment failed. The table also shows that 16 twins, though not tape-recorded, were seen personally by one of us, bringing the total thus seen to 80% of all subjects. A further 9 twins (8%) were seen by other Unit members only—Dr. Slater, Miss Brown, or Miss Harrison—or were interviewed on behalf of the Unit by their GP or by a psychologist in Ireland or a PSW in Scotland. (Included here is a pair where the co-twin was interviewed by the Maudsley registrar.) Only 14 twins, all from DZ pairs, were never seen for the purposes of twin investigation. The

TABLE 3.4

Extent of Personal Investigation

Twins	MZ		DZ		Total
	A	*B*	*A*	*B*	
Interviewed and tape-recorded	21	16	19	19	75
Not taped, but seen by author	1	4	6	5	16
Seen in or on behalf of Unit	2	4	1	2	9
Not seen for twin investigation	0	0	7	7	14
Total:	24	24	33	33	114

reasons why they were not seen were as follows: 4 were dead, including DZ 4*A* and *B*; both members of 2 pairs were abroad (the proband had come to the United Kingdom from Canada or New Zealand, fallen ill, and later returned home); and 6 were unwilling to be seen or trace was lost of them (2 direct refusals, 3 indirect refusals via a relative, and 1 who may have returned to Ireland). Of the 14, however, 3 completed an MMPI by post. In terms of pairs, both twins were seen in 49 pairs (86%), 1 twin only in 4 and neither twin in 4 pairs.

Twins and Family Members Tested Psychometrically

Whenever possible a personality inventory and a test of concept formation were administered to the twins and members of their families. Findings will be presented in more detail in Chapter 7. Here we need only state that 79 twins completed the MMPI. It was completed by both twins in 15 MZ and 18 DZ pairs and by 1 twin only in 5 MZ and 8 DZ pairs. When only 1 twin was tested it was the *A*-twin in 5 cases and the *B*-twin in 8. Those who completed the MMPI were usually also given the Object Sorting Test and *vice versa*. The OST was completed by 72 twins. In only 3 MZ and 6 DZ pairs in the total of 57 pairs could neither twin be tested on either test.

In 20 families—5 families of MZ twins and 15 of DZ twins—at least 1 relative was tested. Thirty-one relatives completed the MMPI—21 parents or adoptive parents, 8 sibs or half-sibs, 1 daughter, and a wife. Seven fathers and 11 mothers, including 2 adoptive mothers, completed the OST. Despite these efforts, it should not be surprising that in a cohort where cooperation is not guaranteed by self-selection, the number of intact families where both parents and both twins completed both tests amounted to only 5—all discordant fraternal twins. Canons of the twin method demand that available DZ pairs be investigated as thoroughly as MZ pairs and we have attempted to abide by them.

Field work in 1963-1965 required chasing the probands and their relatives throughout England and Ireland; we had no subjects in Wales and necessary contacts in Scotland were made by others at our request. Our families occupied all walks of life from London slums to large country estates. The twins were interviewed wherever they happened to be living at the time, sometimes in hospital, most often at home. A few came into the Unit just to avoid a home visit by the investigator. One co-twin was interviewed on his way to the airport after 48 hours emergency leave from a duty station in the Middle East. One proband was seen in concert with his even more psychotic nontwin brother in their recently purchased home. Sometimes a whole day would be spent with a family while both twins and both parents would be interviewed, and tested; sometimes it was difficult to avoid psychotherapeutic interaction with people begging for advice or reassurance. Hospitality of the probands or their families could not be refused in the more remote areas without destroying rapport. Almost all relatives contacted wanted to do what they could in the hope that either their schizophrenic—or future schizophrenics—might be helped by the results of the research.

Status on Follow-up

In order to give the reader a general picture of our subjects at the close of the investigation we shall conclude this chapter with an outline of their status as regards age, length of observation, mortality, morbidity, and current hospitalization. When the time period over which they were selected is borne in mind, a fair degree of variability can be expected. It will be important to bear this variability in mind when evaluating the findings.

The age distribution of the 114 subjects when last observed was as follows: 37 were under age 30; 29 were in their thirties; 30 in their forties; 18 were aged 50 or over. Mean age was 37.5 years, S.D., 11.3.

The last proband to enter the study was DZ 22A who came to the Day Hospital at the age of 20 after a year's illness; with her co-twin she was followed up to the age of 22. All co-twins have been observed at least 3 years from the time of onset of the proband's illness. The longest interval between estimated onset of schizophrenia and the last information on a nonschizophrenic co-twin is 28 years (MZ 16). The youngest twins were MZ 13A and B who were aged 19; they first came to notice in the Children's Department at 15. The oldest twin was DZ 17B, aged 65; the proband in this pair died in 1964 during the period of investigation. The earliest-born pair—MZ 1, born 1893—came to notice when the proband, an ambulatory schizophrenic, attended the Maudsley Outpatient Department at 61; by 1963 both twins in this pair were dead. All told, 9 twins from 7 pairs have died, all of them schizophrenics, 2 of them by suicide. A tenth schizophrenic, MZ 21*B,* is known to have died (?suicide) since the close of the investigation. These observations suggest that studies restricted to living twin samples may underestimate concordance rates for schizophrenia.

By 1965, 8 subjects in addition to the probands were known to have been hospitalized and diagnosed as schizophrenic. Six others had been hospitalized for conditions other than schizophrenia. Of the 70 schizophrenics, 9 were dead, as already mentioned; 13 were in psychiatric hospitals, of whom 9 had been there continuously for the past 2 years; 13 had been discharged from hospital within the previous 6 months; 5 had been out of hospital longer than 6 months but were unemployed; the remaining 30 were working or running a home, but at least 16 of these had gross symptoms. All patients had been treated with the prevailing active therapies for schizophrenia at the time they were hospitalized, including insulin coma, electroshock, social rehabilitation, and, since 1954, phenothiazines. It was largely a group of schizophrenics in a state of remission that was examined. Of the six hospitalized nonschizophrenics, one was recently discharged and the remainder working. How far the remaining co-twins were handicapped will be clear later: three were treated as psychiatric outpatients.

Concordance for Hospitalization and Diagnosis of Schizophrenia

Having indicated how many subjects had been hospitalized and how many diagnosed as schizophrenic, it will be appropriate to present at this stage, without comment, the basic MZ and DZ concordance rates as we found them at the conclusion of the investigation. These are the rates already reported by us as provisional findings (Gottesman & Shields, 1966a,b). In 10 out of 24 MZ pairs (42%) both twins had been hospitalized and at some time diagnosed as schizophrenic by a psychiatrist. The corresponding DZ rate was 3 out of 33 (9%). In our previous report we termed this Grade I concordance. Including psychiatrically hospitalized twins, never diagnosed as schizophrenic, concordance (Grade II) is increased to 13/24 (54%) in MZ and 6/33 (18%) in DZ pairs.

Method of Reporting Concordances

There are different kinds and degrees of concordance and different ways in which concordance rates for twins may be reported, each of which is appropriate for certain purposes. These are discussed in detail in Appendix B. In general we shall report direct pairwise rates as in the above example (N = 24 MZ pairs). Such rates have the merit of simplicity and provide a conservative estimate of the proband method rate. When it appears more appropriate, we shall talk about the co-twins of probands (N = 28 MZ cases). Rates will be uncorrected for age.

First we shall use the direct pairwise comparison of MZ and DZ concordances to establish the case for genetic factors in the etiology of schizophrenia. The ultimate step will be to estimate the lifetime risk for developing schizophrenia for an individual who is the MZ or DZ twin of a schizophrenic. It is such risk figures which should be compared with the base rate for an individual in the general population. For this purpose the proband method is theoretically correct.

It is misleading to think that there is only one meaningful and correct concordance rate for a given set of data. Since we shall be applying different standards of diagnosis to our twins as well as reporting different kinds of rate, there will be a very large number of concordances which it will be of interest to compare both within and between zygosities.

Summary

When the final sample was gathered and followed up in extensive field work in 1963-1965, considerable information collected over the years was already available. At follow-up the schizophrenics were mostly in a state of remission. Interviews with 75 of the 114 twins were tape recorded; altogether 100 twins were seen in connection with the project. The Minnesota Multiphasic Personality Inventory was completed by 79 twins and 31 relatives. A concept formation test was also used. In 42% of 24 MZ pairs and 9% of 33 DZ pairs both members had been diagnosed as schizophrenic and had been hospitalized. Some co-twins had other psychiatric abnormalities. When results are reported in the form of concordance rates, they will generally be uncorrected ones.

CHAPTER SUMMARY

The first section of the chapter described and discussed the sampling of the present schizophrenic twin series and was summarized on page 60. The following section dealt with the zygosity of the twins. Blood-grouping and fingerprinting were among the methods used to decide whether a pair was monozygotic or dizygotic.

The method and extent of our clinical investigation of schizophrenics on the Maudsley Twin Register 1948-1964, their co-twins and their families, was the subject of the final section summarized above. This section also indicated the status of the twins at the time when they were interviewed and tested. The question of final psychiatric diagnosis will be dealt with in Chapter 5. Meanwhile, we present the case histories of all 24 MZ and 33 DZ pairs.

4
CASE HISTORIES

The distilled case histories of the twins summarize the raw data of the study. Rather than relegate them to an appendix, we shall present these basic findings here, together with a final clinical evaluation of each pair, before proceeding to an analysis of the data. Invasion of privacy of our families is avoided by various disguises of facts not crucial to our study wherever we feel that the histories could be too readily associated with our families by potential readers in the twins' communities.

Explanatory Notes

Abbreviations used in the case histories are listed in Table 4.1.

TABLE 4.1

Abbreviations Used in Case Histories

abn abnormal	MH Maudsley Hospital
adm(s) admitted or admission(s)	*Mo* mother
CGC Child Guidance Clinic	MZ monozygotic
DH day hospital	NK not known
diag. diagnosis	OE on examination
DZ dizygotic	OP outpatient
F female	OPD outpatient department
Fa father	OW observation ward
FU follow-up	PSW psychiatric social worker
GH general hospital	PTC phenylthiocarbamide
hosp (d) hospital (hospitalized)	RC Roman Catholic
IP inpatient	Sc schizophrenia
M male	VP voluntary patient

The first section of the case histories provides a condensed synopsis of each pair by tabulating data on age, hospitalization, follow-up, and diagnosis. To explain how we have set out this information, and that on zygosity evidence and family structure, we take as an example the headings of MZ 14 (Table 4.2), a pair of male twins, born in 1943 and classified as monozygotic. The age at which a twin was first admitted as a psychiatric inpatient is designated by "1st hosp" and age at last birthday. It was age 16 in the case of Twin *A* (the index twin) in MZ 14. (For the definition of *A*- and *B*-twins see p. 58. If both twins were probands, the fact is noted in a footnote.) (Sc 16) indicates the age (also 16) when a diagnosis of schizophrenia was first made; this is followed by the number of spells, three in this instance, of psychiatric hospitalization and the cumulative number of weeks spent in hospitals—91 weeks. This excludes information about his hospitalization obtained after the study closed. Attendances at Day Hospitals were regarded as hospitalization, as were also periods in correctional institutions when relevant. Among other conventions adopted was to regard hospital admissions separated by less than 2 weeks as one continuous spell. Follow-up status, FU 21, denotes the age of the twin in 1963–1964 (or at death or when last heard of), whether he was in hospital, a recent hospital discharge, i.e., discharged within 6 months, a home invalid, or working. Presence of gross symptoms in working subjects is noted. *B* below had no history of hospitalization for psychiatric illness. Follow-up for *B* took place at age 20.

The consensus diagnosis of six judges is listed—Schizophrenia or ?schizophrenia, other diagnosis, or normal—with the number of judges agreeing with the consensus in parentheses. When the consensus diagnosis is Sc or ?Sc the number in parentheses refers to the number of judges favoring either schizophrenia or ?schizophrenia (see Chapter 5). The diagnoses made by the two senior judges, Eliot Slater (ES) and Paul E. Meehl (PEM), are given separately. Mean ratings of

TABLE 4.2

Sample Case History Heading: **MZ 14, male, born 1943**

A: 1st hosp. 16 (Sc 16), 3 spells, 91 weeks; FU 21, recent hosp. discharge.
B: ___ ; FU 20.

	A	*B*
Consensus Diag.	Sc (6)	Other (4)
Slater Diag.	A typical schizoaffective	Inadequate personality
Meehl Diag.	True Sc	Pseudoneurotic Sc
Global rating	5.5	3.25
MMPI	278*4601'35-9/	18*7342"690-5/

Zygosity Evidence. Blood, seen.
Background. Social class 4. *Fa* 59, *Mo* 41; 1*A* and *B*, maternal ½-sibs; *M*8, *M*6.

two judges, Meehl and Loren Mosher (LM), on a 7-point scale of severity of global psychopathology (see page 229) are entered for each twin (here $A = 5.5$ and $B = 3.25$) under Global rating. An additional psychodiagnostic device, the Minnesota Multiphasic Personality Inventory (MMPI), is reported in coded form for each twin tested and, after *Background*, for each other relative tested. Details of the MMPI will be given in Chapter 7. The code presents the scores obtained on the ten clinical scales in rank order, highest to lowest, and reveals the extent of departure from a normal control group with a standard score of 50 and a standard deviation of 10 for each scale. Twin MZ 14*A* obtained scores higher than 90 (*) on scales 2(Depression), 7(Psychasthenia), and 8(Schizophrenia), none between 80 and 89 ("), above 70 (') the "danger point" on scales 4(Psychopathic deviate), 6(Paranoia), 0(Social introversion), and 1(Hypochondriasis), above 60 (-) on scales 3(Hysteria) and 5(Masculine–Feminine interest), and above 50 (/) on scale 9(Hypomania). Scales 4 and 6 were underlined to show they were within one standard score point of one another.

Zygosity Evidence. This indicates whether both twins were blood-grouped, fingerprinted, or seen by the same observer, or whether evidence was based only on the history; in this instance, blood-grouping indicated a monozygotic twin pair, and both twins were observed as to their physical appearance by the same interviewer.

Background. Social class 4 above, the first item included under *Background*, refers to the family in which the twins were brought up, and was assessed according to the United Kingdom Registrar General's system: (1) professional; (2) managerial; (3) clerical or skilled laborer; (4) semiskilled laborer; (5) unskilled laborer. The biological father of MZ 14 was a manager, social class 2 in this system. He deserted the family and the twins were ultimately raised in the home of the mother and stepfather who was a van driver, social class 4. Some of these assessments may overestimate the socioeconomic status of the family. The next items under *Background*, the ages of the parents at last information, e.g., father aged 59 and mother aged 41, are followed by the sex and ages of the members of the sibship in order of birth. Parents or siblings regarded as schizophrenic or grossly abnormal are indicated, as are suicides. Psychiatric hospitalization of second-degree relatives is mentioned when known. The information on the premorbid history of the twins is selective. If nothing is mentioned about deviation in health, intelligence, or social relationships, these can be presumed to have been normal. Only religions other than Church of England (the established Episcopal church) have been indicated. Information on occupation, marriage, and children will sometimes be found later in the case history and not under *Background*.

Psychiatric History and *Later History.* These sections are inevitably very condensed; the summaries on which the judges based their diagnoses were "uncontaminated" and (usually) much longer. The only hospitals named are Maudsley and Bethlem: first admission to one of them, after 1948, indicates the point in time at which a schizophrenic twin became a proband.[1] The estimated *onset* of illnesses diagnosed as schizophrenia is always designated. Milestones in the subjects' histories are indicated parenthetically by age.

When Interviewed. This refers to specific interviews done by one of the authors during the follow-up in 1963–1964 and contains verbatim material from tape-recorded interviews. The interviewer's inquiries are italicized.

Comment. This last section of each case history varies according to the circumstances. It includes integration of information reported in the separate accounts of the twins' illnesses; diagnostic considerations, including reference to the opinions of our judges; lessons or suggestions that might be drawn about genetic and environmental influences in the particular pair; and cross-reference to other pairs. Inevitably, the comments will reflect our own viewpoint aided by hindsight.

A synoptic chart of all periods in hospital (Fig. 6.1) supplements the case history information in this chapter with a visual display of patterns.

THE TWENTY-FOUR MONOZYGOTIC PAIRS

MZ 1, male, born 1893

A: 1st hosp. 30 (Sc 30), 3 spells: 51 weeks; suicide 62.
B: 1st hosp. 27 (Sc 30), 4 spells: 201 weeks; died 52 in MH.

	A	*B*
Consensus Diag.	Sc (5)	Sc (5)
Slater Diag.	Mild chronic Sc, hebephrenic or simple	Chronic Sc, hebephrenic
Meehl Diag.	Chronic undifferentiated Sc	True Sc
Global rating	5.5	6.5
MMPI	—	—

Zygosity Evidence. Fingerprints, seen (1931–1938).

Background. Social class 2. *Fa* died 83, *Mo* died 89; 1 *A* and *B*, 2 *M* died 20. *Fa* patent attorney; 20 years older than *Mo*. Spoiled the twins. *Mo* idolized twins, gave them most similar names she could think of; overprotected them in that she did not allow them to mix with other children and did not send them to private school until 8 years of age. Paternal aunt psychotic.

---

[1] First admission as proband is indicated by italics.

Born in Canada where family lived temporarily on account of *Fa*'s business, *A* born first by 5 hr; very difficult birth (*B* had to be turned), 24-hr labor. Birth weight unknown, both "small." The twins were inseparable until 18 years of age (*A* perhaps the leader); they were equally good at school and had an extremely close relationship with their *Mo*. They left school at 18, *A* to a series of jobs—"always wanting change," *B* to employment as a clerk in a business firm. At about 20 both twins claim to have led a fast life to which they attributed their later troubles. At one time *A* related contact with prostitutes 5 times a week and concomitant masturbation without experiencing pleasure. Neither twin married. Their brother (2 *M*) was killed in World War I when twins were aged 25.

Psychiatric History of A. Normally cheerful and careful of appearance. In World War I army (at 22) he was unable to carry on with drill because of palpitations and breathlessness (*onset*). On office job at home discharged (at 23) with "heart trouble." He attended four different hospitals with supposed VD (WR repeatedly negative) and another hospital for this heart condition. He was admitted to Maudsley OPD (at 30); then, an inpatient (IP) with complaints of inability to concentrate, twitching of the legs, feeling as if his head were going to burst, could not look anybody in the face (he avoided main thoroughfares). At admission, he attributed feelings of strangeness and pressure on the head to stagnation of blood in his head, was obsessed by thoughts of VD, and general blunting of intellectual faculties. He was treated medically (iron, removal of septic tonsils, electrical massage) with no change. Fatuously called himself "little bunny"—"little bunny will help you." Strayed into neighboring garden instead of going to work. (Provisional diagnosis was exhaustion neurosis—hard drinking, masturbation, coitus—or obsessional neurosis—fear of crowds. final diag.: dementia praecox). Discharged after 31 weeks, his attendance at numerous hospitals continued (*Fa* died at about this time). As a Maudsley outpatient (OP) (at 35), with a diagnosis of chronic hypochondria with tendency to schizophrenic alteration of personality. Six months later, he spent 12 weeks at MH with hypochondriacal complaints: t.b. meningitis; cloud over his head preventing him from concentrating. The following year (at 35) Maudsley OP—agreed with physicians that he did not have t.b. meningitis; he now believed it was "paralysis of the brain." Continuing hospital attendances (10 hospitals in 14 years) with hypochondriacal complaints while living with *Mo* who had become a spiritualist, he often talked about securing work, which he never did, and about committing suicide, which he did not attempt. No psychiatric information for 16 years; then admitted to *Maudsley* OPD (at 61) becoming proband. *Mo* had died 4 years previously, and he had been steadily employed in unskilled work since her death until 4 weeks before OPD attendance with a neurodermatitis. Complained that his mind was all wrong ("It's due to masturbation, doctor"). Diagnosis: simple Sc. He lodged with an eccentric old lady after the death of his *Mo*. Three months after the Maudsley visit, 8 weeks in MH as VP, he was treated with 8 ECT for involutional hypochondriasis. About a year later, Maudsley OPD (at 62) when found wandering, thinking he had a tumor on the brain, wanting to die. He refused hospitalization; 2 days later body recovered from city lake (suicide at 62).

Psychiatric History of B. B, a clerk, remained with the business firm that employed him after leaving school at 18. He was particular about his appearance—liked to travel to business in spats and light gloves. His head clerk took a homosexual interest in him and, normally cheerful and amusing, *B* would come home from work worried and tearful. His father sent him on a cruise for reasons of health. Upon return (at 21) he joined the armed services but was discharged after 6 months with "heart trouble" (*onset*). He joined up again at 23 (to cure his supposed gonorrhea) but fell or jumped from a roof (circumstances unknown), suffered concussion and ?skull fracture, and was discharged with pension. At 27 he was found wandering, delusional (thought he could make millions by patents); he spent 10 months in MH. Two years later, *B* wandered around home, wished he were dead, continually undid his clothes to inspect imagined discharges, and commented "I can't stand this." At 30, as a Maudsley inpatient (before current Twin Register was begun), dwelled on sexual matters, and some mental deterioration was apparent as did not know date and year. Apathetic, sensitive to noise, believed other patients purposely annoyed him, speech monotonous, thick, and indistinct, he discharged himself after 4 weeks (diagnosis: dementia praecox). At home, increasingly depressed; MH (at 30) where he was morbidly anxious about himself and was regarded as actively suicidal. After 50 weeks, he was discharged to *Mo* (Diagnosis: neurasthenic type of melancholia). At home with *Mo* (*Fa* died about this time) from age 32 to 50, except for 1 week in GH for "gastric ulcer" where he was regarded as "obsessional state in high-grade mental defective." Helping his mother about the house while at home, continually preoccupied with self-abuse, considering that he had thus ruined himself. At one point he could not void, believing he had been given powdered glass. Readmitted to MH (at 50) where he reported he had fractured his pelvis 20 years previously while having sexual relations and that his spinal cord was damaged beyond repair: aurally hallucinated; depressed that he must go raving mad (diagnosis: hypochondriacal state ?schizophrenia). Later disoriented with considerable deterioration, he was diagnosed upon recertification as Sc. The patient collapsed and died (at 52) of bronchopneumonia after 104 weeks in MH.

Comment. MZ 1, our oldest twins, are alike in their history of army discharge as "heart" cases in World War I, suffering from extreme hypochondriasis (VD fixations), with dire effects of masturbation, apathy and lack of drive, repetitive verbal stereotypy, and gradual personality deterioration without florid symptoms.

Taken in isolation this pair should *not* give more comfort to genetic enthusiasts than to environmental ones with respect to weighting the etiological importance of hereditary and intrafamilial factors.

Smothering overprotection by both parents, one of whom became a spiritualist, would almost certainly have made normal social and sexual adaptation difficult. However, the twins were said to be free of neurotic symptoms in childhood, to be sociable, and to have made friends in adolescence. *B* held the same job for 3 years, then worked as a recruiting clerk in 1914, enlisting himself 18 months later.

The influence of one twin upon the other could account for the similar *content* of their symptoms. The less extroverted, more submissive *B* was the first to decompensate and had the more severe psychosis. It can hardly be doubted that both had genuine schizophrenias, rather than *A* having an induced "pseudo-schizophrenia," such as would be implied by the *folie à deux* hypothesis about MZ concordance. Reasoning from a genetic angle, it is easy to see how in this case both parents may have transmitted specific polygenes predisposing to schizophrenia. The mother, in the view of PEM, was not just a "bad parent" but "kind of crazy," while the father's sister was psychotic.

MZ 2, male, born 1915

A: 1st hosp. 31 (Sc 31), 4 spells, 750 weeks; FU 49, hosp.
B: 1st hosp. 33 (Sc 33), 9 spells, 128 weeks; FU 49, DH.

	A	B
Consensus Diag.	Sc (6)	Sc (6)
Slater Diag.	Typical hebephrenic Sc	Chronic Sc, hebephrenic
Meehl Diag.	True Sc	Chronic undifferentiated Sc
Global rating	7.0	6.0
MMPI	87"90 26' 41-5/3	9'7-58026/43:1#

Zygosity Evidence. Blood, seen.

Background: Social class 3. *Fa* died 66, *Mo* died 78; 1 *F* committed suicide at 32 (abn), 2 *A* and *B*, 3 *F* 44.

Fa, skilled laborer, bad temper, "poor mixer," according to *B*, children frightened of father—were "glad when he went out." *B* also said "could never have had a better *Mo*," although described as possessive by psychiatrist who interviewed her. At 68, *Mo* was alert but suspicious about revealing information. 1 *F* had "unfortunate accident" after death of RAF husband during World War II—took overdose of aspirin.

Born "in separate cauls" (dichorionic?) (birth weight: *A* 7 pounds, *B* 3 pounds). *B* was "black" when born, and was not expected to survive. *A* more robust, heavier, and taller as a child, and was sometimes a class ahead of *B* in school, although *B* was considered to have wider interests and to read more. The twin relationship was very close with little difference in the degree of initiative. Equally mother's boys, but both followed their father's trade. *Mo* was the only informant who thought the twins not at all alike, but favored neither. As a child, *A* was broody, touchy, and jealous; *B* had a "sweet disposition." The twins were 36 and 46 when the *Fa* and the *Mo*, respectively, died.

Psychiatric History of A. At 16 *A* suffered "mild nervous breakdown" (*onset*); left his job, went to hospital where he was told nothing was wrong. He was religiously oriented at this time. At 20 broke off his engagement after the announcement party with no reason given. At 31 in the RAF he became engaged to a continental girl (RC). When permission to marry was refused, he left his unit, was picked up because of strange behavior, and was admitted to the RAF psychiatric hospital: apathetic, poorly orientated, preoccupied with thoughts of "truth, his father, his brother, and an incident when he was accused of stealing something." The diagnosis: acute Sc, with hallucinations, periodic negativism, and paroxysms of violence. After 4 ECT and insulin coma, he was medically discharged from service, after 38 weeks in hospital. From that point on he was never really well. He married the girl (at 33); a son was born a year later. First seen at *Maudsley* OPD (at 35) revealing ideas of influence and suspiciousness he was admitted to OW after giving up work and expressing suicidal ideas, then to MH for 4 weeks where he was diagnosed as Sc. In hospital, terrified lest he be going mad again. Treated with ECT, refused insulin. Six months later irritable, depressed, would not eat, would not speak to wife. Maudsley IP for 20 weeks. 4 ECT cleared delusions and hallucinations, but pentothal and methedrine injections revealed schizophrenic thought disorder. (Diagnosis: Depressive psychosis in schizoid personality, changed to Sc.) At 36, attacked wife, admitted to MH where still hospitalized at age 49 (Diagnosis: chronic Sc). At admission, solitary, withdrawn, hallucinated, laughed incongruously, harbored persecutory ideas. A few months after admission, escaped and was brought back. Thereafter he was less agitated, dull, self-absorbed, flattened in mood. By age 37, disinterested, unemployable, remained aurally hallucinated. Visited by wife every 3 weeks.

When Interviewed. At 49, cooperated, seemed "out of it"; smiled, grimaced, whispered to himself (? aural hallucinations). *How are you feeling?* "Alright . . ." *. . . how's your appetite?* "Alright . . ." *. . . how do you sleep?* ". . . alright." Later—*You sleep alright at night?* "Well I don't sleep, I dream." When asked if he had ideas that were different from other people's he suggested that his ideas were not "revolutionary." Described inappropriate laugh as "Habit, don't mind . . ."

Psychiatric History of B. At 10 had "St. Vitus Dance" (head jerking); father mocked him which made him worse. Married at 25, described by wife as silent, unable to make friends, apathetic, and indecisive. Though marriage reported never consummated, later report revealed a son born a year later—"blue baby," died at age 2. Separated temporarily (at 32) from wife by reason of employment; he overworked, slept badly, felt depressed (*onset*). A few weeks later he suddenly began to speak and act strangely—wanted to sleep outside, would not eat—was adm to MH (at 33); was extremely noisy, abusive, resistive, restless, and overtalkative. Mood varied between elation and depression. Good improvement to 9 ECT, disch. after 9 weeks. Two weeks later plagued by disturbing thoughts about his health, talked to himself, became increasingly depressed, and later attended MH OPD. A month after OPD attendance, adm MH: thought disorder schizophrenic in nature—"I seem to be living on my knowledge and memory rather than going forward in a daylight fashion . . . I get coupling of thoughts." Discharged, after 12 weeks and 24 insulin comas, "relieved," and remained out of hosp 11 years. When seen at 36, lacking in warmth, suspicious of fingerprinting and PTC testing, somewhat inappropriate posture and mannerisms. After age 41 began to work less steadily, marital discord due to *Mo*'s interference, increasingly uninterested, careless of his appearance, suspicious and secretive. At 44 first of 7 admissions to MH where depressed with ideas of self-blame and unworthiness. Depresssion subsided after 6 ECT to reveal ideas of influence, shallow affect, unrealistic attitude, and delusions. Discharged after 6 weeks on trifluoperazine (diagnosis: paranoid Sc with depressive elements). The next 5 MH admissions, during the next 5 years, were for periods of 5 or 6 weeks with the exception of one 14-week admission. At each hospitalization he was treated with ECT, trifluoperazine and/or chlorpromazine; each adm was for a recurrence of schizophrenic symptoms—"I know what's wrong with me, it's insanity." From MH to Day Center (at 47) where relapses and domestic stress interfered with persistent and successful attempts at employment. (Diagnosis: uncertain— "very inadequate man").

When Interviewed. In DH, 64 weeks after last MH admission (on chlorpromazine); stated he would like to cooperate and help ". . . with myself it's a sort of worry that goes on in my mind to get clear . . . it's like a battle of wits . . . within myself." . . . *it just seems to come and go by itself?* "Yes, my biggest difficulty is to let myself be free. It's sort of an aversion to paying attention to something . . . your mind seems to want to be on your illness all the time."

Comment. A is one of the most chronic cases in the series. B is also clearly schizophrenic although less severely affected. He succeeded in keeping out of hospital for 11 years after his second admission and continued to work at his trade, even if only "semicompensated at best." Such a difference in severity is not unusual in concordant MZ pairs.

Although the *Mo* was characterized as possessive and the *Fa* was feared by the twins, the intrafamilial climate falls within the normal range. Both twins served creditably during World War II.

The largest within-pair difference in birth weight occurs in this pair. Besides his low weight, *B* may have minimal brain damage, was the inferior in school, and was mocked by his *Fa*. Some would predict that only *B* would be schizophrenic, or, if both were affected, *B* would be the more severe. Neither prediction is borne out. What environmental circumstances might account for *B*'s more favorable course? Perhaps he received more ego support from his wife and *Mo* than did *A* when each was first hospitalized. *A*'s Catholic wife spoke English poorly and was rejected by her mother-in-law. Furthermore, if we are correct in interpreting *A*'s "minor nervous breakdown" in adolescence as a first, mild episode of a schizophrenic process, it may have left him less able than *B* to resist subsequent episodes in his thirties.

In accounting for the case in the context of genotype–environmental interaction, the genotype seems to have played by far the more important part.

MZ 3, male, born 1913

A: 1st hosp. 34 (Sc 36), 5 spells: 13 years 14 weeks; FU 50, hosp.
B: ——; FU 50.

	A	*B*
Consensus Diag.	Sc (6)	Normal (6)
Slater Diag.	Hebephrenic Sc	Normal
Meehl Diag.	True Sc	Normal
Global rating	7	1.25
MMPI	—	712 8349-50/6:

Zygosity Evidence. Blood (ABO, Rh factor only), fingerprints, seen.

Background. Social class 3. *Fa* died 65, *Mo* died 69; 1 *F* 52, 2 *B* and *A*, 3 *F* 42, 4 *F* 33.

Fa skilled laborer. Family essentially normal, both parents lived in the home until twins over 21.

B ahead physically in childhood. Both left-handed, *A* more markedly so. *B* slightly ahead in schoolwork. *A* always quiet and shy, retiring, nervous, frightened, and lacking in self-confidence. Only early neurotic trait was shouting in sleep, age 8–10. *A* only had pneumonia at age 13. *B* exhibited more initiative and was more enterprising. Irritable as a child. Accident-prone on first job as laborer.

Psychiatric History of A. Previously machinist. 1942–1945 Japanese Prisoner of War in Indochina where he developed desert sores, weakness in the legs, trouble with his eyes, and loss of hearing; later diagnosed as nutritionally amblyopic and hard of hearing (perforated right ear drum; however, contradictory answers and quick loss of interest in hearing made assessment difficult); was discharged from services with 100% disability. On return home (at 32) he hardly went out (*onset*), was sensitive about his hearing, and after 5 weeks of work, "just moped at home." Very upset (at 33) with death of *Mo*. Admitted to *Maudsley* IP (at 34), physically as above, no evidence of cortical lesion. EEG: gross excess of bilateral central theta and slow delta focal abnormality in left temporofrontal area suggesting a space-occupying lesion. CSF and skull films normal. IQ Mill Hill 92. Mentally depressed, solitary, at first regarded as an organic case. Fell in love with nurse and became paranoid when his advances were rejected. Absconded after 37 weeks (diag. paranoid state, ? Sc, nutritional amblyopia; deafness). Same month admitted to MH for 14 weeks where he showed no paranoid symptoms (diagnosis: depression). Six months later Maudsley OPD where he admitted "hearing" people discuss him; laughed to himself a lot. All replies preceded by long pause and answers off the point; regarded as a schizophrenic illness. In another 6 months, anxious, withdrawn, unable to concentrate, markedly hypochondriacal; had short periods of agitated behavior as result of hallucinations. After 1½ years insisted on leaving but was returned very shortly because he deteriorated at home; illness described as paranoid hallucinatory state in a very deaf person. Departed after 24 weeks to live with sister where he remained for about 2 years, then brought to OW after causing disturbance (raving, shouting, damaged neighbor's door, tore out wires). Admitted to MH (at 40), staff member observed that "hearing is not so very grossly defective as he occasionally answers immediately when I do not raise my voice much." Illness regarded as Sc. Two series of ECT after which much more cheerful yet sometimes stood around in corridors shouting as if in response to auditory hallucinations and engaged in open masturbation. Started on chlorpromazine (at 45) yelling stopped, within 10 months a "model patient." A few months later ear specialist observed little change, still considered that there was a large functional element in patient's deafness; hearing aid fitted. Age 48, loss in visual fields observed. Still at MH after 10 years, 17 weeks, single.

When Interviewed. At 50, wearing hearing aid, although it did not work, had been vegetating last 6 months; nurse stated that he was an active homosexual on the ward. He was put on 800 mg chlorpromazine daily. Unspontaneous and bland in interview. Questions printed in block letters to facilitate communication. *Tell me something about your times during the war.* ". . . I haven't got much to say . . . I was a prisoner of war . . ." *Tell me about that.* Long pause. *Do you remember it?* "Not much." Believed that he was hospitalized by Ministry of Pensions "to see about me troubles." *Which troubles?* "Me eyes and ears."

Further Follow-up on A. Repeat EEG done (at 53); no longer suggested space-occupying lesion.

Later History of B. Served in army, no front-line action after European invasion. Interviewed for Twin Investigation at 35, described as alert and intelligent, responded to worries by getting "touchy." His EEG showed gross excess of theta rhythm but only on left side. Married (at 37) unexpectedly to a forceful, unconventional, previously married woman; no children. Shortly after marriage, took course at place of employment; in line for promotion as night telephone switchboard supervisor. Described by social worker at 50 as a "slight, shabby, diffident man," shy, monosyllabic, obstinate, occasional sweet smile, bullied by wife. Refused to be interviewed (at 50) by psychologist, but completed 300 MMPI items in 1¼ hours and then said he would not do any more.

Comment. This is the only discordant MZ pair in which *A* is very severe. All judges regard him as truly schizophrenic, ES considering an organic, "symptomatic" schizophrenia to be only a remote possibility. Although the consensus opinion is that *B* is normal if somewhat shy, sensitive, and insecure, our clinical impression is that he is schizotypic and our hypothesis is that *A* is genetically schizotypic too. The difference between the twins must be accounted for by experiences undergone by *A* and not by *B*, acting upon a schizotypic genotype. Already the less outgoing, *A* was exposed to the exceptional stresses of internment in a Japanese POW camp, suffered nutritional brain damage and the psychological isolation of deafness. Originally identical diatheses diverged. Only *A* crossed the threshold. Once ill, hospitalization effects per se probably exacerbated his deterioration.

MZ 4, female, born 1924

A: 1st hosp. 36 (Sc 36), 2 spells: 7 weeks; FU 39, OPD, working.
B: ——; FU 39.

	A	B
Consensus Diag.	Sc (5)	Normal (5)
Slater Diag.	Atypical Sc	Normal
Meehl Diag.	Acute paranoid reaction	Normal
Global rating	4.5	1.5
MMPI	09-2<u>37</u>/5<u>186</u>:4#	98'470-<u>26</u><u>53</u>/1:

Zygosity Evidence. Blood, seen.

Background. Social class 4. *Fa* died 62, *Mo* died 74 (abn); 1 *F* 43, 2 *B* and *A*. *Fa*, laborer, "very nice," happy-go-lucky, worked long hours, therefore not home much. *Mo* mentally defective, (left school at 14; failed 3 times) cheerful, timid, easy-going, devoted to *Fa*, regarded by family as "peculiar," children allegedly neglected and "cleaned up" at local clinics. From age 61 very forgetful, last 6 years of life in MH (diagnosis: paranoid involutional psychosis with senile dementia superimposed on mental deficiency). *MGF* died (at 50) in MH. 1 *F*(MMPI 6-24 03/5179).

Birth weight: *A* 5 pounds, *B* 5½ pounds. Only *A* a breech delivery, "delicate baby, *Mo* wrapped her in cotton wool" (elder sister),? more illnesses in childhood. *A* little behind *B* in school but always in the same class—both "not above average." Childhood happy, yet difficult as twins did some of *Mo*'s work, *A* lacked self-reliance as a child, complying with *Mo* who feared *A* wouldn't live to be 21, even when she was a very healthy 18-year-old. *B* closer to *Fa* (who died when twins 19) than *A*, *B* enjoyed social clubs, *A* tagged along. *A* did clerical work, married at 25 to 21-year-old mechanic, one daughter (9). *B* skilled laborer, married (at 24) widower with child, happily married with two more children (daughters aged 6 and 4). Returned to work shortly after marriage as husband contracted t.b.

Psychiatric History of A. From age 33 complained of tiredness, palpitations, and other anxiety symptoms. Later quit work, further symptoms of sleep disturbance and fear of traffic. GP prescribed Ritalin. At 36 *Mo* died in MH; *A* became abnormally preoccupied with private life of own doctor; felt she was being hypnotized by doctor's wife—puzzled and frightened by experience as "when the color of the fire seemed to draw me into a coma as though I could not fight it" (*onset*). Attended OPD where a schizophrenic episode was diagd. Ritalin replaced by perphenazine, later by prochlorperazine—anxiety symptoms persisted, and "strange feelings," at first relieved, shortly returned. 2 years later she was talking about receiving mental telepathy from the doctor. In the meantime, she had attended *Maudsley* OPD (at 36) on one occasion and been in attendance at DH for 8 days where thought to have a benign Sc developing in a neurotic within one month of *Mo*'s death. Little change for another 2 years, then 1 week before adm to MH (38) became talkative, depressed, agitated, worried that she had syphilis, thought phone was tapped, and was auditorily hallucinated for GP's voice. Deluded regarding death of GP and of husband. IQ Mill Hill 88. Rapid improvement on perphenazine, disch recovered in 6 weeks (diagnosis: Sc; one staff member's opinion: probable affective psychosis).

When Interviewed. (At 39) 3 months after discharged, symptom-free, due for discharge from FU OPD. Was feeling "very well," no longer had "bad heads" . . . "with terrible pressure" which she had when ill. *Do you think you're different now from before?* "Yes I do . . . . I feel much happier." *Do any of those ideas ever bother you . . .?* "No, not at the moment." Later, *Does it bother you to think about the way you were when you were in the hospital . . .?* "Well, it does a bit . . . . because I was hearing these voices . . . well I must have been crackers (laugh)." *Well, you can laugh about it now?* "Yes, I can laugh about it." Within 3 months of interview attended emergency clinic; felt like crying, couldn't concentrate, mildly depressed, warm affect, concerned whether or not she had done the right thing in letting her *Mo* go to MH. Improved when back on perphenazine.

Later History of B. When interviewed (at 39) seemed normal but was extremely unforthcoming and inarticulate. Said took life, with its ups and downs, in her stride, though admitted that it was somewhat of a struggle at times. *What are your usual worries?* "To try and carry on . . . to keep going." She sometimes found it difficult to "sort out problems." *? Feelings* "I hide them . . . except when I'm by myself and then I shed a few tears." *Anger?* ". . . I think I bottle it up inside, really." *How would you describe yourself?* "Well, I like to describe myself as friendly . . . to people . . . . but sometimes that doesn't always do . . . when people don't seem to want to be friendly . . . ." She had "just one or two" friends outside the family and hoped that she was thought of as a "nice person and . . . helpful." *. . . your personality?* ". . . . people do tell me I am very quiet."

Later Information. A year later *A* still well, discharged from OPD (at 40).

Comment. *A*'s short psychotic episodes might be termed schizophreniform in Scandinavia, raising the possibility of etiological heterogeneity for the array of disorders receiving a diagnosis of schizophrenia. *A* was a "good premorbid" with a late onset, and she was one of our least severe cases. *B* is clinically normal with an extremely colorless personality and an abnormal MMPI. A blind reader of the MMPI, noting marked elevations on the Hypomania and Schizophrenia scales, would include a phrase about ruling out a schizoaffective psychosis. Grounds for diagnosing schizotypy in *B* are nonetheless not as convincing as in MZ 3. The psychodynamics of *A*'s illness fit the pattern for discordant MZ's reported at NIMH. *Mo* irrationally feared that *A*, breech-delivered and slightly lighter, would die. *B* was closer to *Fa* and the more dominant. The pattern is not sufficient to cause schizophrenia, but it makes it plausible that *A*, not *B*, should have decompensated shortly after *Mo*'s death.

MZ 5, female, born 1921

A: 1 st hosp. 29 (Sc 29), 7 spells: 123 weeks; FU 42, OPD, not working.
B: ——; FU 42.

	A	*B*
Consensus Diag.	Sc (5)	Other (6)
Slater Diag.	Simple Sc	Reactive depression
Meehl Diag.	Process Sc	Psychoneurosis, anxiety reaction
Global rating	5.5	3.25
MMPI	2864*7''035'1-9/	31''742'89-605/

Zygosity Evidence. Blood, fingerprints, seen.

Background. Social class 5. *Fa* died 67, *Mo* died 67; *A* and *B*.

Fa laborer; rolling stone until marriage, well-liked but irritable, "beneath" *Mo* socially. *Mo* semiprofessional occupation; unconventional, popular, and sociable, domineering, ? stomach ulcers. Maternal grandfather ? alcoholic, ? suicide. Parental rows when twins small. RC family.

Premature birth, both "very tiny," *A*? heavier birth weight, but not expected to survive. *A* left-handed, *B* right-handed. Both grew into fine babies; won baby prize, advertized baby foods. *A* a shy sensitive child, nail-biter, hair-puller, afraid of dark, nightmares, stammered, excused from many school activities because of supposed congenital heart disease, at least after age 9. *B* a nail-biter. Both twins physically active, preferred boy's play, pretended they were boys, wore shorts and boy's haircuts, called each other John and Michael until at least 10. No apparent difference in intelligence or ability as school girls, *B* was ahead in school as *A* often absent or excused for reasons of health. *A* took classical course, *B* commercial in secondary school. *B* was leader in school games. Both twins were close to *Mo*. Traumatic menarche in *A* at age 10; *B* much later, but not so late as to cause anxiety. *B* held several clerical jobs until marriage (at 20) to airman who was killed 8 weeks later (World War II). Later said she "never got over his death, got used to it." Joined WAAF, later (at 25), remarried to schoolteacher.

Psychiatric History of A. Did not get along with other children at school. Tried to learn to write with right hand at age 7, retrospectively said this made her "very ill" and believed stammer began then. Age 9, heart hospital, mild chorea; heart normal except for soft systolic murmur. A year later spent 3–4 months in GH with rheumatic fever, but also ? mute; seen by psychiatrist. Reluctant to wear makeup or feminine clothes during adolescence. Traumatic tonsillectomy allegedly insufficiently anesthetized because of heart condition; became conscious during operation. After (at 16), screaming attacks (*onset*) at night (persisted "for years") when she thought men in white coats were coming for her with knives. Also curious attacks (at 16 or before) over several months when lying down, day or night; would "freeze," accompanied by stiffness and pains, saw stranger who "used to talk." At this age worried about religion, guilty about masturbation and concerned over attraction for other women. A few psychiatric sessions. Indifferent work record (aged 16–28), mostly untrained teaching, interspersed by several years helping at home and in church welfare work. Last employment at 28. About this time began to take larger dosages of barbiturates to combat insomnia. Psychiatric OPD attendance (at 28) with some worries (diagnosis: incipient Sc) followed by many admissions to psychiatric hospitals or neurosis centers. At two of these admissions (aged 33) Progressive Matrices IQ 116, Porteus Maze IQ 121. Rorschach showed "constriction of personality . . . deterioration of ego structure . . . deterioration of ability to use her intelligence . . . signs of paranoid delusions and sinister fantasies, and a tendency to withdraw further into a world of her own." Occasional overdoses mostly for sleep rather than suicide; continued addiction to barbiturates obtained by forgery, stealing, and homosexual prostitution; usually regarded as schizophrenic defect state: thought-blocking, mannerisms, increasingly dilapidated appearance, inappropriate affect; confided in *B* that "they are after her" and will get *B* as well, "they" being things rather than people. *Mo* died when *A* was 36; now regarded as deteriorated schizophrenic whom treatment (modified insulin, psychotherapy, insulin coma, electronarcosis, ECT, modified narcosis, social work) had not helped. Began to neglect self and home to stay in bed if left alone; occasionally suicidal. First seen at Maudsley OPD (at 37) in a residual schizophrenic defect state. The next year saw an onset of seizures. After much investigation and admission to *Maudsley* IP (at 40), diagnosed epilepsy in a chronic schizophrenic (but no evidence of temporal lobe epilepsy or any clear epileptic features on EEG). Concluded that seizures probably secondary to Sc or due to barbiturate withdrawal.

When Interviewed. At 42 just before discharge from DH to OPD: Rational, insightful, hesitant speech, affect flat. *?future* "I haven't got any future." *?Visions* "Yes . . . they used to terrify the life out of me . . . and they used to talk . . . it was a long time ago." . . . *did they happen lately?* "Not for several years . . . except I used to have a voice in my head . . ." *Voice said?* "He never said it, but I knew what it . . . it used to go so fast, but I knew what it was going to say . . . I used to get frightened . . . it slowed down . . . it would say that I killed my mother which I did, but I just didn't want to hear it." *Is this a fact or is this . . .?* "Oh yes." *Does anybody else say that it's a fact?* "Nobody knows the circumstances but me . . ." *? Enjoy* ". . . very little actually, sex and music at the moment," Enjoyed bed on weekends, sleeping by drugs if obtainable. When attempts of *A* and *B*, male friend, and priest failed to persuade consultant to increase dosage of barbiturates, she ceased to attend OPD (at 43) after being IP or OP in at least 12 different hospitals for psychiatric reasons.

Psychiatric History of B. Generally cheerful, alert, capable, active, and got along well with people. Happily married for first years of second marriage; gave birth at ages 26, 28, and 30 (all boys), and miscarried at 35. (Rorschach at 36—normal, sensitive, well-integrated person with some signs of anxiety and immaturity, but these well-balanced by positive personality features.) Porteus Maze IQ 118. Much upset in early mid-thirties by *Fa*'s death, twin's psychiatric history, and developing domestic problems. Partial thyroidectomy (at 36); thyroid tablets prescribed, after complaints for 2 years of weight loss, irritability; physical attraction to pretty women (never before)—"felt sanity hanging on a hair." Marriage became "rather a strain" after miscarriage (at 35) when husband insisted child could not be his because he practiced coitus interruptus.

Many physical problems during year after miscarriage. About this time removed *A* from MH against advice. Regarded by psychiatrist as impossible to talk to, aggressive, paranoid; paranoid traits not in evidence when seen for Twin Investigation a few months later. Loss of parents in two consecutive years added stress. Marriage guidance psychiatrist consulted (at 40), when she began to feel revolted by sexual intercourse; diagnosed as having personality disorder. Continuing complaints of fatigue and continuing marital problems precipitated attendance (at 42) at Maudsley OPD. Opinion was that it was not really a psychiatric problem but a marital one due to differences in religious principle (she was RC, he was not).

When Interviewed. At 42, said her spirits were alright, appetite and sleep not so good; subject to depression, but sustained by her religion. Asked if she worried about becoming mentally ill, she replied "No, but I do worry that perhaps my children might. I think really that's the whole core of my dissatisfied sex life." She was fearful of another pregnancy and worried "a terrific amount" about this. Had not completely lost the tendency to be sexually attracted to women, but the feeling was not strong enough to worry her; also, "I can always think to myself, well this is wrong so pack it in."

Later Information. At 44 *A* admitted to MH; then private treatment.

Comment. In this case no pronounced genetic predisposition to schizophrenia needs to be inferred. The pair may help (less ambiguously than in MZ 4) in differentiating a class of schizophrenic phenotypes with a predominantly environmental etiology. ES noted that *A*'s illness developed insidiously in an anxious abnormal personality. Diagnoses of schizophrenia were hesitant or qualified by the five consenting judges; most commented on the barbiturate addiction itself. *A*'s seizures were generally seen as due to drug withdrawal. PEM guessed that she was so "far out" only because of drugs. The sixth diagnosis was barbiturate addiction in a psychopath.

On the grounds of a ? mild chorea, *Mo* believed *A* had rheumatic heart disease and subjected her to considerable overprotection that may have exacerbated anxiety readiness and initiated a damaged self-concept. *A*'s initial diagnosis was "severe and chronic anxiety with strong guilt feelings" (masturbatory). Barbiturates brought relief and addiction led to escalation via aversive drift to schizophreniclike state with overt lesbianism, thought-blocking, mannerisms, anhedonia, and interpersonal aversiveness. Without the drug induction, she might have remained a chronic neurotic like *B*.

B was also picked on by *Mo* (nagged at and rejected). Still she coped with marked stress. Although it was 5 months since last seeing a psychiatrist, her MMPI was very elevated but not suggestive of a disorder in the schizophrenic spectrum; it was consistent with *A*'s initial diagnosis of chronic anxiety.

MZ 6, female, born 1920

A: 1st hosp. 32 (Sc 32), 3 spells: 50 weeks; FU 43, home invalid.
B: 1st hosp. 27 (Sc 33), 1 spell: 17 weeks; FU 43, working, symptoms +.

	A	*B*
Consensus Diag.	Sc (6)	?Sc (4)
Slater Diag.	Catatonic Sc	Endog dep, inad pers
Meehl Diag.	True Sc	True Sc
Global rating	6.0	5.0
MMPI	18″63 72′540-9/	3412′6870-95:

Zygosity Evidence. Blood, fingerprints, seen.

Background. Social class 5. *Fa* died 67, *Mo* died 50; 1 *M* 45, 2 *B* and *A*, 3 *F* 41, 4 *F* 39, 5 *M* 37, 6 *F* died 19, 7 *M* 34, 8 *F* ? 32.

Fa laborer; "drinker—sometimes heavy." *Mo* nice, understanding. A very united RC London-Irish family of sibs—and very close-mouthed regarding twins' illnesses.

B heavier at birth (birth weights normal). Neither *A* nor *B* exhibited neurotic traits in early life; each left school at age 14 and worked as unskilled laborers. *A* only in World War II Land Army. *A* (at 30) married an unskilled laborer; *B* never married.

Psychiatric History of A. At 32 (*onset*) lost interest in surroundings. Worried by people in streets and shops talking and laughing about her, gave up work. After a few months returned to work but came home depressed, did nothing (didn't eat, looked blank and vacant) later became excited and talkative, made an attack on husband. Subsequent admission to OW, then *Maudsley* IP (at 32) where withdrawn, mute, retarded, blank affect—regarded as characteristic catatonic Sc when seen by ES. Wechsler Bellevue IQ 84. Treated with ECT, discharged after 22 weeks (diagnosis: paranoid Sc, affective features). Baby girl born to *A* about 9 months after discharged from Maudsley. Four months after birth of baby, *A* began to "go quiet." On admission to MH (at 35), mute, no spontaneous movement or gesture—completely passive, facial expression quite blank. Reportedly symptoms prevalent for past 14 months with increasing severity. Regarded as semistupor of schizophrenic type. Treated with ECT, occasionally relapsed—ECT continued. Nervous and tremulous in a schizophrenic way, occasional thought-blocking but no paranoid features; affect lacking but not incongruous. Discharged after 29 ECT, 20 weeks (diagnosis: catatonic Sc). At home during next 3 years but very unwilling to attend OPD or take medication which she needed. Admitted to MH (at 39) dishevelled, mute, retarded, detached, very depressed, apathetic. Improved on ECT and chlorpromazine; after 8 weeks "probably as good as she will be"; husband very pleased with her. (Diagnosis: chronic Sc). One year later seen at OPD in a state of catatonic stupor—but refused MH admission. Later that year husband left, taking daughter. *A* then lived with younger, married sister as psychiatric invalid. Family protected her from anyone's attempts to arrange psychiatric care.

When Interviewed. At 43, appeared to be depressed and probably hallucinated. Extreme difficulty in communicating: *What do you do now that gives you the most enjoyment?* "Oh I can't . . . I don't . . . I can't say that I really do anything . . . that really gives me the most enjoyment . . ." *What worries you the most right now?* "(long pause) . . . What is really worrying me most now really . . . aren't . . . (pause) right now at this moment, I can't tell you what it is . . . worrying me but . . . ." (long pause).

Psychiatric History of B. Attended psychiatric OPD (at 27). Retarded to point of mutism; apprehensive and distressed. History of 3 months amenorrhoea, increasing depression and withdrawal (*onset*). Said to have been "fretting" since death of *Mo* 4 years previously; believed that workmates talked about her. Admitted to MH, on ward entirely self-absorbed; complained only of "pressing feelings on the head." After 2 ECT somewhat overactive and so aggressive that transferred to padded room. After 17 weeks "apparently at her best," discharged as recovered (diagnosis: endogenous depression). Seen (at 32) when visiting twin in hospital (after 5 hr persuasion to make the visit); rarely left home except to work. Described by interviewing psychiatrist as quiet, subdued, monosyllabic, extremely apprehensive—a schizoid "shut-in" personality. PSW who saw her at the same time described her as very vague and muddled: a few months later she was seen for Twin Investigation; movements slow and awkward, had to be told exactly where to sit, during fingerprinting tended to hold fingers in same position until they were moved. Spoke hesitantly in monosyllables in almost inaudible voice. On basis of above two reports, the consultant psychiatrist regarded her previous illness as schizophrenic and considered that she was suffering from a postschizophrenic defect at the time of the Twin Investigation.

When Interviewed. (At 43) apparently much the same as when seen at 32; with same employer after 15 years where she proofread printed checks. Sibs, who did not regard her as ill, treated her like a dependent child. *Any friends?* "No." *Pleasure?* (Long pause, no answer.) . . . *do you read?* "No." *Favorite TV programs?* "No, not any." *Hobbies?* "No."

Comment. Although *B* was only doubtfully schizophrenic according to consensus diagnosis, endogenous depression being the alternative, the pair would not be a convincing example of an MZ pair with one schizophrenic and the other manic-depressive. When ES was "unblindfolded" he commented that when *A*'s certain schizophrenia was taken into account, "it seems impossible to think that *B*'s very similar illness, also leading to partial invalidism, is anything else than schizophrenia too." There are no environmental factors uncovered to account for the concordance. Parents and all other seven siblings were normal. Neither twin seems to have been particularly stressed, and *A* even had the support of a marital relationship before her decompensation. Although *A* had the better premorbid personality, she had the poorer outcome.

MZ 7 *, female, born 1940

A: 1st hosp. 18 (Sc 18), 3 spells: 222 weeks; FU 24, DH.
B: 1st hosp. 19 (Sc 19), 2 spells: 233 weeks; FU 24, hosp.

	A	*B*
Consensus Diag.	Sc (6)	Sc (6)
Slater Diag.	Hebephrenic Sc	Hebephrenic Sc
Meehl Diag.	Process Sc	True Sc
Global rating	6.0	6.5
MMPI	7'89 260-314/5:	24*06'783-519/

Zygosity Evidence. Blood, seen.

Background. Social class 3. *Fa* 47 (abn.) *Mo* 47; *B* and *A*.

Fa barber, discharged from army on psychological grounds after 6 months hospitalization—said he "ran amok"; obsessional, violent temper, wife convinced him that he was to blame for twins' illness because he was "emotionless." *Mo* not psychiatrically treated but vague, muddled, anxious, and intrusive (intervened at school and at twins' place of employment asking that twins not be separated); considered by GP to be disturbed and by psychiatric observer as depressed, negative and almost inaccessible at times. When the parents quarrelled, or *Fa* nagged, twins preferred to stay with aunt; dreaded returning home. *Mo* claimed this was because the twins disliked the long flight of stairs at home. PSW commented that *Mo* excluded *Fa* from relationship with twins. Jewish family.

Premature birth, normal delivery. Birth weight: *A* 4 pounds, 4 ounces, *B* 3 pounds, 4 ounces, *A* taller and heavier as an adult. Menarche within a week of each other, *B* first. Both average scholars until grammar school, where they did not do well and left at 15, *B* brighter. Twins very close—"couldn't make a move without each other." They had few friends, very uninteresting, insignificant personalities, were too quiet, and were extremely dependent on their *Mo* who encouraged their dependence. *B* dominated *A; A* less "touchy" got on better with people. Both worked in same library for 2 years and got on reasonably well with constant support and mutual encouragement until *A* transferred to another library branch.

*Both probands.

Psychiatric History of A. At 17 (*onset*), A became even more withdrawn and self-conscious, couldn't concentrate, thought the man in charge "kept looking at her," finally refused to serve customer, reported sick, and never returned. GH then psychiatric OPD (at 18), autistic, impoverished affect, inappropriate giggling, believed people were looking at her, was almost mute, had recently used words that, according to *Fa*, "were not English" (diagnosis: Sc of 18-month duration). Admitted to MH—"hardly in touch"—absconded after 22 weeks and 7 ECT (diag. Sc). Near stupor, readmitted to MH (at 19); treated with chlorpromazine, trifluoperazine, and after 50 weeks, was disch. to rehab. center, where continued almost catatonic; considered dangerous, persuaded to return to MH (at 21). No change; was transferred *Maudsley* IP where intensive psychotherapy begun. As part of rehab., attempted employment on two occasions but was released after a few days because too slow. (Diag. Sc), transferred MH (age 22) for further rehab. At MH admitted, for first time, to auditory hallucinations: "I walk along and hear voices coming out of me—a woman calling Jesus." Hospital note to the effect that her "personality shows appreciable schizophrenic defect." Treated with phenothiazines and rehab., she showed no change when disch. to DH after 1 year. Regarded at DH as slow but "all right," unemployable.

When Interviewed. At 23 in DH (150 weeks), on trifluoperazine; slow, flat affect, answered questions to the point. *Health?* "I get . . . I get . . . I get . . . brain, brain stoppage sometimes." *Concentration?* "I don't think I have any concentration left . . . I've got a certain amount and then it's no more like . . ." Later, "There's someone saying something to me right at the moment . . . something like . . . please . . . please go off or something . . . it's a kind sort of voice you know."

Psychiatric History of B. Behavior gradually became unpredictable and irrational after transfer to another library and A became ill (at 17); accused *Mo* of causing twin's illness, thought everyone against her; later increasingly withdrawn, didn't return to work after reprimanded by superior (age 18, *onset*). Psychiatric OPD (at 19) with complaint that her mind was too active and would like something to settle it. At OPD observed to grimace and smile inappropriately— later, "Well, that's social, isn't it?" Sometimes mute, evasive, vague and rambling when not so (diag Sc). Almost directly to MH where deteriorated until ECT, then chlorpromazine. Absconded after 21 weeks, but had been "as well as ever" for 2 weeks (diag Sc). Readm MH (at 20) from GH where attended for seizure in which went rigid. Remained uncommunicative and paranoid (thought nurses poisoned her) with periods of slight improvement followed by relapses while on chlorpromazine, trifluoperazine, thioridazine and ECT. After 14 months, absconded, found wandering and returned to MH. *Maudsley* IP from MH (at 21) together with *B*, for rehabilitation: mute, doubly incontinent, immobile, preoccupied and expressionless. Eventually, accelerated response to reserpine diminishing before drug withdrawn. Continued uncommunicative, paranoid, also compulsive and abusive; at times quiet and unresponsive. Still hallucinated for voices–usually at night and most frequently her *Mo*'s. On one occasion "thinking about suicide"–no attempt; another time explained that she felt as if she didn't have a brain–"Sounds silly, doesn't it?" Parents ceased contact after home visit for Christmas holiday (at 22) and almost 160 weeks in hospital. Rift developed when *B* impulsively upset table, then oil heater, causing fire. Transferred to MH for long-term treatment (diag Sc).

When Interviewed. One year later, still hospitalized, on chlorpromazine. Felt "not too well," bored (making paper hats), ". . . miserable sometimes you know . . ." *Do you think you have any kind of mental problem?* "Well . . . not really no . . . can't think of anything . . . . that I should . . . that I should be here you know." *Are you just like everybody else?* "I suppose so . . . I hope so anyway." Denied hallucinations–"not recently anyway."

When Parents Interviewed. At home with *A* in the room: *Fa*, "She (*A*) gets confused, confusions you know . . ." *Mo*, ". . . I think that since she's been at home . . . she has improved quite a lot, really. The only thing is, you've got to have a little patience with her, a little understanding, that's the whole thing really . . ." *Fa*, ". . . that's a minor detail . . . if a person's ill, then there's hospitals for them." *Mo*, ". . . but you can't take a person and stick them in the hospital:" *Fa*, ". . . you can't mind your words all the time can you? Sometimes you've got to have a little tiff . . . it doesn't interfere with other people's children . . . they carry on just the same." *Mo*, "And I want them to get better, you've got to realize that they're home here and we try to make them better without argument, so you have to overlook a lot of things, that's the only way." *Fa*, ". . . mind you. I say this . . . I mean . . . Not because they're my girls, but there's a lot of personality in them, which, eh, through this illness . ." *A*, "I've a feeling I shouldn't be here, somehow (laugh) . ." *Fa*, "No . . . as you know . . . this illness has brought 'em down a lot." *Well, yes, that's the effect of all illness, it brings people down, doesn't it?* Later, *Fa*, "Do you believe in this, that the years I had in the army has made a difference to her life?" . . . *I don't think it could be blamed for the illness. Fa*, ". . . . all of a sudden . . they've got to take orders from . . . like a stranger . . ." *Mo*, "Don't be silly . . the actual fact is, you never really did care for them . . ." *Fa*, . . . "every father cares for their children . . . . I don't get you, to say, to say, I don't really care for them." *Mo*, ". . . I could never leave them with you." *Fa*, "Is there any reason why, even if I cared, it would have . . it couldn't have been appreciated." *Mo*, ". . . yes, it would." *Fa*, "No, no." *Mo*, "Yes, it would . . ." *Fa*, "No, it wouldn't be appreciated." *Mo*, "It would . . ." Near conclusion of interview, *Fa*, "If I could get her over the nervousness . . . if anybody says anything and she thinks that it's goin' to affect her, her security, you know what I mean. Well, if I could get her over that . . . ." *That's sort of minor–that's not the most important thing in the world to . . . . Fa*, "No, what I mean to say, she thinks, eh, the least little things she knows (pause) mind you, my father was like that, very nervous person . . the least bit of trouble begins, you know, sort of shaking, and nervous." *A*, "I don't shake." *Fa*, "You do sometimes . . ." *A*, "No I don't, only when I'm cold." *Fa*, "Yes you do, you do sometimes." *Mo*, "It's only when you upset her." *Fa*, "I don't upset her, no I don't . . ." *Mo*, "Not half . . ." *Fa*, "I try to get her back, try to get her back to stand up to things." *Mo*, "No, that way won't do it."

Comment. Can an intrusive, impervious mother with cognitive slippage and an obsessional, rejecting father with a history of an acute breakdown provide a family climate sufficiently psychonoxious to cause schizophrenia in their children? Since there are no data on the base rates of such "schizophrenogenic" parental behaviors in the general population or in particular subgroups, we cannot say how far they are correlated with schizophrenia. If correlated, the questions remain how far they are etiological rather than merely triggering, maintaining, or augmenting schizophrenia in their offspring; how far they have been elicited by the premorbid and morbid behaviors of the children; or, how far the parents' abnormalities are reflections of genes relevant to schizophrenia and which they have transmitted. Even dual-mating overt schizophrenics have fewer than half their children affected with the disorder.

This East End London family parallels those described in the psychodynamic family study literature of Lidz and others. Such oversolicitous, child-centered patterns of rearing seem to be common among such European immigrant cultures as the Jewish, Irish, and Italian families which provided many of the cases which have found their way into the literature as well as this pair of concordant twins.

MZ 8, male, born 1917

A: 1st hosp. 29 (Sc 29), 3 spells: 45 weeks; FU 46, working.
B: ——; FU 46.

	A	*B*
Consensus Diag.	Sc (6)	Normal (6)
Slater Diag.	Paranoid Sc	Normal
Meehl Diag.	True Sc	Normal
Global rating	5.5	0.5
MMPI	59'6872-013/4:	359-214678/0:

Zygosity Evidence. Blood, fingerprints, seen.

Background. Social class 2. *Fa* died at 68, *Mo* died at 70; 1 *M* 51, 2 *F* 48, 3 *A* and *B*.

Fa hotel proprietor; abrupt, stubborn, strict, dominated his wife. *Mo* "very sweet," understanding. Very happy home except during summer months when children "banished" so guests (paying) could be adequately cared for.

Birth weights unknown. *A* right-handed, *B* left-handed. Very close twin relationship, spoke up together, neither the leader; yet not upset when separated after RAF training. Twins often thought only one child was wanted, not two, and felt as if one too many. Both *"Mo's* boys," treated by family as a unit, other siblings cared little for the twins. Equally good at school. *A* slightly superior at RAF training school (age 16–19), *A* turned down as pilot because of heart murmur (diagnosis: effort syndrome at 26, and as congestive heart failure at 44); *B* graduated at highest level, recommended for cadetship as pilot but turned down (big disappointment) because of heightened pulse rate on exertion.

Psychiatric History of A. Admitted (at 29) RAF hospital for 8 weeks; having tried to tear his penis (*onset*); talkative, restless, grossly hallucinated (heard voice of girl he was in love with telling him to do things). Treated with ECT (diagnosis: simple Sc). Year later (at 30) pressed unwanted attentions on same girl. Some time later (at 33) visited girl's home, fought with her fiance, admitted GH with broken nose. Psychiatric consultation because of insulting, violent behavior resulted in court case and incarceration in prison before admission to *Maudsley* for 33 weeks. At hosp. delusions of bodily change, catatonic features prominent, no insight, treated with insulin coma with good recovery and no apparent deterioration (diagnosis: cat Sc). Sometime later both parents died within a year of each other. Several years later, developed fantasies about a girl being in love with him, compulsion to throw self overboard and felt people looking at him. After half-hearted attempt to cut throat (at 40) admitted to MH where responded well to chlorpromazine, discharged after 4 weeks (diagnosis: para Sc). No subsequent hospitalizations on psychiatric grounds but exhibited "mental confusion" when in GH (at 44) for heart condition. Depression noted, also complaints of ill-treatment by his employers. Trifluoperazine prescribed.

When Interviewed. Single, living alone, employed at less skilled work than previously (later unemployed). Concerned with "getting by, getting by, and getting a job." *How has your mind been working?* . . . "I don't think it works too well actually . . . I don't feel at all like I used to somehow." *If it weren't for that disappointment in love do you think that you never would have ended up in that condition?* . . . "Well, I mean that's something that certainly is . . . because it's passin', but eh . . . say for . . . I . . . I . . . it's in your nature, it's nothing much, it is events and things, disappointments I suppose in your life, but it's eh . . . it's em . . . it's yourself really that em . . . eh . . . is . . . in hospital . . ." Before interview wrote: "I do not consider I am much use reference Psychiatric Research. I do have very much regard for your work for the benefit of others, and certainly if it is felt that my remarks reference a paper-and-pencil questionnaire could be of any assistance—I am on your side! Two years ago I was accepted as a patient at X hospital . . . Since that time I have felt extremely highly of this particular hospital and its services . . . I am not truthfully overkeen on reliving memories by returning to Maudsley Hospital, however. Privately I sometimes wonder how are those small children who were born without hands and legs, etc., getting on. Promotion of good research to me—is so much more good fresh air and good food—agreeable scenery and society—rather than your apparent quest for more different colored pills and tablets? Personally I should recommend anyone to visit X hospital for its wonderful helpful and charming atmosphere. Any assistance gladly." And after the interview: "May I make a truthful confession, Sir? What on earth have those little pieces of wood with nails through, bouncy rubber-balls, and little china dogs—really to do with 'psychiatric and Genetics Research?' Truthfully, I hope you do not mind my saying—which 'side' is really in need of Psychiatric Research. (I admit our landlady is though!)"

Later History of B. Career warrant officer; married at 32, 3 daughters (14, 12, 10). First seen for Twin Investigation at 33; normal impression, cooperative friendly, answered quickly and to the point with no trace of suspicion or reserve.

When Interviewed. At 46 while on leave from isolated base to visit family settled in England. Very interested, since teens, in art; many social activities, does art decor for stage work, oil portraits and cartoons. Believed ". . . it doesn't matter what you are, you've got to suffer in life before you can produce something . . . this beauty you produce is simply an outlet . . . ." Asked if he had ever had any concern or doubts about his mental health, recounted an incident (26) when he associated with a girl from a group living away from England "escaping income tax" and living luxuriously. Marriage seemed appropriate, but he could never talk freely with her because he felt bitter and resentful about the type of people to whom she belonged. So he wrote letters to her for several months telling her about the life he had led and "trying to draw some of the better qualities out of her." Used to get an urge to write, would sit down and "it just used to flow out." Then realized that "there was something a bit queer about me." The association ended, and he had never told anyone about it.

Comment. This pair matches a clearly paranoid-catatonic schizophrenic with a completely normal identical co-twin. The history of *B* above focused on suggestions of pathology, but he has had a successful and satisfying life. The same genotype and good premorbid history have led to quite different outcomes. Our ex post facto explanation for the discordance is that *A* underwent a gross ego insult, the rejection of his love with its implied doubt of his masculinity, that served to release a schizophrenic diathesis. A possible clue to identifying both *A* and *B* as schizotypes is in their failure to contain anxiety as evidenced by their psychophysiological cardiac difficulties. Once *A* had had an attack of his schizophrenia, his resistance to further attacks was diminished. On follow-up *A* seems to have a mild chronic schizophrenia contributing to the correlation between severity and concordance we find in our sample of MZ twins.

MZ 9, female, born 1917

A: 1st hosp. 43 (Sc 44), 3 spells: 40 weeks; FU 47, working, symptoms +.
B: ——; FU 47.

	A	*B*
Consensus Diag.	?Sc (3)	Other (6)
Slater Diag.	Alcoholic Psychosis	Anx state
Meehl Diag.	Acute undifferentiated Sc	Psn, depressive and conversion
Global rating	4.5	3.25
MMPI	4*89"672'0-31/:5#	239760/4851:

Zygosity Evidence. Blood, history.

Background. Social class 3. *Fa* died 65, *Mo* 78, 1 *A* and *B*, 2 *M* 43.

Fa own business, went bankrupt when twins 14; drank heavily late in life; kind but undemonstrative. *Mo* energetic, cheerful, quiet, kind, possessive, strict, dependent, tried to please, "rather nervous"; originally semiprofessional, later when husband ill and after his death provided for family by running guest house. Protestant Irish family. Home life described as unhappy, many parental arguments, especially about money; little show of affection.

A feeble for first 3 months, later overtook *B* in development. *A* left-handed, *B* right-handed. Menarche: *B* first. *A* night terrors as child, also stammering, lisp, occasional tantrums; *B* "temperamental," a lively and active child. *A* above average in school, "teacher's pet"; *B* never studious. Both "ringleaders," *B* more so than *A*; *A* jealous of *B* whom she saw as more successful. No evidence of differentiation in family, *Mo* did not differentiate them at all on Briggs questionnaire. *A* more attached to *Fa*, *B* to *Mo*, each twin saw herself as less favored of the two (denied by others). Essentially *A* seen as the less reliable.

Psychiatric History of A. Left school at 15; 16–21 a pupil, then instructress at riding school where she took over when owner died. At 17 *A* impulsive, irresponsible, and liked to be the center of a crowd, "got in with the wrong people," began to drink, stole money from *B* and possibly from riding school. From 18 on promiscuous, self-induced abortion at this age; from 24 regarded as alcoholic, but served as driver in Women's Service, age 22–30. Fiance died (at 26), "since then I've been lost," never married. Ten years later lived with married man and brought up his children; "they both drank but seemed happy" (*Mo*). By 42, now living on own, was drinking up to a bottle of whiskey a day. At 43, (*onset*) was "influenced by rhythm"—moved in time to music from tape recorder or sometimes tried to resist the music—"I used to look up at the sky and think—I'm being televised . . ." Moved to seaside town because voices told her to. Fined for being drunk and disorderly after falling into sea. Later accused of setting fire to rooming house; admitted to OW, denied history of drinking, improved on chlorpromazine, discharged after 4 weeks (diagnosis: affective illness, hypomania). Continued working (waitress) and drinking, changed jobs and residence frequently. On 44th birthday adm OW after public "scene," transferred *Maudsley*. Recent employer noted personality change within last year in that she was no longer "happy-go-lucky self" but had become irritable, aggressive, resentful, and suspicious. Because of persistence of hallucinations and passivity feelings when not drinking and, at that time, lack of knowledge of the full extent of her drinking, diagnosis of Sc considered. Later accepted half-heartedly that her illness was caused by drink. Wechsler IQ 108. Discharged symptom-free after 29 weeks but already drinking (diagnosis:alcoholic hallucinosis, chronic alcoholism). Within 3 months to MH from OW where expressed belief of control by electricity and of her thoughts being observed by radar. Evidence of thought disorder in clinical examination and in psychological testing. Treated with chlorpromazine, disch after 7 weeks (diagnosis: Sc). Continued at Maudsley OPD, with lapses.

When Interviewed. At 47, in gay mood, frank about everything except her drinking but gin bottles under the bed; laughter merged quickly with crying; any thought disorder manifested in context of flight of ideas. *What gives you the most pleasure?* ". . . if I see a smile in somebody's eye, like the old lady I met down the road, she lives next door, she lives in the basement and she's got about three hundred canaries . . . . . . and she told me tonight . . . she was going to the shops just before eight o'clock, and she said . . . . . . well I suppose, I don't know, I suppose mental people always talk to mental people. I think there's a community with mental people and not with anybody else." *How are your spirits?* "Oh, when my father was stylish, he played the violin, he was a wonderful man, and, I think I'm given my father's name; Daddy was a spirit drinker, Daddy was a whiskey drinker, course he was, but Daddy was very ill for years and years, and my aunt used to give me a little bottle of whiskey every night, and I used to slip up to my father's room, say 'there you are Daddy, aren't you a villain,' and every night I did that . . . . . . I loved my father very much." Talked freely about past illnesses; after the fire, "I was blamed for it, had to get my cat and dog destroyed and that's what sent me cuckoo, it's true."

Psychiatric History of B. Clerical work from 16 until marriage to officer at 23—"it was insincerity . . . inferiority complex that made me get married when I shouldn't have . . ." Marriage eventually dissolved, 2 daughters (22, 20). Variously described as very emotional, a hopeless housekeeper, no intellectual interests–also "like quicksilver." Fond of animals, a dog breeder. At 42, when *Mo*, for whom she was caring, required much attention because of a bronchitis condition, *B* suffered a "nervous breakdown." Symptoms of weight loss, insomnia, neck pain, chest "tightness" and phobias of going out improved when treated by GP with meprobamate, but she remained unwell at home for 3 years, was pensioned off from Civil Service and was unable to care for *Mo* again. First seen for Twin Investigation[1] (at 44), recovering from above psychiatric illness. Wechsler IQ 119. She had never drunk to excess.

When Interviewed. (47): no longer psychiatric symptoms, meprobamate only rarely. Very quiet and restricted social life, enjoyed work with dogs. *How do people regard you?* ". . . I suppose they think I stick too much to one thing and don't get out enough . . ." *To what extent has that mistake [unsuccessful marriage] been rectified?* ". . . I've made my own life in a different way . . . I'm quite happy the way I am at the moment . . ." *How do you think A will face up to the future now?* "I don't know. I think she'll get out of London (laugh) . . . perhaps . . . . Well to be quite honest with you I think if I hadn't had the dogs or something really you know to . . . I've often thought this. If I hadn't had the dogs or something that I really was interested in, I suppose in a way I could have gone the same way, not as bad . . . but I suppose I could easily have . . . . . just as easily as she did."

Comment. The proband represents an example of a symptomatic schizophrenic psychosis, a phenocopy not of genetic origin, in our opinion. All judges commented on *A*'s alcoholism but enough doubt remained about schizophrenia for her to obtain a consensus of ? Sc; as such the pair lowers our overall MZ concordance rate. *A*'s genuine warmth would have been difficult to convey to a judge without his hearing the tape-recorded interview.

MZ 10*, male, born 1939

A: 1st hosp. 23 (Sc 23), 6 spells: 131 weeks†; FU 24, recent hosp. disch.
B: 1st hosp. 17, 2 spells: 4 weeks†; FU 24, working, symptoms +.

	A	B
Consensus Diag.	Other (4)	Other (6)
Slater Diag.	Psychopathic pers	Psychopathic pers
Meehl Diag.	Psychotic character (Sc)	Psn, conversion reaction
Global rating	4.5	4.0
MMPI	84*2"679'13-50/	Not done

*B is proband MZ 12 in Slater (1961).

†Alternatively, counting periods in detention as well as periods as psychiatric IP: *A*'s first admission at 17; 6 spells: 132 weeks; *B*'s first admission at 17; 2 spells: 39 weeks.

[1] We are grateful to Dr. J. W. Patton, Clinical Psychologist for testing and interviewing *B*.

Zygosity Evidence. Blood, fingerprints, seen.

Background. Social class 3. *Fa* 54, *Mo* 62; 1 *M* 32, 2 *F* 30, 3 *A* and *B*, 4 *M* 22, 5 *F* 18.

Fa truck driver (MMPI 09/1436752:8#) unimaginative, inconsistent in treatment of the twins, but also cooperative and tolerant, seen by *A* as a "jolly father." *Mo* (MMPI 51-34802/96:7#) nearly blind; ineffectual, rigid, and passive, also tolerant, sympathetic, sensible, and cooperative—bewildered by twins' difficulties. Parents basically kindly and normal.

Premature birth (birth weight: *A* 4½ pounds, *B* 4¾ pounds). At 2½ only *A* hospitalized 9 weeks for hernia operation. Though smaller, *A* more robust as a child and later. At 6, when *Mo* hosp. after birth of sib, twins cared for happily by relative. Poor scholars, unlike other sibs, *B* slightly better. Different schools from age 11. At 13 WISC IQ, *A* 78, *B* 75. When very young, close twin relationship, *A* the leader; somewhat later would cover up for each other. Neither had close friends. *A* more hyperkinetic; as a child, would scratch and pick self; *B* more lacking in energy, less persistent. Both hypochondriacal (especially *B*), hysterical and egocentric.

Both enuretic till 7, nail-biters, blinking tic, demanded attention, mischievous, a passion for knives and matches. At 7 both referred to CGC for lying, disobedience, stealing at home and school; treatment not considered necessary. Thereafter in different classes at school. *A* occasionally played truant. *A* the first to appear in court (at 12) for stealing a bicycle, put on probation. Four subsequent offenses for larceny by age 15; to Detention Center (at 17). Later, sent once to Borstal and three times to prison. At least 15 convictions by age 24. *B*'s first offense, stealing from fellow employee (at 15). Subsequently several appearances in court; once sent to probation hostel and once to Detention Center, never to Borstal or prison. About seven convictions by age 22. *B* left home at 17, said his parents and *A* picked on him; *A* left home at 18 and maintained a closer relationship with family; each twin had told acquaintances that *Mo* was dead. *B* married (at 23); deserted wife refusing responsibility of son; *A* lived with *B*'s wife for some time, left when considering marriage to an epileptic patient. *A* and *B* each worked as skilled laborer's assistants with periods of part-time work or unemployment—longest job for either was 18 months.

Psychiatric History of A. After first offense, referred by probation officer to *Maudsley* Childrens' Dept (almost 13) much investigation for nose-bleeds. Parents reported him as destructive, noisy, quickly bored, but reported by school as respectful, inattentive, often lost in daydreams. EEG normal (*B*'s EEG record at this time very similar). Seen by psychiatrist as slightly inferior member of an essentially normal family, responded at first to interest and encouragement; had generalized sense of wrongdoing; 60 sessions in 21 months. (Diagnosis: primary behavior disorder, juvenile delinquency in a somewhat anxious personality, prognosis doubtful.)

At 19 referred to Maudsley Adult OPD by prison aftercare service. When recently in prison (at 18) for arson heard voices (*onset*), which told him to set fire to blankets or "saw" man entering his room before going to sleep; investigated for blood in urine, suggestion symptoms self-induced; plausible, evasive, genuineness doubted; chlorpromazine. Two years later reattended, still hearing voices, twice (once in prison) swallowed needles, continued to steal, trifluoperazine prescribed, personality disorder diagnosed but near psychotic at times. At 23 suspected of starting fire in hostel, took overdose of chlorpromazine (instructed by voices), climbed on roof brandishing axe and after skirmish with police adm MH where bland and offhand. Aural hallucinations soon ceased and behavior became "more typically psychopathic"; absconded after 45 weeks (diag. schizôphrenic episode in psychopath). Tried unsuccessfully to secure admission to Maudsley as IP; lived with *B*'s wife in *B*'s absence, told police he was a mental nurse who had had his paycheck stolen; attempted to incriminate an acquaintance as a thief, worked as painter, eventually paid off fine for breaking condition of residence (final Maudsley OPD diagnosis: delinquency).

When First Interviewed. For our Twin Investigation, still in MH, impression of schizoid psychopath who exaggerates symptoms to obtain psychiatric rather than prison treatment. *Reinterviewed* 1 month after absconding, not on drugs, smartly dressed, working, says gets easily depressed, feels fed up and then hears voices "telling me to do these things . . . . To my point of view it is definitely someone speaking to me, you know"; does not believe he has been mentally ill like the other patients in MH who talk to themselves.

Psychiatric History of B. Within 3 days of admission to probation hostel (at 17) and a week after *Fa* collapsed with acute flu, began series of "fits" in which he could "see" his father and temporarily lost use of legs or had spasm of legs, no loss of consciousness; admitted to teaching hospital, findings essentially negative (diagnosis: ?? epilepsy), anticonvulsants prescribed. One month later, pattern of attacks changed, would doze into "coma," fall to floor, attack helpers, no convulsions; admitted to OW, no organic illness (diagnosis: hysteria), disch to hostel and Maudsley OPD supervision where composed, *belle indifference*, regarded as unstable character with unsatisfactory relationships with family and tendency to withdraw into illness. Continued investigations at other hospitals, further delinquency. After injuring wrist (at 20), complained of disability out of all proportion. Referred to psychiatrist in teaching hospital, depressed, self-reproachful, wrist symptoms hysterical, psychotherapy, lapsed. At 22 when in hospital for treatment of t.b. twice lost and regained voice, had attacks of anxiety, hyperventilation and fainting with epileptic movements, severe toothache uncontrolled by tablets, much attention for old thumb injury, psychiatric symptoms regarded as functional, adopted "unreasonable attitude," took discharge before son could develop t.b. immunity.

At 24 unwilling to be interviewed for Twin Investigation; deserted wife and child.

Later Information. A (at 28) continued psychopathic, hysterical behavior, but not epileptic; patient had married, had daughter and son. When wife pregnant, attended hospital for ankle, discovered to be functional; stole from father-in-law, temporarily deserted wife.

Comment. The pair is in the series by reason of A's later hospital diagnosis of "schizophrenic episode in a psychopath". According to the consensus diagnosis neither twin is schizophrenic, but both are psychopathic. PEM classified A as "psychotic character" which the NIMH checklist happens to group with border-line schizophrenia, and commented, "I assume he is a pseudopsychopathic schizophrenia". Since they are asocial and at times heartless they might both be regarded as schizoid. ES, however, when reviewing A's case in the light of PEM's diagnosis did not feel that the main personality deviations were on a schizoid-schizophrenic dimension. He saw both twins as irresponsible, hypochondriacal egoists of dull intelligence and would speculate about brain damage at the time of birth rather than schizophrenia. In view of their stature (both are about 6 feet 3 inches), their intellectual level and their delinquent behavior, dating from an early age and occurring in a basically normal family setting, we are now hoping that circumstances will permit us to exclude the possibility of an XYY chromosomal complement. (Excluded by karyotyping, 1969.)

The twins are extremely alike in the nature of their psychopathy and in many of their symptoms. Since they attended different schools and had little contact after onset, mutual influence cannot be the main cause of the similarity. The more robust A was considerably more troublesome earlier on and only he was treated in a Child Guidance Clinic. Perhaps this fact led to his presenting with such symptoms as hearing voices when he acted out in a hysterical way. B's "hysteria," at its most dramatic took the form of epileptoid seizures. Both have now attended orthopedic departments for functional complaints. It is interesting that pregnancy in their wives triggered off an acute reaction in both. The latest information shows little sign that the twins are stabilizing with age.

MZ 11, male, born 1932

A: 1st hosp. 23 (Sc 23), 3 spells: 31 weeks; FU 32, working.
B: ——; FU 32.

	A	*B*
Consensus Diag.	Sc (5)	Normal (6)
Slater Diag.	Atypical Sc	Normal
Meehl Diag.	Chronic undifferentiated Sc	Normal
Global rating	5	1.75
MMPI	Not done	Not done

Zygosity Evidence. Blood, seen.

Background. Social class 4. *Fa* 62, *Mo* 60; 1 *B* and *A*.

Fa self-employed scrap-dealer; dull, unobservant; *B* very contemptuous of him; *Fa* removed *A* from hospital against medical advice. *Mo* "emotional nagger," nervous, highly strung; stood in the way of Twin Investigation by opening letters addressed to him and answering in his behalf. PSW described her as weepy, confused, and evasive; also, querulous, full of complaints, and uncooperative.

B born first (birth weights: *A*, 7½ pounds, *B* 7¾ pounds.) *A* weaker in early development, *B* physically very fit as a child. *A* left-handed, *B* right-handed. Twins always together as children—*A* followed and depended upon *B*. *A* rejected for army service at 18—upset at not being called up with *B*. *B*, by own description, nervous and excitable, at least during adolescence.

Psychiatric History of A. Numerous job changes, mostly unskilled factory work, ambition to be bookie's assistant. Attack of giddiness (at 22), acute infective labyrinthitis diagnosed in GH—never well after that; felt something moving in his head and his left eye rolling about; anxious, tense, frightened of people, uninterested, at times "quite dead inside" (*onset*). A year later admitted to MH (23). Psychologist's report: IQ VS 112 but PS 88; disturbance of concentration, severe blocking. Treated with insulin comas but tendency to relapse. Complained only of lack of energy, not interested in rehabilitation, left hospital against advice after 18 weeks (diagnosis: para Sc). Three months later at GH OPD gave long story of his deceiving doctors at previous hospital; no work since discharge: readmitted to MH more frankly schizophrenic, discharged to *Fa* after 4 weeks (diagnosis: cat Sc). Two months later (at 23) at Maudsley OPD, complaining of brain fatigue, "all my feelings have gone, they're dead," Sc diagnosed. Readmitted to MH (at 24); gave rambling account of illness with twitchings of the face, thought his "nervous system is coming alive," bewildered, frightened. Treated with chlorpromazine, after 9 weeks removed by relatives who had "no insight into the seriousness of his illness."

At 31, *Mo* would not allow him to be interviewed for FU, as he was "nervy and not right yet." *Mo* "does everything for him." Working regularly as bookie's assistant, engaged to marry, and then did (at 32).

Later History of B. Refused scholarship as thought only sissies went to grammar school; therefore did factory work until army at 18, then became technical clerk. Later studied accountancy and at 31 worked as accountant, having married at 29; had at least one child (daughter 6 months). Described as "very abusive about male staff, angry, excited, no insight" when visited *A* in MH. Seen for Twin Investigation (at 24); unnaturally jaunty in manner, seemed affected and a little lacking in warmth. Contacted again (at 32), much the same; agreed to interview, then later wrote, refusing, saying that his wife was opposed to his cooperating.

Comment. The pair illustrates many of the factors involved in evaluating MZ discordance. *A*'s unexpectedly good outcome caused three judges to express doubts about the diagnosis of schizophrenia. Slater noted that no characteristic schizophrenic phenomena of a delusional or hallucinatory kind were recorded. On the information given, he thought that *A* could merely be an inadequate personality, which would fit the eventual remission. In diagnosing "schizophrenia, atypical and benign", however, Slater gave weight to the fact that all doctors who saw him when ill considered *A* to be schizophrenic. At most, therefore, *A* seems to have had a mild schizophrenia.

There are also doubts about the extent of *B*'s normality. Slater's blind opinion was "within normal limits, ? a bit schizoid in that he shows peculiarities in social relationships." He was clearly much involved in trying to secure the best treatment for *A* and was highly critical of much of the treatment he actually received. His attitude led some doctors to consider he was "going the same way" as *A*. (We may note that such an opinion did not influence our consensus diagnosis any more than it did in the case of MZ 5*B*.) Of equal or perhaps greater significance may be *B*'s reluctance to cooperate further on follow-up. He cannot yet be considered entirely out of risk for developing schizophrenia. On the other hand, far from becoming schizophrenic, *B*, whether we regard him as schizoid or not, is the only MZ co-twin to have risen socially, eventually qualifying as an accountant.

Given a discordant pair, it is not surprising that *A* is the schizophrenic. Perhaps more than any of the others, the pattern fits that described by the NIMH Twin and Sibling Unit. *A* was marginally lighter at birth and was regarded as the weaker; is said to have been slightly favored as a child by the overemotional mother and to have been more upset than *B* by parental disagreements. He did less well at school. As a child, *A* used to follow *B* everywhere. It is not clear why *A* was rejected for the army at 19. It could be that he was already suspected of being schizophrenic. *B*'s call-up and subsequent removal from the family milieu may well have protected him as much as anything else from

developing schizophrenia too. It is tempting to conclude from the relatively
good outcome in *A*, despite his failure to cooperate in treatment, that these MZ
twins may be less genetically predisposed to schizophrenia than others in the
series. If so, this would make discordance more likely.

MZ 12, male, born 1934

A: 1st hosp. 16 (Sc 17), 2 spells: 33 weeks; FU 25, working.
B: ——; FU 25.

	A	*B*
Consensus Diag.	Other (5)	Normal (6)
Slater Diag.	Adolescent anxiety neurosis	Normal
Meehl Diag.	Acute undifferentiated Sc	Normal
Global rating	4.5	1.5
MMPI	Not done	Not done

Zygosity Evidence. Blood, seen by GP.

Background. Social class 3. *Fa* 48, *Mo* 45; *A* and *B* (order ?)

Fa shopkeeper, strict, insisted on absolute obedience, reserved but welcomed
visitors to his home, well liked by many, took much interest in family; thought
A's illness his own fault, grossly uncooperative with psychiatrists. *Mo* kind,
affectionate, sensitive, and easily hurt, active, appeared happy. Allegedly *Mo* a
beautiful girl who married against parental opposition; the marriage failed
sexually, and she went away with another man for 3 months when twins aged
14, but returned, and maintained that *Fa* was a good man. GP observed that
Mo's "attempts to compensate have left her with a curiously distorted personality."

Twins only children of the marriage. Both attended private school, both
athletic but *B* overtook *A* at sports at 13, after that *Fa* favored *B*. *A* described as
"more sensitive"; *B* as "more happy-go-lucky." *B* also noisy, active, and popular
in teens and *A* shyer, quieter, more submissive.

Psychiatric History of A. At 13, felt certain he had chest disease, not
diagnosable, this about the time that housemaster died. Three years later
increasingly depressed (*onset*), pressure on top of the head, fantasies of strangling people, guilty, fearful, feeling of unreality. Admitted to GH (at 16) transferred to *Maudsley* IP but removed 2 days later by *Fa* (probable diagnosis:
anxiety state). Subsequently seen by consultant psychiatrist and clinical psychologist, then to private MH (at 17) where diagnosis uncertain but Sc suggested—
numerous fears expressed without emotional display, depersonalization, rather
negativistic, vague, bizarrely hypochondriacal. Improvement with insulin comas,

discharged after 30 weeks. Joined the Merchant Navy where psychiatric history stood in the way of promotion. At 25 examined by GP on behalf of Twin Investigation, no evidence of instability, excessive introversion or lack of affect. Had worked for past 3½ years in one position with an airline. Did not like to be reminded of his past illness. Not traced for final Twin Investigation.

Later History of B. Interviewed at 25 by GP for Twin Investigation; described as well, alert, cheerful, self-assured. Happily married, no children, cycled for pleasure and for keeping fit. *Mo* reported that he had been "difficult about his jobs" in that he "thinks he knows all the answers." At time of interview working with *Fa* after fighting Mau Mau in Kenya. Attempts to contact for final Twin Investigation failed.

Comment. Influenced by the outcome and the original diagnostic uncertainty, rather than by the fact that he was treated with insulin coma therapy, five judges rejected the view that *A* was schizophrenic, preferring a diagnosis such as "adolescent crisis with mixed features." In the absence of more detailed information, one can suppose that discordance with respect to this condition was associated with the quieter, more submissive twin's having been overtaken in athletic prowess by his brother, by his being severely criticized on this account by their rigid father, and by the sudden death of the housemaster with whom *A* would well have had a closer relationship than *B*.

MZ 13, female, born 1945

A: 1st hosp. 14 (Sc 15), 6 spells: 91 weeks; FU 19, recent hosp. discharge.
B: 1st hosp. 15 (Sc 15), 4 spells: 53 weeks; FU 19 hosp.

	A	*B*
Consensus Diag.	?Sc (3)	Sc (6)
Slater Diag.	Inadequate personality	Schizo-affective, atypical
Meehl Diag.	Chronic undifferentiated Sc	Schizo-affective
Global rating	4.5	5.5
MMPI	9*8″67′51-<u>04</u>/<u>23</u>:	89*<u>67</u>″4′12<u>35</u>-0/

Zygosity Evidence. Blood, fingerprints, seen.

Background. Social class 3. *Fa* 40 (Abn), *Mo* 39; 1 *B* and *A*, 2 *F* 18; paternal ½-sib: 1 *F* 9; maternal ½-sibs: 1 *NK* 16, 2 *NK* 3.

Fa (MMPI 4*8<u>5</u>″6 <u>12739</u>′0-), civil clerk, unstable from early age, at times elated, irregular sex life followed by remorse, discharged from army because of "moods." Maudsley OPD (at 25) considered him psychopath pers with depression, hospital recommended, not admitted. Similar attacks at 27, at 32 admitted to OW and MH for 6 weeks as hysterical psychopath; at 35 as OP. Inconsistent in attitude to his children but twins fond of him. Both marriages, at 21 and 31, ended in divorce. *Mo* 19 at marriage, deserted family when twins aged 2 years.

Unwanted pregnancy, preeclampsia; birth weight, *A* 5 pounds, 3 ounces, *B* 5 pounds, 8 ounces. *A* more childhood illnesses; *B* slightly ahead developmentally. Nurseries and foster homes 2 until 5½; grandparents 5½ until 9; residential school until 9½; *Fa* and vindictive step-*Mo* until 13, then with *Fa*. Both left-handed, *B* totally, both poor students, occasionally in separate classes but at same level. *A* only hospitalized at age 3 with dysentery and treated for enuresis 5 until 12; *B* occasional enuresis, *A* rebellious, erratic, moody from "nervous attacks" from 12–15; *B* lively, undisciplined, temperamental, daydreamer, poor school behavior. Both afraid of dark as children, both nail-biters. *A* (at 15) described by *Fa* as the more sympathetic and responsible and *B* as more talkative, less irritable, and usually the leader.

Psychiatric History of A. Juvenile court (at 14) for sexual activities, anal intercourse established (diary contained lurid passages thought to be fantasy), remand home 1 week, then probation (*onset*). Referred to *Maudsley* Children's Dept. (at 15), moody, apathetic, almost mute, IQ WISC short form, Verbal 86, Performance 50 (uncooperative). (Diagnosis: primary behavior disorder). A few months later, ran away after row at home, subsequently at remand home (twice), with aunt at school until admission to MH as unsuitable for training school for delinquents because of IQ, emotional disturbance, and sexual promiscuity (still virgin). At MH marked mood and thought disturbances—said that in Classifying School she had been classified as a ballet dancer or figure skater, but she preferred science, "study the world, the moon, the stars, Jupiter—all that sort of thing that is geography," some persecutory ideas. Regarded as diagnostic problem. Treated with chlorpromazine with improvement. Discharged to *Fa* after 23 weeks (diagnosis: Sc). Admitted to MH month later in a "restless, irresponsible state, ? Sc.," reported induced abortion (untrue); heard "nice voices," contemplated suicide for "getting pregnant" (untrue). Improved on chlorpromazine; sent to after-care home but returned because of sexual promiscuity and disturbing influence to older patients. (Diagnosis: behavior disorder). Readmitted to MH (at 17) after disturbance at home and suicide threat, evasive, emotionally flat (diagnosis: relapse of Sc). Treated with thioridazine; discharged after 24 weeks. Employed at tourist resort for 1 week then employed as cinema usherette. MH again (at 18) after 6-week depression, giggled incongruously, lacked insight, absconded twice (one diagnosis: a relapse of schizophrenic illness, chiefly an affective disturbance; another opinion: previous diagnosis of Sc not supported by present state, ? personality disorder; final diagnosis affective disorder). Discharged (20 weeks) at *Fa*'s request to work in fish and chip shop.

When Interviewed. 6 months later, working irregularly, admission to hostel anticipated. Described feeling depressed ". . . you just don't feel right doing anything, just like sittin' down all the time . . . ." When asked how she felt at the time of first MH admission, replied . . . "kept thinkin' I wanted to kill myself . . . take an overdose . . . When I'm out of work I still feel like taking an overdose you know." *How does your mind work right now . . . solve problems and concentrate?* "Oh yes, I can concentrate alright. But when I was in the hospital . . . I was all muddled, I couldn't think straight or anything . . . it was just as if you'd lost your memory . . . (laugh) in a road accident." *What kind of a person would you say you were? . . . Are you happy or sad?* "Sad (laugh), well, I've never been happy; you know, really . . . ." Commented about an American movie star "but she's dead now isn't she? Took an overdose, they all go the same way (laugh). I think I would have gone that way ages ago . . (laugh)." *How is it you can talk about these sad things while smiling about them?* "I don't know (laugh) I'll laugh at anything." *Do you ever cry?* "Cry, yeh, a lot . . . I sometimes don't know what I'm cryin' about . . . (laugh)."

Psychiatric History of B. At 15 became disturbed and progressively excited (*onset*) after row at home which coincided with *A*'s running away; excited, restless, and elated, persecuted by drugs and poisoned milk, deluded about needles and sterilization of blood; affect warm and hypomanic (diagnosis: schizoaffective psychosis, probably hypomanic reaction). 94 Mill Hill IQ. Gradually improved on chlorpromazine; discharged symptom-free after 9 weeks. Hospitalized again in less than a year after four short-term jobs, all of which she lost because "a woman went around talking about her and taking part of her brain away" (diagnosis: schizoaffective psychosis). Treated with chlorpromazine, haloperidol, and lithium citrate; discharged on chlorpromazine. Readmitted about 1 year later on account of hypomanic attacks; paranoid, hallucinated, disorder of affect; memory and judgment intact. Absconded once, discharged after 27 weeks for summer employment at tourist resort—was ready for discharge earlier but *Fa* refused her a home. Later worked as cinema usherette while living with grandmother. Admitted to MH (at 19) (Diagnosis: relapse of Sc with paranoid ideas) to locked ward as aggressive and unmanageable; multiple complaints, i.e., pregnant although denied sexual intercourse, argumentative, accused family of stealing things.

When Interviewed. While still hospitalized; said had a mark on head that "proves" she's "mental," later "could have caused me to go mental," still later "it's just a birth mark" and "I'm not mad . . . ." Admitted to "a temper . . . an unnatural type of temper . . ." *Do you start to throw things?* "No I don't, I'm not a schizophrenic." Said she used the term for interviewer's interest and that her doctor said she had ". . . imagination , . . debility . . ." *I'm trying to find out what actually you think about.* "I like music. I don't like people who bash me across the 'ead like what I've had to put up with for years and years. That's what has brought this all on me . . ." Expressed fondness for dancing, American TV, and Americans.

Comment. Despite *A* being only ? *Sc* on consensus and the possibility that her diagnosis of schizophrenia was "contaminated" by being seen in the same hospital where *B* had been so diagnosed, Slater concluded (when blindfolds removed) that they both suffered from one kind of illness, without doubt schizophrenic. They have quite similar disorders in the face of premorbid personality and physical differences: *A* was shy, sympathetic, persevering, fatter, and unattractive, while *B* was histrionic, undisciplined, impulsive, and good looking. That their pathology is not just an exaggerated adolescent personality disorder caused by the twice-broken home and inadequate mothering is supported by the normality of their sister, one year younger, reared similarly. *Fa*'s MMPI is quite comparable to the strikingly identical profiles of the twins although differing in sex and age; his lower *Hypomania* score could be age-related since he had been described in his youth as unstable, bursting with self-confidence and a ragtime piano player. We could speculate that the "familial" schizoaffective picture shown here is segregating in a Mendelian way and running true to form; if correct, it would add to the pedigrees in the literature where a dominant gene is implicated.

MZ 14, male, born 1943

A: 1st hosp. 16 (Sc 16), 3 spells: 91 weeks; FU 21, recent hosp. discharge.
B: ——; FU 20.

	A	*B*
Consensus Diag.	Sc (6)	Other (4)
Slater Diag.	Atypical Schizo-affective	Inadequate personality
Meehl Diag.	True Sc	Pseudoneurotic Sc
Global rating	5.5	3.25
MMPI	278*4601'35-9/	18*7342"690-5/

Zygosity Evidence. Blood, seen.

Background. Social class 4. *Fa* 59, *Mo* 41; 1 *A* and *B*, maternal ½-sibs: *M* 8, *M* 6.

Fa, manager, not married to *Mo* whom he left when twins 18 months old. *Mo* (MMPI 1–39/256048:7#) 18 years younger than *Fa*; quick-tempered, excitable, dogmatic, dominating personality, sociable. *Mo* migrainous, becoming "vacant" during attacks, therefore "fearful." She did not visit *A* during first two hospital admissions as locked doors upset her, later did not visit when no locked doors. Also, did not acknowledge his 21st birthday though professing concern about him. *Mo* married when twins aged 10. Step-*Fa*, van driver, morose, irritable, bad-tempered. Maternal grandmother helped raise twins; kindly, affectionate, permissive, an "anxious worrier." Maternal uncle assisted *Mo*, unmarried, helpful and supportive, understanding, opinionated, and generous with advice. *Mo*'s DZ twin sister had "twisted glands," heavy facial hair, exophthalmia, was obese; treated surgically (unsuccessful). Jewish family; homes in walking distance.

Said to have been overdue, *A* forceps delivery (birth weight: *A* 3¼ pounds; *B* 4½ pounds). *A* left-handed, *B* right-handed. Milestones for both slow; walked at age 2½, did not talk well until age 7. Residential nursery ages 1–3; *Mo* worked there as nurse; foster home (at 3–4), then with maternal grandmother, and maternal uncle (at 4–10) when *Mo* married; 11 on with *Mo* and Step-*Fa*. Only *A* hospitalized for surgery (cyst removed from right eye), (at 5–6). *A* better student—he completed commercial course; *B* "keen on sports," no secondary education; IQ's (at 16): *A*, Progressive Matrices 109; *B* Mill Hill equivalent 83 (low in vocabulary). *A* regarded as introvert by *Mo*; also, less mature sexually.

Psychiatric History of A. Poor work record after leaving school; heard voices (at 15) saying "I was dead, seriously ill, mad," couldn't think so quickly (*onset*). Also, numerous somatic complaints, hypochondriacal fears, impulses to harm self (swallowed pins), auditory hallucinations and ideas of reference; affect incongruous. Referred psychiatric OPD by GP (diagnosis: Sc), then to GH (at 16) and *Maudsley* IP for 6 months where often tense and elated (with accompanying compulsive thoughts of strangling), expressed hypochondriacal fears for which he sought reassurance—"I have the feeling that I'm going to die . . . These moles on my arms, are they cancer?" . . . expressed feeling of body not belonging to his head (diagnosis: adolescent depression with obsessional features). Remained reasonably well, living with maternal grandmother, maternal aunt, and uncle, and working from 17–20 when complained of tension, urges to kill, rapid mood changes, recurrent thoughts; was walking streets with knife in pocket and touching women to relieve tension. Admitted to MH (at 20), (diagnosis: schizoaffective). Within week of discharge, admitted to Maudsley IP; hallucinated for own voice, recurrent thoughts . . "You've got to kill somebody . . you've got to kill yourself" . . deluded that someone had "taken over his mind." IQ dropped 10 points, normal EEG; treated with trifluoperazine, chlorpromazine and supportive interviews (diagnosis: depressive illness in abnormal personality, although three consultants favored Sc). After 8 months, to hostel, where thought warden tried to blackmail him; heard own thoughts spoken by others.

When Interviewed. Third visit, second interview (at 20): Spoke in short phrases interspersed with numerous pauses. . . . *are you looking forward to something with happiness?* "No . . . ." *Are you able to enjoy yourself?* "No . . . I'm not . . . . em . . . at the moment, . . . I feel fairly rough at the moment ... I em ... (mumble) ... he said you'll have your bad days and your good days .... nothing physically wrong with me. It's the ... it's my mind that's making me feel .... reaction of the mind that's makin' me feel like this." *How is your ability to concentrate now?* "... not all that well. I don't ... if I read a paper, I don't read it for more than about two minutes ... eh ... em ... I don't read any books. I wander about quite a bit ...." *What kind of a person do you want to turn into?* "Em .. I would like to turn into a happily married .. happily married rich man (laugh). Em ... I will say something about yourself ... I have noticed that you have been touching your head while you've been talking to me ... everybody has some habit or other, you know .. but, em ... as I say ... There's an old sayin' that goes ... There's more out than in .. (laugh) ... I just wonder at times ... whether I could have worked .. worked with this illness ... I don't think I could have done ...." Had made no plans for the future as injured hand (had put fist thru window) and medication were "holding me back no end."

Later History of B. Continued to live with *Mo*, poor work record from age 16, often unemployed and drifting, a "terrible gambler" (because of this *Mo* took over his savings for safekeeping); no girl friends. GH IP (at 18) 3 weeks (diagnosis: postdysentery arthralgia); innumerable physical complaints, "seems mentally maladjusted." Various and continued investigations, all negative.

When Interviewed at 20. Very hypochondriacal ... "every time I start work I seem to be ill" ... thinks he had an ulcer and takes milk for it. *Mo* no longer cooks for him because of his unpunctuality, inconsiderateness, and her disapproval of his gambling. *What kind of person would you say you were?* "... I'm just an ordinary worker and a gambler, in a sense, you know ..." *What does it mean to say that you're a gambler?* "Well, all the spending money I do get goes on gamblin'." *Successful?* "No, that's why I'm trying to cut down on it. I want to be successful." *Do you have any difficulty concentrating?* "Yes, I do in a sense. I start to concentrate and something else comes across my mind and I've got to check myself to do it then."

Later Information. A admitted to MH within 8 months, then OPD until another short admission at age 24.

Comment. To call this pair discordant does an injustice to the similarity we see in their underlying personality structure. Meehl called *B* a pseudoneurotic schizophrenic, and the MMPI would be read blindly as that of a schizophrenic with somatic delusions. *A*'s illness defies easy subtyping but it certainly is atypical in many respects. Both twins have been exposed to disruptive moves as children and to a series of unhealthy psychic events, primarily an intrusive mother and a harsh rejecting stepfather. *A*, additionally, suffered the forceps delivery and was 1¼ pounds lighter at birth than *B*. The mother here as in MZ 7 may merely be acting according to her subculture Jewish stereotype and not as a specific causal agent; *B* has been and continues to be more exposed to her influence but he, while still very much at risk (age 20), has not been made overtly crazy by his "schizophrenogenic" mother.

MZ 15*, male, born 1916

A: 1st hosp. 36 (Sc 37), 4 spells: 503 weeks; FU 47, hosp.
B: 1st hosp. 34 (Sc 34), 2 spells: 54 weeks; FU 47, working, symptoms +.

	A	*B*
Consensus Diag.	Sc (6)	Sc (6)
Slater Diag.	Inadequate pers, ? Sc	Para Sc
Meehl Diag.	True Sc	Acute para reaction
Global rating	5.5	5.0
MMPI	7"82'54-30 9/16:	8*56'071294-3/

Zygosity Evidence. Blood, seen.

Background. Social class 3. *Fa* died ? 32, *Mo* died 52; 1 *B* and *A*.

Fa, technician, killed in World War I before birth of twins. *Mo*, unconsummated marriage for at least 2 years until last 2 days of *Fa*'s leave before going to France in 1916. *Mo* had ambitions to become a singer; supported twins as well as assisting grandparents and aunt.

Premature birth by 2 months; birth weight: *A*, 3 pounds; *B*, 4 pounds. *A* right-handed, *B* left-handed. *B* better at school, although little difference and both in same class and considered "very bright." *A* perhaps more aggressive and more dominant. Both won places in school choir. Neither could succeed in exams and failed school certificate. *A* left school sometime before *B* because of seizures—aura of little things hurrying past him, similar to recurrent nightmares previously. EEG as adult was normal. During and after childhood *B* suffered from frequency of micturition. Only *B* married.

Psychiatric History of A. Difficulty fitting in at early jobs; then apprenticed for trade until "singing bug got him." After *Mo*'s death (at 18) joined army hoping to save money for singing lessons, but discharged within 9 months as unsuitable. Employment usually temporary but continued to study and to sing; in professional choir, and sang with army during World War II (only happy time of his life). Engaged twice, never married—"I've had little love life and would rather not talk about it." Treated for headaches (at 29), unable to manage teacher's training after World War II (*onset*). Sporadic employment, occasionally in chorus of shows where he often broke contracts feeling victimized. At this

---

*Belmont Hospital proband.

time had "the most amazing experiences of the heights and depths of human experience," possibly related to hopes and failures of various auditions, also anxiety symptoms and possibly derealization—"had no sense of contact with the world." On one occasion, sacked from job for staying off work, believing he had had successful audition for professional opera company whereas in fact they had not accepted him. Also about this time (at 35) turned out of musicians' club for "misunderstanding its amenities." Soon suffered an "emotional collapse" after failing another audition; also now (at 36) described by aunt as "like a child." *Belmont* Hospital (at 36) for 24 weeks, "on top of the world"; very immature emotionally; Rorschach suggestive of Sc. Group therapy somewhat successful (diagnosis: inadequate psychopath, near psychotic). Two short admissions to MH (diagnosis: Sc—then psychopathic personality). Expressed opinion that he was much happier when hospitalized—"I can't stand money anxiety. As soon as I'm back in my routine here I am a different person." In MH hallucinated, vague, preoccupied, incoherent—"I'm fine . . . I never have been here. I'm an incarnate spirit"; ideas of influence and control. Treated with ECT. At next admission (55 weeks) deep insulin treatment. Expressed much conflict—"to sing or not to sing," also hypomanic. IQ (at 38) Wechsler Verbal over 120, Full Scale 106. Tried working outside the MH on various occasions, using MH as hotel. Readmitted to MH (at 38), agitated, on point of tears, threatening suicide, discharged after 69 weeks. Within 2 months, readmitted, talk "floridly schizophrenic." Treated with chlorpromazine and occupational therapy. Very successfully organized hospital Christmas concerts during last MH admission.

When Interviewed. Still in MH (at 47) (355 weeks), working in laundry, singing in choir, and chairing social committee. *Plans?* ". . . I mean I haven't got . . . I haven't got any plans at all. You ask me what my plans are, I'm afraid one feels very negative about that. My plan is to keep regularly going to the choir and going to my work in the laundry (laugh) . . . and out 'round the social world . . . 'round the social life . . . I had comparative use for that . . . I can really be useful at that because I do . . . I seem to be able to do that alright." *How is your ability to concentrate?* "Not very good I'm afraid . . . I skip lines when I'm reading . . . I lose interest very quickly . . . ."

Psychiatric History of B. By own account, nervous, cowardly, bad-tempered as a boy; later passion for singing and Shakespeare. Joined RAF (at 19), remained for 10 years. Experienced overwhelming "urge to work at singing" and said to have been "on threshold of big career" when transferred to another station. Later strong interest in philosophy and psychology. Married at 26, had 2 sons (16, 13). About this time felt his "natural personality was being killed by overwork." Soon compelled to study elocution and dramatic art, with reasonable amount of local success in performing. Selected for special armed services training school where working hours were so loathsome to him that his "reason and speech" were affected; sent overseas where he suffered "a major rebellion of the unconscious mind" (at 28, *onset*) with the return of his voice and the possession of tremendous energy "which could only be directed toward voice production." Admitted to MH (at 34), followed siege of influenza; talked of "revelations"–elated, full of self-adulation, amplified account of his life with irrelevancies (showing thought disorder rather than flight of ideas–no rapport); felt that he had acquired Nijinski's legs and that his heart was outside his body. Treated with deep insulin, discharged after 22 weeks (diagnosis: Sc). Sudden relapse about a year later, readmitted to MH (at 36)–"Sullivan told me yesterday that Gilbert was Ivan the Terrible and Judas–that's interesting information." At this hospitalization, epileptic seizures reported, treated with phenobarbital. Discharged after 32 weeks, insulin therapy (diagnosis: Sc). Spent 2 years touring (at 36–38) with professional company chorus; left after "humiliation" (wanted a big part). Described by wife at this time as "like a grown-up child"; also grandiose, unrealistic, kind, undersexed. At 45 employed in a factory, the most recent of four jobs held since last discharged 10 years previously. He was well known and tolerated in the village; allowances were made for his odd ways–coming home singing in the middle of the road and getting at odds with people; participated in local operatic productions; could not manage except for sympathetic, understanding wife.

When Interviewed. Still with factory (at 47) and involved in local opera company. . . . *I'm interested in how you feel* . . ". . . I've got a splendid house and I've got eh . . . my wife is a . . . excellent . . . wonderful worker, wonderful worker you see with lots of connections, expert cook, highly expert cook and, eh, always in demand for doing different jobs; very lucky in that respect." *Enjoyment from opera work? Appreciated?* "I do get enjoyment out of it, naturally, I do, but there is an enormous responsibility in it . . . I'm appreciated not only by people that don't know but by some people that do know . . . they've got vast experience. I worked with him before with the opera company . . ." . . . *nervous breakdown?* "Well, I think I was completely out of my mind, I think so . . . Well em . . . I was conscious of being . . . I was conscious of being em . . . of possession to be a certain extent . . . of be . . . of doing things completely outside of my own control . . ."

Comment. The great difference in outcome for this concordant MZ pair is misleading. *A* had been hospitalized for nearly 10 years, *B* for only one. However, their overall clinical state is quite similar. *B*'s symptoms are perceived as eccentricities by the community and he has a more severe degree of pathology showing in his MMPI than *A*. Only *A* was on phenothiazines at the time of testing. *B* was better premorbidly in that he was able to hold a routine job for 10 years and get married to an accepting spouse. *A*, unmarried, admitted to using the nearby hospital mainly as a hotel, conveniently located in his brother's village. These factors, *in toto*, may have given *B* a self-concept not possessed by *A* that enabled the former to stay gainfully employed after his own break-down—9 years without a day's absence. The twins are remarkably alike in their psychopathic, histrionic—rather than schizoid—personalities. There is also a hypomanic tinge in both which (Slater's opinion) may have kept them socially organized and friendly.

MZ 16, male, born 1919

A: 1st hosp. 17 (Sc 30), 3 spells: 89 weeks; FU 44, in analysis.
B: ——; FU 44.

	A	B
Consensus Diag.	Sc (3)	Other (4)
Slater Diag.	Severe obsessional	Paranoid, schizoid
	personality	personality
Meehl Diag.	Pseudoneurotic Sc	Schizotype
Global rating	4.75	2.5
MMPI	862 45*7″0′-39/1:	Not done

Zygosity Evidence. Seen.

Background. Social class 1. *Fa* 80, *Mo* died at 68; *A* and *B* (?order).

Fa banker; pedantic, domineering, narrow-minded, no use for arts, lacked imagination . . . "Mr. Heretofore Notwithstanding" (*A*). *Mo* chronic neurotic, "highly strung," overprotected *A'* who was "tied to her apron strings" as a boy.

A probably born first, birth weights unknown. Both right-handed, only *A* left-footed. Twin relationship never close; dominance not noted but strong rivalry suspected. *B* had university degree, service World War II, became partner in *Fa*'s bank.

Psychiatric History of A. Did well at prep school. At 14 anxious about going away to boarding school where he was homesick and totally unhappy; while there, gave up competing with *B*. Two years later, asked to be taken from school, worried about masturbation (*onset*), referred to psychotherapist who recommended school change, but *A* changed his mind and returned to the school. About that time he stole from *Mo*. Broke down again a year later, sent (at 17) to "rest farm" under psychiatric supervision for 13 weeks; same worries, feelings of depression, remembered running over fields to relieve tension and looking around because he thought he heard his mother calling his name. Read economics, then law at university for 2 years, but refused to return because he thought he had failed an exam which, in fact, he had passed. Registered as a conscientious objector during World War II. Psychoanalysis begun (at 23?) after an encounter with a girl; very slow progress because of his "severely obsessional temperament." For 12 weeks stayed at a psychiatric hostel in the country as he was "inert and severely inhibited in speech and action." In analysis 1 year, 3 times a week, analyst in touch for 4 years. Two years later unable to work because of extreme fatigue, insomnia, weight loss, depression, obsessional ideas and action (e.g., pulling out eyebrows) and fear of developing t.b. Admitted to MH (at 30), treated with insulin comas, followed by psychotherapy, where discussed without the slightest inhibition a series of paintings in which he disposed of his whole family by slaying them. Psychotherapist noted marked thought disorder, incongruity of affect, inability to concentrate. Improved long before discharge and had to be forcibly discharged after 64 weeks (diagnosis: anxiety state, Sc). Eventually found work as publisher's clerk in a firm specializing in psychology. Treated by two other analysts before seen at *Maudsley* OPD (at 34) requesting analysis under the Health Service as his money had run out; felt "guilty about life, feel I am a failure; feel others notice it . . . I'm not expressing my true personality"; once, when talking to a certain girl, he had felt free from problems for a few seconds. No sexual experience. Consultant's opinion: a partially recovered schizophrenic; really, nothing can be done. Treatment with analysts resumed, one of whom considered *A* to be one of his patients least likely to improve. Later analyst of opinion that, although he was friendless and affectively flattened, there were no features that would "classically justify the diagnosis of schizophrenia."

When Interviewed. Still single (at 44), seeing analyst weekly. Continued to be employed in position as clerk where there was "no possibility of making a mistake." Regarded himself as an artistic, literary intellectual, had written an epic poem in cockney without obvious schizoid characteristics. Lived a "hermit's life, I'm shamed to admit," attended movies "to escape," overate "as a compensation," gave up smoking (cigarettes "a sex symbol"). *How is your ability to concentrate?* ". . as I told you I read practically nothing at all . . . the Sunday papers . . . I can't apply myself to an intellectual job . . ." *How do you feel about the future?* ". . . I'm running away from things . . . I daren't think about it . . . I carry on from day to day . . . as Freud himself says . . ." (later) "I feel my life's all sort of confused you see . . . I don't regard myself as a twin . . . my brother doesn't exist. The feeling I have . . . . I'll learn to do better than my brother or else wipe myself out completely . . . so to speak you see. Which is what I've done recently."

Later History of B. Married, son (8); did not get on well with "bossy" wife. No adm to local hospital. Did not reply to letter sent him about twin research (at 44). When contacted over the telephone, he immediately became antagonistic saying he couldn't possibly give an hour of his time for such a project; offered 10 minutes at 2:30 that afternoon, then arrived at 10 minutes past the appointed time. He refused to sit down in his office, was evasive, uncooperative, demanding, and rude. Near the end of the interview he wondered, with little apparent sincerity, how he might help his brother. He left an overwhelming impression of a paranoid personality. Despite an expressed willingness to do the MMPI, as well as to answer questions submitted in writing, he did not complete the forms and did not reply to a further letter.

Comment. A's schizophrenia typifies the focus of the controversy between European and American views of what may properly be diagnosed in this category; he is a standard Hoch-Polatin (1949) pseudoneurotic schizophrenic and is still in analytic therapy. After being asked to comment on Meehl diagnosis, Slater said schizophrenia would "be quite possible." Without the field work, *B* would have gone on record as merely discordant for schizophrenia; as it stands, we now know him to be personality disordered and well within the spectrum. The different outcomes may be traced to the unique relationship *A* had with his neurotic mother who tied only him to her apron strings. The twins are an interesting example of negative identification, i.e., they tried to be as different as possible from each other. For example, *A* was a C.O. while *B* served on active duty; *B* went into *Fa*'s bank, *A* became an "intellectual." Blood-grouping was not done, but they were seen within a few months of each other and were as like as two peas in a pod.

MZ 17*, male, born 1934

A: 1st hosp. 22 (Sc 22), 3 spells: 327 weeks; FU 29, hosp.
B: 1st hosp. 22 (Sc 22), 3 spells: 339 weeks; FU 29, hosp.

	A	*B*
Consensus Diag.	Sc (6)	Sc (6)
Slater Diag.	Hebephrenic Sc	Hebephrenic Sc
Meehl Diag.	True Sc	Chronic undifferentiated Sc
Global rating	6.0	6.5
MMPI	Not done	Not done

Zygosity Evidence. Blood, fingerprints, seen.

Background. B and *A* illegitimate. Separated at birth. See pedigree (Fig. 4.1). Social class of adoptive parents: *A* 4, *B* 2, *6*, mechanic; *7*, although never responsible for *A* and *B* seemed to favor *A*, married American, *8*, during World War II; *Fa* and *Mo*'s relationship a casual encounter, *Mo* 19 when *A* and *B* born.

*Previously reported as Sm P4 in Shields (1962). Case MZ 11 in Slater (1961).

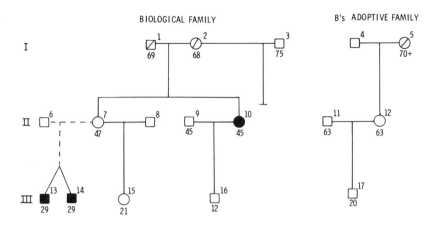

BIOLOGICAL FAMILY

B's ADOPTIVE FAMILY

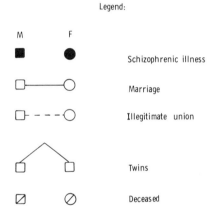

Legend:

M   F

Schizophrenic illness

Marriage

Illegitimate union

Twins

Deceased

Fig. 4.1. Pedigree chart for MZ 17 family. I, II, and III indicate generations. Arabic numerals above symbol represent individuals. Arabic numerals below symbol indicate age at last information. Key persons:

1 Maternal grandmother, Chinese
2 Maternal grandmother and *A*'s adoptive *Mo*
3 Maternal grandmother's second husband and *A*'s adoptive *Fa*, Chinese
6 Biological *Fa*
7 Biological *Mo*

9 Jamaican negro
10 Maternal aunt, adopted by 4 and 5
11 *B*'s adoptive *Fa*
12 *B*'s adoptive *Mo*
13 *A*-twin
14 *B*-twin

2, reserved, undemonstrative; *1*, deported for law infringement; *3*, skilled laborer, little interest in *A*. *2* did not raise her own children (*7* and *10* reared in hostel, later *10* adopted by *4* and *5*, *11*, business manager, responsible position in professional organization, keen gardener, fatherly interest in *B*; *12*, talkative, histrionic, warm-hearted, perhaps inconsistent, possibly intrusive; lenient and protective toward *B* who idolized her. *B* also idolized *5* who was somewhat stricter than *12*. *10*, of normal personality, developed a paranoid schizophrenic illness (at 43) 5 years after *A* became ill.

Born after 2 days labor; *A* forceps delivery, lighter weight and more delicate baby. *A* had three accidental falls as a child in which he injured himself (probable concussion at age 10). *A*, after being kept in hospital after birth, was in residential nurseries until age 4½, then to *2* until outbreak of World War II (in 9 months) and evacuation to country town; *B* to foster homes (three in all) until evacuation, where twins first met (5 years). They did not get along and were later separated for this reason, with little further contact. At about 6½ *A* returned to *2*; at 7½ *B* left his evacuation home but *2* refused to accept him; instead, she made arrangements for *B*'s care with *11* and *12*, the family who had raised *10*, her own youngest daughter. *B*'s adoptive parents later had a son of their own, *17*, to whom *B* was always very attached. School and education both subnormal, *B* attended a special school for the backward from 10–14. *A* committed to Approved School for a year at age 15, IQ 87. Both saw national service in the Pioneer Corps for dull soldiers. At 22 both twins illiterate; IQs: *A* Wechsler 76; *B* Wechsler 75. Low IQs perhaps due to dyslexia and disadvantaged education, as both twins had a good speaking vocabulary and were very sensitive about their illiteracy. *A* raised in low social class, *B* in relatively high one. Both had number of unskilled jobs, were fastidious about appearance premorbidly and not interested in girls (single).

Psychiatric History of A. As schoolboy truanted and stole, enuretic until 12, talked a great deal in his sleep; from age 7½ set several fires. On probation for stealing (at 15), described by consultant psychiatrist as "a boy with a poorly formed character, self-absorbed, impulsive, and subject to periodic depression in which he is careless about what happens to him." Two months after *Mo*'s visit to England from America (at 22) began to neglect appearance, grimace, and laugh to himself; interpreted sound of passing cars as aerial activity indicating the imminence of war (*onset*). Eventually sent home from work, referred self for treatment; to OW where thought to be schizophrenic, then to *Maudsley*; symptoms of Ganser's syndrome, more than one consultant thought he was schizophrenic (classified hysterical pseudodementia, borderline intelligence); absconded after 8 weeks. Ten months later admitted to OW (at 23) having developed spells of aggressive behavior (struck maternal grandmother), ideas of reference, thought blocking, and neologisms. Seen for Twin Investigation: spoke of lack of education, puzzlement of word meanings seemed to go beyond simple ignorance—"Words don't mean what they say." Answered "don't know" and "can't remember" about his past history. Transf MH (8 wks), improved on chlorpromazine. 4½ months later to OW, then MH (at 24) after smashing television set at home (diagnosis: Sc). No essential change over the years.

When Interviewed. At 29 still hospitalized, treated with trifluoperazine, very schizophrenic impression: "Things have been worrying me ... people been talking ... 2 million camels ... 2 hundred Queen Marys ... 10 million taxis ... Father Christmas on the rebound." *What was that?* "Father Christmas on the rebound ... 20 million Father Christmases ... there's so many things that have been worrying me ... it sounds silly ... and ... 1000 ships ... 20 ... 200 stadiums, and there's all those things that worry me ... because ... it's very dangerous. Not only for me ... it might 'urt somebody else, mightn't it? ... to get on all right ... you have to, 'ave to, 'ave someone ... and you 'ave to ... 'ave to talk very straight to them, but it's 'ard really, because when I'm ... I'm fairly brainy but I–can't spell ... it's very 'ard like when you ... er ... 'ave thoughts like this coming in to your head ...." [verbatim passage, no omissions]

Psychiatric History of B. When adopted age 7½ sent to private school where had to be removed because of rowdy behavior. Child guidance clinic at 10, enuretic until 14, from age 9 set several fires. On a few occasions after death of 5 (at 15) he said he heard her talking to him, warning him about something. Convicted of stealing (first at 17), probation hostel; described by probation officer as "a strange, detached boy." Left first job after National Service because thought other workmen were spying on him. The evening of the day he was taken to visit *A* at home (at 22), at the onset of *A*'s illness and heard of biological *Mo*'s visit, he was found crying; the following day seemed distracted, made clicking sounds with his tongue. Two weeks later developed strange, unintelligible talk (*onset*); felt "special powers" left him when cigarette packet discarded–later smashed bathroom window with china dog– "the devil was there and it was either him or me." To MH (at 22) where inaccessible with occasional overactive spells. First seen for Twin Investigation: answered "don't know" and "can't remember" about his past history, spoke of lack of education, puzzled over meaning of word "fitzoscrene" to describe schizophrenic twin. Released to adoptive parents after 14 weeks and treatment with tranquilizers (diagnosis: Sc). Readmitted a few months later (at 23) after cutting cheek with razor blade to "remove a beauty spot." Little change but released to parents after 79 weeks only to be readmitted 3 weeks later (diagnosis: Sc); Episodes of unpredictably violent behavior (not controlled on chlorpromazine) and occasional absconding continued (once went to London where pulled out of Thames–"wanted a swim"); flat affect.

When Interviewed. In hospital (at 29) after 246 weeks continuously, very schizophrenic impression (on trifluoperazine). Said when he first came into hospital "I was paralyzed, and I can't remember a lot about it. The only conclusion I could think of was because my head wasn't working properly. But now my brain is working normal. It enables me to use my body the way I wish to use it, and if all goes well I hope to leave hospital and er .... make as much money as I can .... and er .... eventually marry and settle down by the seaside." *How would you describe yourself?* "Well, I mean, er ... I'm inoffensive, inoffensive, inoffensitive, jokative, likable, er . . . . . (very long pause), carefree, um, (another *very* long pause) very constrengthative."

Comment. In the 7 years since the twins were described (Shields, 1962) they have remained chronic patients in different hospitals, although some attempt was made by *B*'s adoptive parents to look after him at home. When first treated at the Maudsley *A* was given the benefit of the doubt and officially described as hysterical, but there is now no doubt that they are both genuinely schizophrenic.

The main point of interest is, of course, that the twins were reared apart from birth. They were together for only about 1 year at the age of 5, when they did not get on with one another. After different institutional or foster home experiences, one was adopted as an only child by an elderly, reserved woman in a half-Chinese home in the East End. The other was brought up with a foster-sib by a younger, highly demonstrative woman in a suburban higher-class home. Cultural level, family structure, and the age and personality of the parents differed markedly, even if neither upbringing could be described as ideal. Nevertheless, extremely similar abnormal patterns of intellectual and personality development unfolded, culminating at the age of 22 in psychoses that are as similar as any in the series. Hospitalization of the twins occurred within 3 days.

The similarity and time of onset may be due to both twins experiencing similar stresses at the same time, related to their unusual background about which they were sensitive. The arrival, for the first time since his adoption, of the biological mother at *A*'s home early in October preceded his manifestly odd behavior by about 2 months. *B* was taken to visit his sick twin on December 22 in the belief that this might improve *A*'s condition. Instead it appears to have precipitated *B*'s illness, for the same evening he was disturbed and on January 5 was admitted to hospital, just before *A*.

MZ 18, female, born 1927

A: 1st hosp. 30 (Sc 30), 5 spells: 137 weeks; FU 37, working.
B: 1st hosp. 20, 2 spells: 53 weeks; FU 37, working.

	A	*B*
Consensus Diag.	Sc (4)	Other (5)
Slater Diag.	Schizoaffective	Inadequate personality
Meehl Diag.	Schizoaffective	Pseudoneurotic Sc
Global rating	5.0	3.75
MMPI	56/234 809 71:	Not done

Zygosity Evidence. Blood, fingerprints, seen.

Background. Social class 4, *Fa* died 68 (abn), *Mo* 77; 1 *M* died inf. 2 *M* 54, 3 *M* 50, 4 *M* 48 (abn), 5 *M* 46 (abn), 6 *F* 45, 7 *F* 43, 8 *M* 39, 9 *B* and *A*, 10 *M* 32, 11 *M* 30.

Fa junk dealer with cart, often out of work, heavy drinker, aggressive when inebriated, so ill-treated *Mo* that police were called in. Had seizures, stuttered, possibly hypochondriacal; in MH (at 66) a few days with headaches, hissing tinnitus, anxiety, insomnia, dreams, weight loss, poor appetite, and thoughts of suicide of 4 years duration (? cerebral arteriosclerosis); died the following year. *Mo* very kind, "phobia about blood," went off to Australia to visit daughter at 70. 4 *M* MH for 6 months after being torpedoed during World War II, possible second MH admission with delusions of persecution (?). 5 *M* MH for 6 months with delusions of persecution (?). 3 *M* a male mental nurse, difficult to get on with; had ulcers. Roman Catholic family, Irish descent.

B always bigger (birth weight: *A*, 5½ pounds; *B*, 6½). Both left-handed. *A* had more childhood diseases, hosp. at 2 (scarlet fever) and at 7 (facial abscesses); although of equal ability at school, *A*'s illnesses caused her to lose one year which she later regained. *B* dominated *A* but took *A*'s part in disagreements. *B* got own way with parents, tried to dominate the family. Both shy as children, dependent on *Mo* (*A* more so), afraid of dark—*A* had night terrors with screaming; *B* sleepwalker. Both irregular work records. At 18 *A* and *B* competed for affection of *A*'s future husband, *B* eventually marrying his friend whom *B* had unsuccessfully tried to pass off on *A*. *A* married (at 21) to bus conductor: 2 sons, 13 (treated at CGC for "habit spasms" at 2) and 6; *B* married (at 20) to caretaker: 3 daughters (14, 9, and 6).

Psychiatric History of A. Menarche at 11, a shock: at 11–12 imagined hands swelling (had recently put hand through window pane), thought she was getting thinner, would stop breathing, and die; would not go out for fear of fainting; seen at teaching hospital OPD, improved with 3 month convalescence. One day before planned wedding day (at 20) seen at *Maudsley* OPD with complaint of "silly idea I might be going mad," feeling brain had "turned over," denied worry about marriage but "not enthusiastic," improvement with sedation and post-ponement of marriage date (diagnosis: anxiety state). Hospitalized (at 26) for acute nephritis, no psychiatric symptoms recorded. At 30, toxemic pregnancy, overanxious, sensitive; Caesarian delivery. Four months later (at 30) suddenly ran around naked and uncontrolled (*onset*), 3 days later admitted to MH, sleepless, aural and visual hallucinations of *Fa*, vaguely paranoid. Same day miserable, wept copiously. Later predominantly manic, then unreality feeling, "muddled" thinking, occasional ideas of reference or depression (ruminated about slashing wrists). Treated with chlorpromazine then with reserpine. Discharged after 11 weeks (diagnosis: Sc). About 4 months later began to slip back; Maudsley OPD (diagnosis: depression) refused DH vacancy. In and out of MH as well as OPD treatment during next 2 years; would wander away from home, leaving baby unattended, was deluded and hallucinated (light in sky means she's Joan of Arc), exhibited regular mood changes. Treated variously with ECT, insulin coma, chlorpromazine and sodium amytal (diagnoses: Sc; depression; hypomania; schizoaffective disorder and again, hypomania). Absconded several times, also home and coping well with housework and children for short periods. Made one suicide attempt after receiving insufficient medication from GP (attempted to cut throat with razor). At 24 successful discharge to OP supervision; on reserpine, had never felt so well.

When Interviewed. Still felt (at 36) "perfectly well ... capable of doing anything"; on reserpine and sleeping tablets—seemed to be coping well. Slightly affected (voice, athetoid gestures—mannered gait had been reported by age 24). *Strange ideas when ill?* "No none at all, no. I just seemed to go berserk, really. I mean I wouldn't have harmed the children, you know, or anything like that because I haven't got that sort of nature you know what I mean." *Odd thoughts or ideas come into your head?* "No, none at all .... you know I never heard anything like that, I mean I did have the strange idea once or twice that I was Joan of Arc, I did have that sort of thing come over me and I didn't hear voices of any kind you know sort of thing." She was considering part-time employment, only concern was that she may become ill again—"I must admit I do really hate MH. . ."

Psychiatric History of B. "Nervous breakdown" at 11 following collapse of brother-in-law, his death of peritonitis shortly afterward and wife's (*B*'s favorite sister) hysterical behavior at the time. *B* ill for a month with stomach pains, weight loss, screaming attacks, and fear of imminent death. Treated with 2 weeks rest in convent where cried for *Mo* the entire time. Second "nervous breakdown" (at 16) over boyfriend's lies and affairs with other girls—conflict in sexual activity with boy friend and religious scruples. To teaching hospital OPD with fears about eating, pains and vomiting after meals, and screaming attacks at night (diagnosis: anorexia nervosa). Then Maudsley IP but left after 1 day (diagnosis: hysteria of visceral type). Adm Neurosis Unit within a few days; fear of stopping breathing, disappointed with married life (intercourse since 15—now no libido, much guilt, fear of pregnancy). Treated with sedation, individual and group psychotherapy (psychodrama). Discharged after 51 weeks (diagnosis: hysterical hypochondriacal phobia: an acute exacerbation of long-standing anorexia, nervouslike state in lifelong neurotic). Complaints persisted, to OPD again; also gynaecological unit where pregnancy terminated (at 23) because of her instability. Symptoms continue—unwilling to talk about anything but self and symptoms. Two years later thyrotoxicosis suspected, unsuccessfully sought termination of pregnancy. Continued unhappy with husband and within 2 years of last delivery again unsuccessfully sought legal abortion. OPD visits continued, continued irritable, anxious, made unreasonable accusations of husband's relatives and later had a row with neighbor and threatened litigation. Employment, first part-time, then full-time, alleviated some symptoms. Classical migraine developed (at 33). At age 36 still attending OPD (diagnosis: hysteria), symptoms "consisted mainly of anxiety and neurotic-type depression."

When Interviewed. A few months later (at 37) no longer attending OPD; coping better. Asked *why?* "Oh I mean, I don't say I've mastered them (symptoms) completely, I mean I do come up against them, but then I try to sort of sit down and sort of think about them all, you know, try to work them out a bit that way, you know. And also, I think I have got a bit harder, which does help, I used to be terribly sensitive." . . . . *Plans for the future?* "To get a house eventually perhaps and garden, and watch the children grow up . . . you know . . . nothing very drastic, really . . ." *Except for your occasional difficulties with your husband, you've been getting on well for most of your married life?* "Well, my husband, I mean quite honestly he bores me stiff . . . he's just . . . rather boring at times, but I just have to pretend he's not (laugh) . . . *What do your friends think of you?* ". . . a bit outspoken, I think, as much as anything . . . I sometimes come out with things that offend, you know." Expressed pleasure in coming home from work and "settling down for the evening or something like that," or for a really good time "if it's nice weather, to go down to the beach." Despite reminders did not complete MMPI.

Comment. This MZ pair is remarkable on account of the unexpected occurrence of severe illnesses of different kinds in the two twins. *A* had a recurrent schizoaffective illness with paranoid, manic and depressive features, which improved with the nonphenothiazine, reserpine; *B* had a chronic illness, including a year in hospital with anorexia and hypochondriacal phobias. Hers was a long illness compared with those of our other hospitalized but nonschizophrenic co-twins. Although *B* was described as near-psychotic on admission, Meehl is the only judge to consider her as lying high on a schizoid dimension. She is generally diagnosed as an anxiety state and hysterical personality disorder. The twins are indeed concordant in respect of anxiety neurosis early on. *A*'s psychosis seems more difficult to account for on the basis of the early history than *B*'s chronic neurosis (but cf. MZ 4 where a schizoaffective psychosis also developed in the more protected twin). We may note, however, that our male schizoaffective proband MZ 14*A* (though his illness was of much earlier onset and more chronic than that of MZ 18*A* and he was never manic) is matched with an inadequate hypochondriac. In the present family, attempts to discover more about the illnesses of the twins' brothers unfortunately met with resistance.

MZ 19*, female, born 1926
A: 1st hosp. 31 (Sc 31), 3 spells: 283 weeks; FU 38, recent hosp. discharge.
B: 1st hosp. 31 (Sc 35), 2 spells: 56 weeks; FU 38, working, symptoms +.

	A	B
Consensus Diag.	Sc (6)	Sc (6)
Slater Diag.	Paranoid Sc, or hebephrenic	Paranoid Sc
Meehl Diag.	True Sc	True Sc
Global rating	5.5	5.75
MMPI	14*8"356'79-20/	Not done

Zygosity Evidence. Blood, seen.
Background. Social class 4. *Fa* NK, *Mo* 70; *A* and *B*.
Fa unknown. *Mo* unmarried, unintelligent woman from closed and inbred rural Irish community, where suspicion of outsiders and harsh attitudes to unconforming members are normal.

B always taller and stronger. Total convent institutional upbringing with occasional summer day spent with *Mo*. Both poor scholars. IQ scores (age 31) while hospitalized: *A* Wechsler 73 (VS 76, PS 73); *B* Wechsler 76 (VS 85, PS 69); probable underestimates for both. Both stubborn children; later, as adolescents, backward "but nice"; regarded as moody, suspicious, "keeping together." Dominance varied in that *B* was dominant when they were well and *A* when they were ill. *B* (at 26) left Ireland and went to London to better self financially; *A* followed shortly and they worked together as domestics at same hotel. Both remained single.

Psychiatric History of A. Allegedly paranoid with attack of Asian flu at age 30; two months later (at 31) accepted transfer, without *B*, to a branch hotel; when the day arrived suddenly felt persecuted (by Japanese), rambled on about food being poisoned and tried to jump through a window (*onset*). *B* joined in this wild scene, became equally disturbed, and talked about attempts to murder *A*; admitted to OW together. Remained suspicious, resentful; deluded: thought married to a Sikh and grandfather was King of America (provisional diag "situation reaction," Alternative diagnosis: Sc). Transf *Maudsley* IP with *B* from whom she was inseparable; they usually sat huddled together and were upset by any suggestion of separation. Symptoms persisted, eventually transferred (after aggressive outburst in which twin joined) to a different branch of the hospital without *B*; visually hallucinated for first time (black serpents jumping off curtains). Treated with chlorpromazine, 4 months later refused further medication "because it is poison." By this time there were long intervals when

*Both probands.

she functioned reasonably well; however, soon became more disturbed (final Maudsley diagnosis: paranoid Sc). Transferrred to OW (at 32), then to MH where a bit more alert and cooperative. Treated with chlorpromazine and insulin coma; said ideas she had were pure imagination, discharged after 93 weeks on chlorpromazine. Almost a year later found wandering; markedly deluded (she was a secret agent, Moscow paying salary), to MH where treated with phenothiazines, to work after 11 weeks (diagnosis: Sc). Returned to Ireland after working a few months where strange behavior precipitated MH admission and where she remained 179 weeks (diagnosis: hebephrenia). While hospitalized was "nice enough" until questioned, then resentful and hostile; if Russians or Germans mentioned "would attack."

When Interviewed. Working successfully in laundry (at 38); cooperative but evasive in interview... *Strange ideas?* ".... when I was run down I used to have awful dreams, sort of nightmares I used to have ... just dreams, but I hadn't any ideas." ... *Concentrate?* "I can concentrate, it depends on what it is, but not learn. I found it difficult to learn especially." *What was it allowed you to get better and to get out of the hospital the first time you were in?* ".... courage I suppose, I picked up courage .... I didn't like hospital ...." *What would you like to tell me that we haven't talked about?* "... I seen the other night on the screen .. do you think we will get to these rockets that they're flying, because it's awful unfair to the farmers in Ireland having this bad weather and I think the rockets that's doing it ... the farmers should have better weather in Ireland because it's unfair to them ...."

Psychiatric History of B. Considered odd by other workers at hotel: always rambling about leprechauns, dying, going to be murdered, and everyone being pregnant. No symptoms until *B* joined "wild scene" when *A* became disturbed (*onset*): felt persecuted, thought food poisoned, tried to jump through window; became equally as disturbed as *A* and talked about attempts to murder *A*. Together admitted to OW (at 31) where wept and wailed continuously; delusions of marriage (married to a Sikh) and of persecution identical to *A*'s; rapid mood changes, vague and evasive (provisional diagnosis: Sc in a dull patient). Transferred to *Maudsley* IP; delusions of grandeur and of persecution (grandfather was a king ... parents were Sikhs), thought to be tortured by English (they should torture their own blood. It's because I'm a redskin); treated with chlorpromazine. Socialized only with twin, upset by any suggestion of separation, once responded by immediately breaking vase, allegedly accidentally, deeply cutting wrist. *A* was eventually transferred to another branch of the hospital after an aggressive outburst in which both twins joined, culminating in a wild scene—hot tea was thrown in nurse's face and *B* put arm through a window; thereafter *B*, though missing *A*, gradually improved, discharged after 35 weeks (official diag. sensitive psychopathic personality; later regarded by consultant in charge as genuine paranoid Sc). Worked satisfactorily for at least a year and a half, although still deluded (reported weekly letters from *Mo* which always contained news of her nonexistent husband and twin's nonexistent husband and children). Seen when *B* hospitalized second time (at 34); withdrawn, tense and belligerent, refused to look at interviewer or to see twin as twin was "not her sister." Returned to Ireland (at 35) where admitted to MH after police found her wandering. Improved with ECT, discharged after 21 weeks (diagnosis: hebephrenia). Returned to England and worked as hotel maid.

When Interviewed. At 35, uncooperative, denied having ever been in hospital; later, "did visit (twin) in a hospital once." No longer deluded about marriage, stating that neither twin had ever been married; also "(twin) will never get married" and "I'll never get married because I know all about men." Querulous, suspicious attitude . . . "I don't like to be pressed"; more aggressive and agitated than *A*; general paranoid demeanor. Refused repeatedly at beginning of conversation to allow tape recorder to be turned on and later wanted to be reassured that it had not somehow or other been turned on.

Comment. It is reasonable to raise questions about *folie à deux* and culture shock in this pair. Clearly their twinship was responsible for their shared delusions and simultaneous onset of psychosis. At first low intelligence was thought to be a major cause of their inability to cope with life; later interviews when they were in remission proved that they were of at least normal IQ and suggested that the observed deficit had been the result of schizophrenic thinking disorder. The fact that both had breakdowns when in Ireland and separated from each other devalues the importance of the first two points above. At follow-up *A* surprisingly had a boyfriend and was talking about the possibility of marriage and was actually pleasant unlike *B* who, about the same time, retained her paranoid armor.

MZ 20, female, born 1918

A: 1st hosp. 43 (Sc 43) 3 spells: 36 weeks; FU 46, working.
B: ——; FU 46.

	A	*B*
Consensus Diag.	Sc (4)	Normal (6)
Slater Diag.	Organic psychosis (symptomatic Sc)	Normal
Meehl Diag.	Acute paranoid reaction	Normal
Global rating	4.5	1.5
	4'-31 8679/25:0#	93/58170 264:

Zygosity Evidence. Blood, seen.

Background. Social class 1. *Fa* died 52 (abn), *Mo* died 80; 1 *F* 54 (abn), 2 *B* and *A*.

Fa company director; quiet, serious, "old-fashioned, a good father"; died from accidental gunshot wounds, probably not suicidal. Three years before death "nervous breakdown," attributed to business worries, treated at home (necessary to keep knives off his trays, later on holiday wanted to jump overboard); thereafter, periods of depression and worry. *Mo* "a dear," "old-fashioned," inclined to be ill. Maternal grandmother MH at menopause. Maternal uncle, alcoholic, 1 *F*, 9 MH admissions between age 34 and 56; depressions or, more often, mania. At FU, 56, hypomanic, presented herself in good light.

Equal birth weights; *A* right-handed, *B* left-handed. *A* the more mischievious, *B* quiet, solemn. Both secretarial training after boarding school, no difference in intellectual ability or in achievement. Spent one year in Paris after completing school-divergent personalities thereafter. *A* married (at 22) to graduate engineer by whom already pregnant, son (age 21) born later; irresponsible, drank excessively; second pregnancy terminated at request of husband. Divorce (at 40), reluctantly let husband keep son rather than have him fought over. *B* married armed services officer; stable, affectively warm (daughters 14, 12, son 10).

Psychiatric History of A. Seen at teaching hospital (at 37) diagnosis of thyrotoxicosis made, after treatment nervous tension persisted—headaches, acute anxiety about her job, and a feeling of inadequacy; also suicidal tendencies. Complete hysterectomy performed (at 41) because of menorrhagia; after hysterectomy difficult, overtalkative, mainly about religion (*onset*). Toxic goiter developed; thyroidectomy performed (at 41). A few weeks after surgery became "much the same as after hysterectomy," to OPD "in a very agitated and toxic state," admitted to teaching hospital (at 43). Psychiatric opinion—acute schizophrenic reaction, possibly triggered off by hypothyroidism. Treated with trifluoperazine, transferred to *Maudsley* with catatonic symptoms, paranoid regarding family, hallucinations, ideas of reference and influence (by deceased *Fa*'s ghost). Discharged to *Mo* after 8 weeks (diagnosis: acute schizophrenic reaction). Almost a year later aggressive, verbose, hypochondriacal,—in "touch" with deceased *Fa* and recently deceased man friend, *Mo* was forced to leave residence to protect herself. Readmitted Maudsley (at 44), haloperidol and benztropine mesylate used with some improvement. Serum PBI: 5.6 mg/100 ml. Discharged to friend after 7 weeks (diagnosis: paranoid Sc). Holiday abroad after discharge, very depressed. Feelings of unworthiness and depression continued after return to England. To OPD, then readmitted to Maudsley for 21 weeks (diagnosis: depressive illness). *Mo* died suddenly during this hospitalization. With medication improved gradually, returned to work and lived in own flat.

When Interviewed. Working (age 46), generally free of psychiatric symptoms, mood and affect appropriate, described illnesses as always the same—"excitable and suicidal," now gives self "a talking to" if becoming tense and excited—"I just sit down and say, well you must calm down. . ." Belonged to several clubs, played golf, still enjoyed men's company but "I don't sleep with them any longer. . ." Saw no connection or similarity between older sister's illness and her own.

Later History of B. Suffered from "thyroid trouble" for 2½ years (at 30) no psychiatric symptoms; successful radioactive iodine treatment.

Interviewed (at 46), rather heavy growth of facial hair and a low voice. Very happy family and marriage, "wonderful," helpful and understanding husband; many shared hobbies and interests (including building of rockets) and also good health and many friends. *Do you consider yourself to be an emotional person, do you show your feelings?* "No, I don't think I do. . . I'd be rather sentimental perhaps but I'm very easy-going and calm natured. . ." *You don't have a temper?* "No" *How is your appetite?* "Quite normal, I think. . ." *What kinds of things do you do that give you the most pleasure?* "I just love gardening . . . spend the whole time in the garden. . . I just love to see land looking well looked after. But I love to garden. . ." *Do you have a circle of close friends?* "Yes, I have a tremendous amount of friends . . . one can't keep up with them all. . ."

Comment. The consensus diagnosis of schizophrenia for *A* is mitigated by a number of factors: PEM voted acute paranoid reaction which the NIMH-Danish checklist we used categorized under acute schizophrenia; Slater saw her as a symptomatic schizophrenia; LM's schizophrenia rating was used instead of his diagnostic impression of involutional depression. Our own opinion, admittedly somewhat contaminated, is that *A* had a thyrotoxic symptomatic schizophrenia. The Maudsley diagnosis was changed on her third hospitalization from schizophrenia to depression independently of our views. *B* is quite normal despite her own thyroid disorder which was not, however, treated surgically. The affective disorders in the father and older sister are noteworthy, as is the greater number of stresses undergone by *A* compared to *B*'s good luck.

MZ 21*, male, born 1923

A: 1st hosp. 17 (Sc 17), 13 spells: 107 weeks; FU 40, recent hosp. discharge.
B: 1st hosp. 30 (Sc 29), 8 spells: 246 weeks; FU 40, hosp.

	A	*B*
Consensus Diag.	Sc (5)	Sc (6)
Slater Diag.	Atypical Sc, defect state	Paranoid Sc
Meehl Diag.	Acute undifferentiated Sc	True Sc
Global rating	4.75	6.0
MMPI	2* 5<u>87</u>''04'<u>96</u>-3/1:	8*2<u>74</u>'' 6' <u>50</u>139-

*Both probands. *A* is proband MZ 1 in Heston and Shields' (1968) study of homosexual twins.

Zygosity Evidence. Blood, fingerprints, seen.

Background. Social class 3. *Fa* died 77, *Mo* died 80, 1 *F* 57, 2 *M* died 33, 3 *F* 52, 4 *F* 42, 5 *B* and *A*.

Fa skilled laborer, "drinker," gastric ulcers. *Mo* favored *A* as a boy, "weaker"; when *A* became ill, preferred *B*. Paternal uncle general paresis; paternal aunt chronic mental hospital patient from early adult life.

Equal birth weights. As adult both over 6 feet tall, *B* more robust. *B* received slightly higher marks in school because *A* missed more school than *B*. *B* more dominant in twin relationship which was never close. *A* alleged homosexual seduction by priest at age 10; brain concussion 11–12, OP treatment only. *A* clung more to *Mo*, neither liked *Fa*. Parents did not speak to each other for 30 years before *Fa*'s death when twins were 32.

Psychiatric History of A. Change in character noted (at 14)–dreamy, pretended to go to work but did not, sat for hours scribbling nonsense (*onset*). Maudsley OP (at 17) before start of present twin register, felt people could read his thoughts MH (4 months) where received insulin therapy; manneristic, smiled with but one side of face (diagnosis: Sc). Psychiatric discharge from armed services (at 18) after 6 months. Unsettled work record principally as clerk, wanderings from home, four alleged suicide attempts. About this time, increasingly homosexual. *Maudsley* OPD (at 24) treated with psychotherapy— wandering, and homosexual behavior continued. MH (at 25), "obviously hallucinated," 43 insulin comas, discharged after 4 months (diagnosis: Sc). Seen for Twin Investigation (at 28); very normal impression, frank about his homosexuality, verging on total recovery from schizophrenia (Slater), working. Two years later Maudsley IP for 1 month (at 30) following attachment to man at work–anxious, depressed, suicidal (diagnosis: anxiety state with depression). Regarded at teaching hospital as a severe obsessional. ECT at another MH brought some relief, but returned to obsessional rumination on disch: refused leucotomy (at 32). Readmitted to MH (at 32) after various temporary jobs; unable to get to office on time because of obsessional counting of all the books in a well-known bookstall. At hospital, obsessional counting continued unabatedly, organized other patients to join in. Discharged (at 32) after 2 weeks on death of *Fa* (no affect). (Diagnosis: obsessional symptoms intercurrent in a fairly well adjusted chronic schizophrenic).

In next 7 years sought OPD treatment when necessary (e.g., depression after homosexual incident). There were five hospital admissions, two to DH, where consistently diagnosed as obsessional symptoms with anxiety or depression in chronic Sc. Treated with chlorpromazine and perphenazine. Attended Maudsley OPD at least 99 times.

When Interviewed. Age 40, about to attend DH, just out of OW. *What would you say worried you the most right now?* ". . .the fear which is predominant is a sort of homosexuality. . . See I was . . . I had been involved years ago with homosexuals . . . and I always think . . . in case anyone ever finds out. . ." *Is it frightening to think about sex with a female?* "Well I can't perform it with a female, I've tried . . . and I couldn't. . . . I was going to take my life, I didn't. . . I was very close to it. . . ." *. . .How are you right now?* "Oh no I still carry on with the. . . I still carry on with the numbers. I've done so for years. . . It was always . . . it was number four at one time . . . brushed my teeth four times, washed four times and . . . always you know . . . and em . . . shaved four times, and then the number went down to three, and now it has gone up to five. . . . I don't want the tension to come back again. . . ." *Where does the tension come from?* ". . .It's all sort of . . . head hurts . . you feel like scratching someone, hitting someone, it's terrible, it's terrific, terrific tension. . . ." *How would you say your mind works?* ". . .my mind is working . . . is not so fast . . . can't remember so quickly. . . ."

Psychiatric History of B. Head injury at 12, treated in hospital, 1 month convalescence. Always impractical, could not concentrate, sporadic work record (various unskilled positions). ?acute anxiety attack (at 20) after running into a German booby trap, 2 days fugue; court martialled, hospitalized with? psychoneurosis. From 24 increasingly slow, lacking initiative (*onset*), panicky feelings, could not get enough sleep. Seen for Twin Investigation (at 28), unemployed, no energy, no confidence, extremely hesitant, schizoid impression. *Maudsley* OPD (at 29), getting slower and slower, encephalitic Parkinsonism suspected (final diagnosis: Sc). MH (at 30) recent indecent exposure; believed everyone knew about him; spied on him; heard nurse say "there goes that sexual maniac," thought he might be turning into a woman. Received insulin comas, discharged after 37 weeks (diagnosis: Sc). Incarcerated for larceny (at 33), again (at 34) after attempt to gas his entire family; spoke of spine influenced by machinery and voices coming through the wall, transferred MH (85 weeks), compulsion to expose himself "because Jimmy Hobgoblin told me to," treated with chlorpromazine and stilbestrol (diagnosis: Sc). Five further shorter admissions by age 39. Persistent absconder and at one time worked for 9 months as a messenger— before asking readmission (diagnosis is consistently schizophrenic). Further indecent exposure and attempt to gas sister. Ideas of reference and auditory hallucinations persisted, thought disorder sometimes so severe that he was unintelligible, but "an interesting and likable chap."

When Interviewed. At 40, while again in hospital (72 weeks). Able to hold rational but limited conversation; still controlled "to a certain extent" by spiritual influence, which controls his "moods and nervous system." Twice during conversation emphasized that parents provided children with material things, but provided no affection. *How would you describe yourself?* "I don't like . . . too . . . too much . . . too many people, I prefer to remain solitary and alone and live in peace and quiet and so on, and no . . . sort of . . . irritations or noise or anything like that. I . . . I always mixed with people and talked to people and may befriend a few, but I just don't make the effort . . don't trouble too much about friends ... because I found out they can ... (mumble)". ...*how do you occupy yourself?* ... "used to go to the opera and ballet ... and theater ... concert hall at South Bank it's not too expensive to go into ... some of these places, in the gallery...." *Do you have any talent along these lines yourself, artistic talent?* "...I used to draw... I've got some poems with me now, doctor.... I've written a half-dozen odes ... lyrics...." Read poems written about Victorian poetess whom he admired, who had reached the age of forty ... "going on half-way through life...." The poem, Solemn Autumnal, had been written but 2 weeks previous to interview.

Later Information. A, after interview, lived in mission-run dormitory; readmitted MH at 41 and for 18 months at 42 and again to Maudsley at 44.

B absconded 4 months after interview; four subsequent admissions. One month after leaving hospital, *B*'s body was found on the side of the highway; death was due to exposure (at 44).

Comment. The pair is of interest for the different though eventually converging courses of the twins' illnesses. Of the fully concordant MZ pairs this one shows the biggest difference in age at first hospitalization—13½ years. Both twins could be regarded as schizoid premorbidly—they were the odd ones out in the family, did not mix well, not even with one another, developed unexpected interests in religion (particularly *A* when young) and literature (*B* wrote verses). At work *A* had the reputation of excessive reserve and dignity; a relative described him as going through life with his eyes shut; *B* has been described as unpractical, painfully shy. *A*, premorbidly the weaker and more dependent of the two, fell ill insidiously at about 14 with a schizophrenia of a kind that normally carries a very poor prognosis. But at about the age of 26, some time after his third hospitalization, he made a remarkable improvement. *B*, previously regarded as well, was noticeably handicapped only from about 24 when he became increasingly slow. When they were first seen for Twin Investigation at 28 it was surprisingly *B* rather than *A* who looked as if he had had a severe schizophrenic illness. For several years *A* continued to function at the level of a chronic neurotic with anxiety, obsessionality, and homosexual problems. Psychiatric interviews uncovered a wealth of history of sexual maladjustment, but he made no progress in psychotherapy. Around this time his condition fitted the concept of a pseudoneurotic schizophrenia. At about 35 there was some social deterioration. Meanwhile *B* had fallen acutely ill and never made such a good remission as *A*. When interviewed at 40 the twins were as alike as they ever were (cf. particularly their MMPI profiles), though *B* had been a year in hospital and *A* was functioning precariously outside. Since then *B* had died (under circumstances suggesting suicide) and *A* has had a lengthy period in hospital.

It is of additional interest that *B*, unlike *A*, never developed obsessional rituals in the course of his illness. Furthermore, though *B* exposed himself indecently and at one time had the delusion he was turning into a woman, he was never an overt homosexual. Could A's obsessionality and sexual outlets have in some way held the process of his schizophrenia at bay (Stengel, 1945)?

The father was strict and unsympathetic, and the mother is said to have dressed both twins in a girlish way when they were young. The fact that a sister of the father was a chronic MH patient and the mother is described as "very quiet, hardly ever talks to the neighbors" would lend support to the theory that genetic factors from both sides of the family are of etiological significance.

MZ 22*, female, born 1935

A: 1st hosp. 19 (Sc 19), 5 spells: 228 weeks; FU 28, recent hosp. discharge.
B: 1st hosp. 19 (Sc 19), 6 spells: 130 weeks; FU 28, working.

	A	B
Consensus Diag.	Sc (5)	Sc (6)
Slater Diag.	Recurrent Sc, hebephrenic	Recurrent Sc, hebephrenic
Meehl Diag.	Acute undifferentiated Sc	Acute paranoid reaction
Global rating	5.0	4.75
MMPI	86*742'0391-5:	80237-6491/5:

Zygosity Evidence. Blood, fingerprints, seen.

Background. Social class 2. *Fa* 70, *Mo* died 35; 1 *F* 35, 2 *B* and *A*. Adoptive *Fa* 78, adoptive *Mo* 64, adoptive sib: *M* 43.

Fa, tradesman. *Mo* died of postpartum hemorrhage. Twins adopted at 6 weeks into well-to-do family, adoptive *Fa*, director of large company, then 50, adoptive *Mo* then 36. Adoptive *Fa* kept in background at home. Adoptive *Mo* extroverted, obsessional, intrusive, undemonstrative, unpredictable, dominant, also kind and sincere (MMPI 468-7231/509:); Object Sorting Test (at 64) showed poor concept formation, obtaining a score of seven by saying, for example, that items in three of the set groups belonged together because they had a polished or shiny surface.

Birth weight: 7 pounds each. *A* always a little weaker as a child, on one occasion had temperature of 103° when *B* had none at all; *A* did not talk properly until age 3½. Both temper tantrums and night terrors as children. After many changes of schools, private boarding school 13–17 where poor students. IQs (at 21): *A* Wechsler 95; *B* Wechsler 94. *A* passed three subjects and *B* five subjects for school certificates; both began secretarial course which only *B* completed (*A* transf to domestic science). Twins fond, as well as jealous, of each other, *B* dominant, *A* more docile and passive. Both affectionate, very attached to adoptive *Mo*, day dreamers. *B* extravagant, impulsive, and always optimistic.

*Both probands. Case study of response to drugs reported by Benaim (1960).

Psychiatric History of A. Complained (17½) of sexual attraction exerted on her by some of the girls at secretarial college, fell behind in work, left school, and began domestic science course. Again homosexual feelings, felt she was going mad and after a year (at 18) failed exams (*onset*). Later irritable, vague, "muddled," "no sense of time," frequent job changes; consultant psychiatrist suspected schizophrenia (at 19). Admitted to private psychotherapeutic community in dissociated, disoriented state (diagnosis: hysteria); received analytic psychotherapy, turned against adoptive parents. Psychotherapist of the opinion that *A* would have developed sufficient maturity and insight to enable her to live a satisfactory life away from her adoptive parents had it not been for the uncertainties introduced by the inconsistent attitude of her parents and the interference of the psychiatrist whose second opinion was sought by the parents. Then to *Maudsley* (at 20) where untidy, unkempt, sullen, petulant, inappropriate laughter, episodes of screaming (once caused patients to leave lounge in a panic), incoherent (unable to complete a sentence), thought disorder, episodes of muteness, disoriented for time; improved slowly on reserpine: discharged after 76 weeks (diagnosis: acute Sc). At subsequent hospitalizations [ages 21, 23, 28] treated variously with chlorpromazine, reserpine and trifluoperazine. Exhibited thought disorder, was withdrawn, had occasional paranoid ideas; no definite delusions or hallucinations. Much of the hospitalized time was spent in hostels; there were frequent job changes and attempts at special training. Symptoms controlled by supportive OPD treatment between hospitalizations. Voracious appetite and obesity a continual problem. At 24 found to be pregnant by former MH patient (diagnosis: Sc); pregnancy terminated; 2 years later married another former MH schizophrenic, a factory worker, at first happily.

When Interviewed. Twice seen (at 28) in 2 months; appeared remarkably well at first interview; MMPI then 0'284-69<u>73</u>/51:. *How does your mind work?* ". . .I don't know. . . I think of purely practical things the whole time now. I don't sort of sit and think. . . I try to live on the surface. . ." *Is that because you're afraid of anything other than the surface?* "Yes I suppose I am. . . I don't quite trust myself. I mean nobody is infallible . . . and yet I crave for excitement you know, and think wouldn't it be lovely if something different would happen. . . I can't explain really." No specific fears, was looking forward to the future. Had relapsed and was attending DH at time of second interview: Described feeling like a "completely different person" from what she was at first interview; felt "out of control" and as if "going to be terribly ill again." Related incident of imagining she was in love with boy at work when relationship in reality was a perfectly innocent friendship. Of marriage stated "I don't feel like a real wife."

Psychiatric History of B. "Crushes" on older girls at school, many superficial friendships, later ones heterosexual. At 15, experienced episodes of crying, was quarrelsome, argumentative, irritable. Failed to hold various secretarial posts (longest 8 months) because of poor personal relationships. At 19 became especially difficult, developed peculiar gait, slow and hesitant speech, was unable to complete sentences, and lost use of hands. Occasionally would sit staring straight ahead (*onset*). Shortly admitted to private MH: withdrawn, deluded (pregnant but no sexual intercourse), inappropriate laughter, temper outbursts, complained of last employers "prying on her"; later tense, moody, resentful, suspicious—not in a real hospital with real doctors, detained for some unpleasant purpose she would not specify. Treated with insulin comas and occasional ECT with slow improvement; when free of schizophrenic symptoms, markedly immature; discharged after 19 weeks (diagnosis: Sc). Art school (1 term) then, 4 months after *A, Maudsley* IP (at 20) as would sit staring for as long as 9 hours a day; later very disturbed behavior, treated with reserpine. Discharged after 28 weeks (diagnosis: Sc). Within 4 months again Maudsley IP, on reserpine, discharged to OPD after 30 weeks (diagnosis: Sc). Dismissed from secretarial positions for incompetence; in and out of MH four more times in next 4 years (consistently diagnosed as a schizophrenic). Much OPD supervision between hospitalizations. Treated with reserpine, chlorpromazine, trifluoperazine, chlordiazepoxide HCL, and sodium amytal. Remained reasonably well and married (at 28) a 43 year-old attorney who had received private psychiatric treatment (inadequate personality).

When Interviewed. At 28, affect warm, adequate, and appropriate; most objective about early years—"unreal sort of idea what life is all about . . . very immature," . . . *marriage?* "Life's much more fun . . . I've got somebody to share it with. . ." *Adjustment*. . .? ". . .much happier . . much more settled . . for the first time in my life I know what it is to really feel well." A little concerned about resentment feelings against *Mo*, mother-in-law, and previous boyfriend— "since I seem to sort of be unkinder about other people. . . You know when you're very unhappy yourself you can't afford to be unkind to other people."

Later Information. At 31 after birth of own twins, *A* "feels unreal," "nothing makes sense." *B* emigrates with husband; at 30 twice hospitalized for paranoid Sc; has one child.

Comment. At 19 the twins were first admitted to hospitals in different parts of the country, *A* 7 weeks before *B*. Slater described their remarkable preservation of personality despite recurrent attacks of schizophrenia as "a triumph for the control of symptoms and relapse by treatment." Premorbidly, in her psychiatric history, and in her choice of marriage partner, *A* has done less well throughout. In this pair we see that features in the differential history which in some discordant pairs are associated with the affected twin here merely make for a difference in severity. It is possible that in the course of time the difference in severity may even itself out. It is of particular interest that the adoptive parents show many of the characteristics claimed by psychodynamicists as schizogenic. The father is very much out of the picture and the parents' marital relationship might be described by Lidz as schismatic. The twins' relationship was close and both were attached to the mother, *A* more particularly. The mother has been seen as interfering, lacking in understanding for her less intellectual, less tidy-minded adoptive daughters' needs and as herself being emotionally disturbed and showing a pathological degree of thought disorder. On interview she was indeed controlling and likely to interrupt with diversionary material. Her object sorting protocol was abnormal and her MMPI profile, though not elevated, is psychotic according to the Meehl-Dahlstrom rules. Such observations do not, of course, exclude a genetic etiology, though they are consistent with environmental contribution, at least in this case, of the kind hypothesized. Studies of other adoptive parents of schizophrenics (Wender *et al.*, (1968) do not support the view that they are usually of this kind. To help to decide whether such patterns of parental behavior can be a sufficient cause of schizophrenia, one would wish to know the incidence of schizophrenia in the children of a representative group of such parents not selected by the presence of a known schizophrenic child. Our view is that such patterns of rearing are nonspecific for schizophrenia and commoner in some cultural settings than others (cf. *Comment* to MZ 7).

In this pair we have good evidence that the mother-child relationships were unsatisfactory for both parties. However, positive aspects of this relationship must not be disregarded. The mother and the twins have largely resolved their differences. The mother has accepted the disability of the twins without rancor or censure and has doubtless been a support to the twins financially as well as emotionally, which may have contributed to the relatively favorable course.

MZ 23, male, born 1938

A: 1st hosp. 22 (Sc 22), 1 spell: 6 weeks; FU 26, working, symptoms +.
B: —— ; FU 26.

	A	*B*
Consensus Diag.	Sc (6)	?Sc (3)
Slater Diag.	Acute paranoid Sc	Schizoid pers, ?insidious Sc
Meehl Diag.	True Sc	Chronic undifferentiated Sc
Global rating	5.0	3.75
MMPI	64'895-720/1:3#	46835-917/20:

Zygosity Evidence. Seen, no doubts.

Background. Social class 2. *Fa* 52, *Mo* 50; 1 *F* 31, 2 *A* and *B*, 3 *F* 17.

Fa sea captain; silent, moody, little affection, much time away from home for work, usually drunk (then violent) when home; ? homosexual who "interfered with" *B* and possibly *A*. *Mo* shop manager; sociable, inconsistent—erratic care and discipline. *Fa* and *Mo* eventually divorced.

Birth weight: *A* 7¼ pounds, *B* 7 pounds. *A* right-handed, *B* ambidextrous and left-footed. Twins were first boys in *Mo*'s family in three generations and she "doted on them." Very close twin relationship, "did everything together," and were often mistaken for one another. No neurotic traits in childhood, both left grammar school at 16 where both "bright," *A* slightly ahead. *A* was dependent on *B*, *B* fought *A*'s battles for him. Both many jobs, National Service (both discharged with ear trouble), interest in philosophy, psychology, and Eastern religions since adolescence. *A* married twice, three children (daughters 6, 4; son 2); *B* married twice, daughter (6) and son (1).

Psychiatric History of A. At 20 forced to marry woman he had known for 5 years, only to leave 3 months later for another girl; untreated depression followed (4-month duration). At 22 became intensely interested in an oriental religious cult. Suddenly left mother of his second child to join the religious headquarters staff where he hoped to succeed the retiring director and become a new Messiah (*onset*). He had a special sense of mission and was given training in "psychic alignment" and in "thought radiation," came to believe he had telepathic powers and could detect fragments of the thoughts of people around him. Admitted to *Maudsley* (at 22), described "odd experiences" of past few years, "spoke in tongues," still believed that he was a radiating telepath though his range had been decreased; trifluoperazine begun, discharged after 6 weeks (diagnosis: Sc).

When Interviewed. At 26, working well at two jobs, day and night, well-dressed, now adherent of another religious cult, thought self well-balanced; only treatment since Maudsley discharge had been ten sessions of treatment by a practitioner of this cult who had himself been a patient in a psychiatric hospital. Spent much of interview talking about religious cult and its special treatment by means of a galvanometer, and how it had been used to solve his sexual problems. Described some of his strange experiences since discharge, ". . .working overtime . . . thinking . . . nothing significant . . . but all of a sudden I changed my point of view . . . and regarding myself as I am, I was something entirely different. Em .. more like . . . well . . . electric lightbulb, that's a white one, not these transparent things . . . but without any of the stem at all, just a round ball virtually of white light and from me, although this was a sort of em. . . I was still here, but as I said I wasn't particularly conscious of my body. . . I sat there, well I couldn't have said, well there's my leg just at that particular time. It was an experience as such. I experienced myself as this, as this broad white light . . . since I left the hospital various states like this happen. .,. It just sort of keeps on, you might say." *Have you ever been in any direct communication by words say with anything else, or anybody else?* ". . .I had another experience. . . I've got very good reason to believe . . . that I've been rendered . . . what I did formulate then as a radiating taper, that means that seven people have been connected to

me, or our mind is locked .... they can hear me think...." *This is mutual...?*
"No, I find it's very selective..." *Oh?* "...I can hear them think ... it's like
turning the dial over on the radio, or turning it on, virtually. I turn my own
thoughts off just to listen to other people's virtually...."

Psychiatric History of B. Age 15 episode of loss of memory—later found to
have been contrived to divert *Mo*'s attention from the late hour he returned
home. Early unsuccessful marriage (at 19), deserted wife before birth of child.
At 22 (*?onset*) became follower in same oriental religious cult to which twin
belonged, lived with *A*'s wife while *A* ill to "open" her as practiced in religious
cult. Family much upset by his fanatical embracing of the cult and its unortho-
dox attitudes, e.g., refused to let child consult dentist when in great pain and
considered that "a visit to the dentist was the beginning of all his inhibitions."

When Interviewed. At 26: In 1 hour he spoke at great length of the many
insoluble problems of his religious, emotional, and sex life, of his grandiose
ambitions. About to leave secure job as accountant for part-time clerical work
and further training at religious cult headquarters. *Is the philosophy of (your
religious cult) compatible with earning a profit?* "Yes, because ... earning a
profit is only another way of saying condensed energy...." Food relatively low
in his scale of pleasures, and he did not particularly enjoy sleeping—"I'm cutting
this out." *Feeling of aggression and hate?* Uses the "sit-and-grin" technique ...
if you ignore the anger "flowing" to you from someone, it bounces "back to
them...." *What things in your childhood would you say were most influential
in making you the kind of person you are today?* "...Well, there's quite a lot of
things .... my father was a drunkard, among other things ... he wasn't the sort
went happy drunk-style ... he used to go violent... I've 'ad to stand up to him
and belt him one to stop him hitting mother anymore .... observing my parents
in action, in this way, produced problems ... and everyone I observed was
destructive, no solutions.... I think my religion did the trick ... made me
realize this difference in level of people ... what makes a man able not to do all
these destructive manifestations. I mean, for instance, you're faced with the life
of Christ, which is an example, a real example ... that there is somewhere a
difference in level, and that He can do the things you want to do.... Well I
have, from time to time, read lives of Saints ... All this contributed to give me
this ability to view problems from these different levels—that there was a way to
go...." *Have you noticed any similarities or parallels between your life and say
the life of Christ or any of the Saints?* "...well I have traveled the same road ...
and there must be a parallel ... but I mean you take people like Paul of
Tarsus—he started off in many respects a similar way to what I did ... and we're
fortunate enough to be able to see him after he died ... review his whole life,
me I'm just in the middle.... The things we do could be similar because we use
the same data ... I think... I'm in a better position ... we've got a lot better

information available than they possibly could ever have had in Christ's day and age ...." *What would you say has been your most vivid mystical ... experience?* "...I picked up one or two past track items—that is, previous lives. The other thing is, possibly an exteriorization [sic] .... I had the impression, let's put it this way, of being outside my body...." *Ever been insane?* "Depends on level," never insane "on same dynamic level" as twin: "I'm too careful a cuss." At one time tendencies toward homosexuality "...after I'd been married and I was what you call ... disenturbulated [sic] —um, that is, came back up top and there the problems got less and so on and so forth, and I managed to resolve some things and my living situation got better...."

Shortly after interview gave up self-supporting job as planned and moved with family to the headquarters of the cult. Reported (*Mo*) as "not too well." Did not reply to letters.

Later Information. A (at 27) 6 weeks in MH, paranoid Sc. *B* still no news.

Comment. This is our only consensus concordant pair where the *B* twin has never been hospitalized. We have no reason to disagree with *B*'s consensus diagnosis of ? Sc. The almost mirrorlike similarity in their thinking styles is quite striking on interview even though *A* was on medication; in fact *B* obtained a pathological score on the Object Sorting test but *A* didn't. Their MMPI's are very similar in shape but different in elevation. The difference in severity might be explained by some combination of differences in ego strength and *A*'s exposure to a "consciousness expanding" mystical experience which *B* did not pursue to the same degree.

MZ 24, female, born 1923

A: 1st hosp. 33 (Sc 37), 3 spells: 38 weeks; FU 40, working.
B: 1st hosp. 31, 1 spell: 4 weeks; FU 40, working.

	A	*B*
Consensus Diag.	Sc (5)	Other (6)
Slater Diag.	Paranoid Sc	Reactive dep
Meehl Diag.	Acute undifferentiated Sc	Psychotic dep reaction
Global rating	4.5	3.75
MMPI	02"86'4371/:#59	4381-2607/95:

Zygosity Evidence. Seen, no doubts.

Background. Social class 1. *Fa* died at 50, *Mo* died at 72; 1 *M* 50, 2 *F* 49, 3 *M* 45, 4 *F* 43, 5 *A* and *B*.

Fa clergyman; inconsistent and violent, died when twins 13; *A*, who was very fond of *Fa*, was only family member who cried at his death. *Mo* active, domineering; "indifferent" to twins and favored their older sister; *A* fought with

Mo while *B* accepted the situation. Later *Mo* convert to Christian Science. Family to England from New Zealand when twins aged 10. Many family disagreements.

Birth weight 7 pounds each; *A* left-handed but forced to use right hand. *B* right-handed. As infants only *Mo* could tell them apart; as children often mistaken for one another. Boarding school at 10 where lived in separate houses. *A* usually first in class, *B* second; attended 1 year college. *A* sensitive, moody, subject to temper outbursts in childhood and adolescence, more dominant; later the more vivacious, attractive twin. *B* more submissive, had more feeling for other people. At 35 *A* had benign t.b. lesion (routine chest X-ray); *A* was 1 inch shorter and 7 pounds lighter at time of last hospital admission (at 37). *A* successful secretary (until 30); engaged for many years, eventually broke engagement. *B* various secretarial positions then marriage (at 22) to financially successful but boorish and domineering man, two sons (16, 9).

Psychiatric History of A. At 30 disappointed in love affair, developed paranoid ideas (*onset*). On return to work thought she was being victimized and ridiculed in subtle ways, resigned job of 5 years; again felt victimized in new job, also thought elder sister was trying to make her mentally ill. Admitted to *Maudsley* (at 33), 8 months after death of *Mo*; still no insight; ideas of persecution ceased after a fortnight, complained of exhaustion; received psychotherapy, discharged after 17 weeks (diagnosis: paranoid reaction in a sensitive, dependent personality). Six months later readmitted in much the same state as before; also talk hesitant, interrupted, vague, strong sexual guilt feelings, multiple ideas of reference; symptoms cleared on large doses of chlorpromazine. Private psychiatrist and psychotherapist seen between first two hospital admissions and occasionally for 2 years following last hospitalization: considered very paranoid, twice refused ECT for depressive symptoms. Third hospital admission (at 37) treated with ECT, chlorpromazine, modified insulin (diagnosis: paranoid Sc). About this time made an effort to get out more—dancing, rambling, made unsuitable choice of men friends.

When Interviewed. At 40, living on own, steadily employed (same position 2 years). Had friends but did not confide—"There's no time to have a lot of friends!" Somewhat depressed, gets very tired and low, feelings of faintness which attributed to not eating enough. *How is your mind working?* "...(laugh) How's my mind working, how do you mean, do you mean have I got any of the symptoms that I used to have?" Recognized from reading article that she had had symptoms of schizophrenia and found article "very, very helpful." *...one of the things I was interested in was whether or not you were able to concentrate...?* "...I can't concentrate as well as I would like to, I know." Very pessimistic about the future—"If you're not clever enough to get into one of the professions and be really good, I think you (women) should get married when you're seventeen ... life's a matter of achievement really...." Not content with what she's achieved, but not bitter. Talked about her family and Christian Science; considered herself an atheist which led to some sex problems—"If you're not religious, life's only biological, isn't it?" MMPI not returned until 5 months after interview, at which time she was in poorer psychological condition.

Psychiatric History of B. Eight weeks after birth of second child (at 31) admitted to GH for 4 weeks with weakness and faint feelings; thought she was about to die (*onset*); responses slow and concentration poor. Showed immediate improvement when away from stress at home but improvement not maintained and consultant psychiatrist called in, diagnosis psychoneurosis (obsessive concern with bodily well-being), ECT prescribed. Refused politely but coolly to be interviewed for Twin Investigation at age 35. Three years later not contacted on the perhaps biased advice of elder sister that *B* was very sensitive and might be upset by an interview.

When Interviewed. At 40, made a very normal, friendly and relaxed impression. Seemed satisfied with activities as housewife and mother; only worries were about children, toward whom her attitudes seemed mature. *Have you ever worried about anything like this happening to you that happened to A, just because you're twins?* "Yes, I think I did a bit when I had a nervous breakdown, that was about 10 years ago." *But that didn't last long?* ". . .after I got over that I was alright." *What do you think other people think of you?* "I don't know, I don't much care, frankly (laugh)."

Comment. The psychiatric abnormality in *B* is not genetically related to schizophrenia in our opinion. The difference in outcome is not easy to analyze but the fact that *B* escaped the possibly malignant influence of a deviant mother and older sister by marriage at 22 to a strong male is probably important. Only *A* was hurt by the lifelong rejection of the mother, being, for unknown reasons, the more sensitive and having the greater need for achievement of the two twins. *A*'s illness is not the typical kind of paranoid schizophrenia in that there are affective and warm elements in her personality when in remission; she is mildly schizophrenic on our criteria and her twin is expectedly discordant. Disappointment in love, another traumatic rejection experience, precipitated *A*'s onset of schizophrenia at age 30, although she was not hospitalized until age 33 after her mother's death.

THE THIRTY-THREE DIZYGOTIC PAIRS

DZ 1, male, born 1920

A: 1st hosp. 28 (Sc 29), 3 spells: 73 weeks; died 30 in hosp.
B: ——; FU 33.

	A	B
Consensus Diag.	Sc (6)	Normal (6)
Slater Diag.	Paranoid Sc	Normal
Meehl Diag.	Acute undifferentiated Sc	Normal
Global rating	5.0	1.0
MMPI	Not done	Not done

Zygosity Evidence. Physical description by adoptive *Mo*, i.e., *A* taller, *B* shorter and darker; never mistaken for each other.

Background. Social class 3. *Fa* NK, *Mo* NK; 1 *A* and *B* (? order), Adoptive *Fa* 65, *Mo* 63; adopted at 3 months.

Fa Australian, unknown; *Mo* "highly strung, anemic." Adoptive *Fa* machinist, "straightforward, active trade unionist," adoptive *Mo* "kind, uninformed."

Birth order and weight unknown. Twins got on well together. *A* got on less well with adoptive *Fa* while *B* "followed in father's footsteps." *A* anxious, worried, sensitive about illegitimacy, rich fantasy life, effeminate, moved in homosexual circles, impotent, much sexual conflict. *B* healthy, successful, married with one child (4), good husband and father. *B* "good with hands and better with hammer and nails than at studying", became machinist and active trade unionist. *A* quarreled with adoptive *Fa* who prevented his attending University, at 27 received government grant to study English.

Psychiatric History of A. At 27 for a few days believed he was in telepathic communication with people next door and that he could control the frequency and strength of the contact (*onset*). Six months later, after a sexual experience that aroused "intolerable anxiety," he climbed on a roof intending to jump but changed his mind and was subsequently admitted to *Maudsley*. On admission (at 28) there were no delusions or hallucinations and he thought his former feelings of telepathic communication were "probably delusions." Mill Hill IQ 138. After being attacked by another patient he took up boxing which, together with psychotherapy, produced change in his sexual life in that he took women out, behaved aggressively. Discharged after 11 weeks (diagnosis: anxiety state). One month later at OPD distressed with much sexual conflict. In 3 months (at 29) admitted to MH (diagnosis: Sc) with "terrific, endopsychic tension." While hospitalized exhibited stereotyped mannerisms, thought block almost continuous: when asked to repeat alphabet "....um A.. A..... (22 seconds).. no sorry that's .. B's .. (29 seconds) ... C.." then complete block. Talked rapidly, precisely, and without block after first insulin coma. Inoperable carcinoma of the bladder diagnosed, treated with deep X-ray, then home. Within 10 months, admitted to MH (at 30), confused, retarded, hallucinated, flattened affect (diagnosis: Sc). When asked about suicide threat stated that he had a reason to commit murder, but not suicide ... he killed himself long ago.

Leucotomy recommended but never performed. Physical condition deteriorated, died of carcinoma of the bladder.

Later History of B. No psychiatric history. Family unwilling for him to be contacted for Twin Investigation.

Comment. DZ 1 sets a pattern that will recur frequently in DZ pairs. *A* has been diagnosed as schizophrenic by all our judges and *B* as normal, again unanimously. The pair differs from most in that neither twin was seen by us. When first followed up in 1953, *A* was already dead. Though the adoptive *Mo* gave sufficient information for us to be able to include the pair, other members of the family stood in the way of further investigation at that time. In 1964 *B* was untraced.

A, sensitive and intelligent, is the odd one out in this working class home. Though diagnosed conservatively at first as an anxiety state, he proved without doubt to be a schizophrenic. Interpersonal factors may well have contributed to his problem but they seem to be an insufficient cause for the schizophrenia.

DZ 2, female, born 1913

A: 1st hosp. 46 (Sc 49), 1 spell: 16 weeks; FU 51, working, symptoms +.
B: ——; FU 51.

	A	*B*
Consensus Diag.	Sc (5)	Normal (6)
Slater Diag.	Paranoid Sc	Normal
Meehl Diag.	Acute undifferentiated Sc	Normal
Global rating	4.75	2.25
MMPI	6*8<u>231</u>″704′9/5:	94<u>231</u>-0587/6:

Zygosity Evidence. Blood, fingerprints, seen.

Background. Social class 4. *Fa* 89, *Mo* died 68, 1 *M* died 10, 2 *F* 54, 3 *M* 52 (Abn), 4 *A* and *B*, 5 *F* 48.

Fa nurseryman, deaf, alcoholic, violent. *Mo* dressmaker, sensitive, somewhat nervous, domineering, unsociable; 2 *F* ? "nervous breakdown," age 41, treated at home; 3 *M* involutional depression (hypochondriacal and with paranoid features) treatment at MH, age 49; son of *A* diagnosis: anxiety state at 19, acute Sc at 22 and chronic Sc at 23 with paranoid symptoms.

A heavier as an adult. *A* and *B* were never especially close, *B* the leader. *A* favored by *Mo* who rejected *B*. *B* therefore turned to *Fa* and was only family member in contact with *Fa* after separation of parents when twins 9. *A* average scholar, *B* "not so bright"; at age 49 *A* scored higher on short intelligence test. *A* known as "the quiet one," anxious, tense, unsociable, few and only superficial friends, resident domestic for 14 years; married age 26, husband laboratory technologist, odd, shy, MH after 13 years marriage (diagnosis: alcoholism and inadequate personality). Two sons, ages 23 (see above) and 17 (asthmatic). *B* nervous, quick-tempered, "good natured," kind, apprenticed as hairdresser, then worked as civil service clerk; married shop assistant at age 26, one daughter (24). Twins 38 when *Mo* died.

Psychiatric History of A. OPD at 40 treated for depression attributed to domestic tension and husband's alcoholism. At 44 again depressed more continuously. At 45 sexual intercourse, on one occasion, with decorator who did apartment; suicidal thoughts, ideas of reference (*onset*). After one year had to quit work because of delusion that people made references to her immorality. Admitted to *Maudsley* (at 46) (diagnosis: reactive depression with paranoid trends). While hospitalized exhibited hysterical symptoms and made histrionic suicidal gestures; improved slightly with 8 ECT and phenothiazines. Attempted to work at husband's place of employment but co-workers became involved in her delusions and she had to leave. At 49 first seen for Twin Investigation (Dr. N. Parker); confined to home as paranoid delusions had persisted and made it impossible for her to appear in public for long. Diagnosis: paraphrenia by Slater (blind).

When Interviewed. Fifteen months later, delusions and ideas of reference generally unchanged. In explaining why she remained a recluse said ". . .it's because I hate people saying things. . . . I know that they're talking about me . . . I think that it is the people at the end of this block . . . that have spread things . . . it's sort of gone all over the place." *Do you feel as if anybody had been following you?* "I think they have." She expressed the opinion that her husband was "getting sick" of her "always talking about it" but that he believed and trusted her. Domestic stress was present in that the schizophrenic son came home from MH on weekends; his unpredictable behavior demanding constant supervision. Also, the younger son was not doing well in school. *A* on medication without which she was "terribly depressed."

Later History of B. "Nervous breakdown" (at 21) when discovered boyfriend, her first, was married; breakdown described as a "blackout" during which she lost weight and spent a week in bed, demanding someone to be with her constantly; recovery complete. At 46, developed migraines which coincided with *A*'s psychiatric illness. Three years later regarded as normal but "difficult to draw out," described herself as "on the nervy side."

When Interviewed. After repeated attempts to avoid appointments, stated that she enjoyed her work. . . . "I'd absolutely be bored stiff at home." *What kind of a person would you say you were. . .?* "Well, I don't know, I suppose happy-go-lucky . . . there's only one thing, I wouldn't like to be left alone, I wouldn't even be left alone in this flat, I wouldn't like to be, I just like friends around. . . ." Later ". . .I like old people better than young people. . . I get on with everyone, but I . . . prefer their company." When asked if her "nervous breakdown" was "all part of the past" replied "Oh, yes, yes. I think what a fool you know, I was—there you are." Her health was excellent but she got slightly depressed when physically ill. Snobbish, sanctimonious, insincere, proud of daughter's place in social circles.

Comment. As in DZ 17, *A* is a late onset "paraphrenic," previously regarded as depressive, who comes into the series by virtue of follow-up information and a blind diagnosis of schizophrenia by Slater in connection with the unit's study of neurotic and personality disordered twins. Though *B* had a mild reactive depression in her early twenties and developed migraines later on, she was regarded by all judges as within normal limits, and the MMPI also failed to reveal significant pathology. *B* remained normal despite rejection by the mother and, unlike *A*, having only the violent, alcoholic father to turn to. *A* was the more intelligent and the more introverted. She would rank as a "premorbid good" on the Phillips scale. Her paranoid illness followed the stress of extramarital sexual relations; sexual relations cannot be regarded as "causing" schizophrenia, even in introverts. Without the incident one can speculate *A* would not have revealed her genotype.

The illness of the brother illustrates an association sometimes noted between late onset depression of a hypochondriacal kind and schizophrenia (cf. DZ 16). The illness of the son illustrates the more generally accepted view that paraphrenia and early onset schizophrenia belong to the same group of disorders. That they occur in the same family suggests that they at least have significant genetic elements in common, rather than being completely distinct in genetic etiology. We do not know what was the basis of *A*'s husband's alcoholism, but it may be a "spectrum" disorder, and so make the appearance of chronic schizophrenia in their son more likely than it might otherwise have been. His symptoms of mental illness occurred 4 months after his mother's discharge, showing in this instance a close link in time with an environmental precipitant.

DZ 3, male, born 1927

A: 1st hosp. 30 (Sc 30), 2 spells: 49 weeks; FU 37, working, symptoms +.
B: ——; FU 37.

	A	*B*
Consensus Diag.	Sc (5)	Normal (6)
Slater Diag.	Hebephrenic Sc	Normal
Meehl Diag.	Chronic undifferentiated Sc	Normal
Global rating	5.25	0.5
MMPI	Not done	Not done

Zygosity Evidence. Confirmed by clear descriptive material by 2 independent sources as to differences in physical appearance, i.e., *A* is tall and thin while *B* is shorter and of a stockier build. Also, *A* is almost bald and *B* showed no baldness by age 32.

Background. Social class 3. *Fa* 63, *Mo* 66; 1 *M* died inf., 2 *F* 38 (abn), 3 *A* and *B*. Canadian family. *Fa*, aircraft mechanic (former mortician), seclusive, suspicious, bigoted, hypochondriacal, allowed no one to enter locked attic room full of "personal treasures," feared by children. *Mo*, overprotective, hypochondriacal, yet active and energetic, "worshiped by children."

A hated father and "let everyone know about it." *A* and *B* both able intellectually, *A* less forceful and less successful. *A* observed by *Mo* to be different from age 7 when he would lie on bed staring at ceiling and smiling; always quiet, reserved, uninterested in environment, had few friends. *A* completed auto mechanics course, then learned insurance appraisal (16–20), numerous positions, never satisfied or successful. *B* completed auto mechanics course, worked as auto wrecker then began own business as insurance appraiser for insurance company.

Psychiatric History of A. Left Canada to get away from *Fa*, broke down (*onset*) after arrival in London (at 30). Went to Greece to investigate death of friend by drowning, drank water against advice contracting severe dysentery. On return, weak physically and preoccupied with his experiences in Greece, walked about at night, often moaning; would lie on pavement certain he was about to die and got up only with encouragement. Adm *Maudsley* OPD (at 30) (diagnosis: anxiety state), treated with sodium amytal with instructions to consult psychiatrist upon anticipated return to Canada. Three months later, in Canada admitted to MH (diagnosis: simple Sc). He admitted hearing voices 3 years previously but not at admission where he showed depersonalization, depression, and flattening of affect with inappropriate smiling and laughing to himself. After insulin coma therapy he showed some improvement but remained "quite schizoid" and soon relapsed. Within 3 weeks symptoms were severe enough to necessitate readmittance to MH (diagnosis: chronic undifferentiated Sc). On reserpine, developed side effects; on chlorpromazine with little change; less agitated and depressed but very passive and dependent after insulin therapy. Discharged to parents (at 31). At 37, working, inappropriately telephoned friend (nonreciprocated homosexual attachment) in London on Canadian employer's phone and wrote lengthy letters. (FU information provided by parents of above mentioned London friend with whom *A* had roomed.)

Later History of B. Married at 27, son born 3 years later. At 32 described in *A*'s hospital chart as a short, fat, rather aggressive man, forceful, and socially successful. No psychiatric illness.

Comment. Despite a family that some might describe as "crazy," *B* seems to have remained remarkably unneurotic, let alone schizophrenic, without any obvious attempt to escape from the family environment.

DZ 4, female, born 1915

A: 1st hosp. 27 (Sc 27), 5 spells: 324 weeks; died 40 hosp.
B: 1st hosp. 34 (Sc 36), 4 spells: 56 weeks; died 45 hosp.

	A	*B*
Consensus Diag.	Sc (6)	Sc (6)
Slater Diag.	Recurrent catatonic Sc	Recurrent catatonic Sc
Meehl Diag.	Acute undifferentiated Sc	Schizo-affective
Global rating	6.0	5.0
MMPI	Not done	Not done

Zygosity Evidence. Blood.

Background. Social class 3. *Fa* died 65; *Mo* died 49; 1 *M* 63, 2 *F* 60, 3 *M* 57, 4 *F* 49, 5 *A* and *B* (? order), 6 *F* died NK.

Fa railway clerk and salesman, responsible, steadily employed. *Mo*, died when twins 7 years old, step-*Mo* after age 16.

Normal pregnancy and birth. *B* bronchitis from infancy; *A* no illness and "easy to manage." Pleasant childhood, happy with *Fa* and elder sister after death of *Mo*; unhappy after remarriage of *Fa*. Children disciplined by *Mo* only. *A* "mother's girl; *B* "father's girl." *A* a "little shy," friends of both sexes; *B*, no boyfriends. Neither twin married. Neither dependent on the other.

Psychiatric History of A. Sudden *onset* (27) following scandalous talk about *A*'s relationship with married man (? true) wandered at night, expressed delusion that she had killed her father. Admitted to MH (diagnosis: catatonic Sc) stuporose, restless, impulsive; improved with ECT and within year recovered. Remained well 17 months until 3 days before second MH admission (at 29) when she became confused, restless, overactive and hallucinated for her step-*Mo*'s voice. Treated with triazol convulsions, ECT and insulin. Acutely disturbed phases interspersed with mannerisms, fragmentary responses, and "schizophrenic preoccupation," discharged 10 months after admission, "recovered." Eleven months later, third admission to MH (at 31) in a state of catatonic excitement. Leucotomy performed; within 2 weeks pleasant and well-behaved. A few months later, well-behaved with marked memory defect and intellectual deterioration. She remained well for a year and a half when she again became restless, sleepless, said "strange things," and "thought the marks in her sister's dress were germs." On admission to MH (at 34) required spoon feeding, said she had come back to MH because she had "let the medical profession down." (diagnosed first, as recurrent depression, then mixed psychosis, later Sc). Also, recurrent rheumatoid arthritis (acute). Transferred to *Maudsley* (at 36) after 27 months for hormone therapy investigation. Moderate arthritic response to oral cortisone also increased vigor in behavior and talk and well-sustained elevation of mood; return to pretreatment state when treatment discontinued, no change on placebo. After 3 years, to halfway house where she remained surprisingly well(?) for another three years when a sudden relapse necessitated admission to MH (at 40), (diagnosis: Sc). Within a few months, while arrangements were being made for transfer to halfway house she became physically ill and died of bronchopneumonia, emphysema (at 40).

Psychiatric History of B. At age 29, described as pleasant, shy, vague; also as independent and schizoid. She recognized the symptoms of *A*'s fourth mental breakdown and for a week before *A* was hospitalized, *B* (at 34) was restless, slept little, and worried about "whether she had done the right thing by her sister" (*onset*). Five days later; *B* became confused, repeatedly flushed toilet, became agitated, and eventually violent and was admitted to MH (at 34). At admission described as catatonic, required tube feeding and heavy sedation, and was "in grave danger of exhaustion." (Diagnosis: mania, *folie à deux*). After treatment by modified ECT and insulin, she recovered except for bronchiectasis and returned home. She did not return to work as a clerk because of her physical health. Two years later following a visit with *A* at Maudsley, *B* had to be taken from a subway station to OW in a state of catatonic excitement. Modified ECT failed to produce any improvement, but she showed dramatic improvement after three unmodified ECT and was recovered in 3 months. Her third psychotic attack also followed a visit with *A* (at 38). At admission to MH she spoke slowly and was unable to complete a sentence; she believed her food and everything she touched was contaminated (diagnosis: Sc paranoid reaction). Her mental condition improved with modified insulin treatments following which she nearly died of "chest trouble." Nine months later she had recovered physically and mentally. Final admission to MH (at 45) followed 3 months depression with hallucinations and delusions (diagnosis: ?Sc associated with asthma and chronic chest infection). Within 6 weeks, after three ECT with dramatic improvement of her mental condition, she died of bilateral bronchopneumonia and bronchiectasis.

Comment. The occurrence in both these DZ twins of very similar recurrent catatonic illnesses, while the rest of the family are normal, suggests Mendelian segregation, the illnesses in gene carriers running true to form. Both twins had died in hospital by the time of our follow-up. Though reliable information on early personality differences is lacking, there is good evidence of precipitation of three of *B*'s attacks by contact with *A* when sick. However, *A*'s death did not evoke a psychosis in *B*. Like DZ 7, the pair shows that it is not only in concordant MZ pairs that one can sometimes trace an influence of one twin on the other.

DZ 5, female, born 1921

A: 1st hosp. 31 (Sc ?16), 3 spells: 8 weeks; FU 43; working, symptoms +.
B: 1st hosp. 41, 2 spells: 15 weeks; FU 43, recent hospital discharge.

	A	*B*
Consensus Diag.	Sc (6)	Other (6)
Slater Diag.	Paranoid Sc	Chronic anxiety state
Meehl Diag.	Acute paranoid reaction	Psn anxiety reaction
Global rating	4.5	3.5
MMPI	Not done	Not done

Zygosity Evidence. *B* stated that *A* had gray hair from age 17 and was taller than *B*; *B* congenitally missing pectoralis major; never considered to be other than fraternal twins.

Background. Social class 1. *Fa* died, 70; *Mo* died, 70; 1 F 48, 2 *M* 46, 3 *M* 45 (Sc) 4 *A* and *B*, 5 *F* 38.

Fa an academic, strict disciplinarian, unemotional, generous, and fair. *Mo* emotional, dramatic, obsessional. 3 *M* developed Sc, age 18, MH 21, unchanged to 39, then some improvement on chlorpromazine. Irish, Roman Catholic family.

Normal pregnancy and birth, *B* the "weaker" twin, breast fed, normal milestones. No intelligence differences noted. *A* and *B* got on well together; *B* got on especially well with *Mo* who was particularly solicitous when children were ill. The *Mo* removed the children from a school in which they were enrolled because it was "dirty"; subsequently taught by governess, then boarding school in Ireland and England. *B* more placid of the twins but with more childhood fears (dark, storms, spiders, dogs). *A* never married and was not encouraged to work until age 38 when she became a sales assistant. *B* married at 33 to busy lawyer, after working for 2 years as saleslady. Financial and sexual difficulties in marriage; for first 7 years the husband rarely spoke in the evening. Husband overprotective of daughter, first of two children (daughter 7, son 5). Daughter seen by psychiatrist for night terrors.

Psychiatric History of A. Details lacking—variously stated to have had between four and ten attacks, the first (*onset*) about age 16 (uncertain whether diagnosis was Sc), as well as two short hospitalizations at a private MH (first at 31) and one probable subsequent hospitalization before age 40. At each attack she exhibited sleeplessness but refused to take medication for fear that spies would take her to a concentration camp while she slept. She appeared terrified and unkempt, was occasionally overreligious, had delusions about civil war in Ireland, but was never hallucinated. Each episode lasted from 10 days to a few weeks. At each MH admission, treated successfully with modified ECT. Diagnosis: paranoid Sc by *Maudsley* consultant at 40, believed chosen by God for her prophetic qualities, enemies trying to prevent her telling the world of the awful fate that was approaching, refused hospitalization ("If left alone I will get out of it"), and did not take the prescribed chlorpromazine. Recovered in a short time and (at 43) still well and caring for *B*'s children, but not seen for first-hand account of her status and said to be taking chlorpromazine.

Psychiatric History of B. Efficient, capable perfectionist with variety of interests, cheerful but anxious, tense and sensitive but "very controlled," "pretended to be more calm and assured" than was in reality. Developed panic attacks (at 41) in waiting rooms, shops, at parties, accompanied by somatic symptoms (*onset*). Had difficulty going out unless accompanied. After nearly one year, admitted to GH psychiatric unit seeking reassurance that she would recover. (Diagnosis: endogenous depression in a somewhat obsessional, retiring personality). Major response to three modified ECT but remained tense and anxious while attending OPD for next few months, then admitted to *Maudsley* (at 42) in anxiety state. Remained tense, evasive, suspicious, and uncooperative; considered neither schizophrenic nor schizoid but "snobbish" (Mill Hill 101, Progressive Matrices, 108). Refused to fill in MMPI or cooperate in any other way in Twin Investigation. Admitted to being more relaxed out of the home but refused parttime employment because "you can't work and run a home . . . I can't have my children brought up to have meals in the kitchen and potatoes served straight from the saucepan." Improved on amitriptyline hydrochloride and diazepam and, after 11 weeks in hospital, with OP supervision, able to shop and meet people but unable to travel alone (at 43).

When Interviewed. At home prior to Maudsley admission, reluctantly gave erroneous information about herself (denied any psychological problems) and would not give *A*'s address. She and husband forbade the turning on of the tape recorder and, after studied politeness and tea, ended further questioning. They thought inquiries of Irish hospitals about *A*'s admissions would compromise the family's reputation.

Later Information. A continued to attend OPD for a further 2 years.

Comment. Information is lacking because *B* and her husband refused to give a complete history and stood in the way of our contacting *A* who had previously been seen only once at the Maudsley for a reluctant OP consultation. Nevertheless, there is no serious doubt that *A* is schizophrenic. She appears to be a recurrent paranoid, while her elder brother is a chronic hebephrenic. On a polygenic theory, one would suppose that they had inherited different combinations of schizogenes. It would be tempting to see *B* as standing at the lower end of a schizophrenic continuum. However, we do not regard *B*'s mild chronic travel phobia as a borderline condition. The reluctance of *B* and her husband to permit us to seek further data for fear of embarrassing the family in Ireland is likely to be more socially determined than a latent, genetically determined aspect of paranoid schizophrenia.

DZ 6, male, born 1939

A: 1st hosp. 20 (Sc 20), 6 spells: 172 weeks; FU 26; home invalid.
B: ——; FU 26.

	A	*B*
Consensus Diag.	Sc (6)	Normal (4)
Slater Diag.	Hebephrenic Sc	Normal
Meehl Diag.	True Sc	Normal
Global rating	6.0	2.5
MMPI	8716*23''5'940-	4'9-783/5162:0#

Zygosity Evidence. Description from parents and sister. Childhood photo of *A* and *B*: No doubts.

Background. Social class 2. *Fa* 60, *Mo* 53; 1 *F* 27, 2 *B* and *A*, 3 *F* 24, 4 *M* 21 (abn). *Fa* (MMPI 7''08'2-64159/3), civil service manager; described variously as "as near to a split personality as you can get" (*Mo*); inconsistent in child-rearing practices, "perfectionistic," stern, unsympathetic, stable, a "somewhat dominant personality." *Mo* (MMPI 948/37165-02), ex-nurse "more understanding" than *Fa*. Disharmony in the home in early life; later family provided special environment for *A* by moving from urban area to rural area where he could be more overtly psychotic without hospitalization. 1 *F* provided role of nurse and confidante for *A*. 4 *M* grand mal epilepsy. Irish, Roman Catholic family.

Premature birth (about 4 weeks); *B* heavier in build, more dominant, "brighter in school." *B* did more "A level" work in school, however *A* received higher scores on a few tests. *B* a ring leader, *A* a follower; *A* walked earlier, neither had unusual childhood illnesses or problems. As children, *A* was bullied by *B*; as adults, *A* was more normal when in the presence of *B*. In late adolescence *A* and *B* emigrated and lived together in London, using assumed names and changing residences often to avoid payment on incurred debts.

Psychiatric History of A. Course in engineering begun but soon changed to arts with no academic progress, left University (at 19) and drifted (*onset*). One year later became difficult at home, admitted to MH. On admission was auditorily hallucinated, exhibited thought disorder, incongruity of affect, and expressed passivity feelings (diagnosis: Sc), treated with trifluoperazine. Absconded from the hospital after 5 months, little improved, ceased taking his medication and went to England where he had six different unskilled jobs in 4 months (at this time he filled notebooks with senseless poetry). He consulted a psychiatrist and was sent to *Maudsley* OPD Emergency Clinic (at 22) where "residual schizophrenic defect" was diagnosed. In the next 4½ years he was hospitalized almost continuously (diagnosis: schizophrenic reaction, hebephrenic type). He had auditory and visual hallucinations, ideas of influence and was unable to concentrate; also, he was occasionally depressed and at times mildly aggressive. At home, 6 months after release from the hospital, he was extremely overtly psychotic but accepted in the rural community as a kind of "village idiot" who spent his time sleeping or at the pub where he sang songs for the local inhabitants who would then buy him beers. In the village he was considered to be harmless and good-natured.

When Interviewed. At home, he was in a state of excitement over the impending interview. In response to an inquiry as to how he was getting on and what medication he might be on he replied "I'm a saint.... I am taking Stelazine and Largactil..." Then "...I've got no brains at the moment. I'm schizophrenic you see. It makes me do things (mumble) ...I've got no control over my brain, I'm waiting for the day it comes back...." *What led to your illness?* "I heard a voice ... tell me to eat my movement and I did, I hear voices now ... now I see visions." *What kind of things give you the most pleasure?* "Smoking, sleeping... I masturbate every night, 2000 since I left the hospital. ...I get no pleasure out of anything." He mentioned twice that he was "in hell," then stated that he thought about the future "with enthusiasm."

Later History of B. After 2 years at the University studying engineering he gave up his studies before eventually going to England. As a "drifter" he was "rather wild" and inclined to drink too much. At age 26 he was unmarried, unsettled occupationally although employed as a door-to-door salesman, and described as a spendthrift. Plans for him to be seen personally for Twin Investigation did not materialize, though he completed the MMPI by post.

Comment. A was out of hospital at follow-up only by virtue of his parents having moved to a village habitat; he was one of the most disturbed patients in the series. B is judged as normal but his global pathology rating of 2.5 together with his MMPI and history show definite sociopathic tendencies; we cannot say whether B should be viewed as "spectrum" or not. The father is described as perfectionistic and a "split personality" by mother and both are reflected in the MMPI which could be read blindly as one from a schizoid-obsessional character. A's illness is compatible with polygenic theory or dominant gene theory.

DZ 7*, female, born 1936

A: 1st hosp. 22, (Sc 22), 5 spells: 144 weeks; FU 28, hosp.
B: 1st hosp. 22 (Sc 22), 3 spells: 38 weeks; died at 24 in hosp.

	A	B
Consensus Diag.	Sc (6)	Sc (6)
Slater Diag.	Hebephrenic Sc	Hebephrenic Sc
Meehl Diag.	True Sc	True Sc
Global rating	6.5	6.0
MMPI	86*97 13''420'-5/	Not done

*Both probands.

Zygosity Evidence. Blood, appearance.

Background. Social class 5. *Fa* 59, *Mo* 56 (Sc), 1 *M* 34 (Sc), 2 *M* 31, 3 *B* and *A*.

Fa, dock laborer, made fun of term "mental illness" and repeatedly removed *A* and *B* from MH although they had been certified. *Mo* paranoid Sc, at least from age 34; 1 *M* chronic Sc, at least from age 29 with mild depression and hypochondriasis earlier, received psychiatric discharge from armed forces. 2 *M* removed from home by court order for truanting when twins age 9. Roman Catholic family.

A ahead in early development, later "more intelligent," and "took the lead" in everything. *B* had pneumonia and German measles as a preschool child when *A* did not. Only *B* afraid of dark as a child and developed "coughing habit" at 7. *A* completely left-handed, *B* right-handed for writing only. *A* "lively", *B* "quiet." Constant companions as school girls and later, employment with frequent changes together. Parents separated when twins were 3, wartime evacuation, and return three or more times. Spent childhood variously in children's homes, with *Mo* (already paranoid) and, most often, with *Fa* and mistress with whom *Fa* had set up household when twins 4. Attended RC schools and classes for millinery and from 11–14 mostly in training schools for delinquents; both before Juvenile Court at least three times for truanting. 1 *M*, already abnormal, lived in home from time twins 12; when twins 20, 1 *M* a chronic, deteriorated, paranoid schizophrenic. Eventually, after *A* and *B* both psychotic, *Mo*, now deteriorated, also joined household which still included *Fa*'s mistress.

Psychiatric History of A. At 21, complained of bad skin, gave up work (*onset*); went to GH for skin almost nightly; later ran away from home to care for children in Roman Catholic orphanage. Attended hospital with various physical complaints but did not return because she thought the physician had followed her; then to psychiatric OPD requesting ECT as brother had received. One year later (at 22) requested ECT at *Maudsley* OPD complaining of inability to work, converse or concentrate, hysterical pseudopsychosis suspected at first. Refused hospitalization, but eventually admitted to Maudsley IP (at 22), remaining but 2 days (diagnosis: Sc). Within next 6 weeks admitted twice, each time removed from MH after few days by *Fa* (diagnosis: paranoid Sc). Throughout exhibited delusions of sex changes, violence, kidnapping and persecution; at times hostile and abusive; occasionally fatuous and elated. Treated with trifluoperazine. Over next 2 years there were several similar MH admissions. At each admission either her *Fa* removed her from the hospital or she absconded before treatment was concluded. Again admitted (at 26) after writing wild letters on religion, beheading and cosmic experiences (diagnosis: chronic Sc) again treated with trifluoperazine.

When Interviewed. Still hospitalized, seemingly happy with fantastic delusions about sex and religion. Said she committed adultery with a policeman because she was God, had been in hell where the devil gave her a cigarette and no light, was watched by the Americans on TV, Jesus Christ was writing a book about her which she wanted dedicated to research into mental illness. "I've been in hell you know. Have you got a cigarette for me doctor?" *Yes.* "Thank you." *How was hell?* "I wasn't frightened down there. The devil set me up in the corner and he said. . . The nurse said she tried to kill me . . . she said that to Jesus." *How do you know?* "Because Jesus looked at my legs and then she was crying and saying, we tried to kill her . . . and Jesus lifted his head and looked at her you know, and then he gave me an injection."

Psychiatric History of B. At 21 (*onset*—about 10 months after *A*'s onset), took *Fa*'s iron tonic for run-down feeling, thereupon thought she could feel her face changing, believed she was getting hairier and heavier, might be changing sex. Seen at OP, inappropriate affect, refused hospitalization. One year later (at 22) to *Maudsley* OP (5½ months after *A*) with somatic delusions; arm twisted, periods stopping, changing into a man since drinking *Fa*'s medicine. Admitted as IP next day after suicide attempt by gassing during night (diagnosis: paranoid Sc), discharged self within 2 weeks when insulin therapy suggested. Three months later admitted to MH (diagnosis: hebephrenic Sc). Expressed same delusions as previously with inappropriate affect; also believed she had caused *A*'s illness through drinking *Fa*'s medicine. No improvement on trifluoperazine, improvement on new drug until *Fa* offered to take her home. She then became restless and agitated, but was removed from hospital by *Fa*, against advice. Six months later admitted to MH with same delusions; voices told her she was the Virgin Mary; incoherent, disjointed speech, at times noisy and quarrelsome. Died unexpectedly (at 24) of acute pneumonitis.

Comment. With a schizophrenic mother and brother, this is one of our most heavily loaded families. The concordance of the DZ twins fits with a polygenic hypothesis, the risk for a co-twin increasing with the presence of other affected relatives (cf. DZ 10, another concordant DZ pair in a genetically high-risk family).

The chaotic environment, to which the illness of the mother probably contributed more than any other factor, is reflected in the partially institutional upbringing of the twins, the many changes of home, and the delinquency of Sib 2 and both the twins. At one time four overt schizophrenics, the mother, Sib 1, and both twins were living in the same house along with the father and the "stepmother." The influence of one member on another can be seen more convincingly in the content of their ideas than in the form of their illnesses. It was from her brother that *A* had the idea she needed ECT. *A*'s delusion of sex change was doubtless influenced by *B*, who was the first to express such an idea after swallowing father's medicine. Both twins expressed religious delusions but never the same ones at the same time. When *A* first said she was the Madonna, *B* did not believe her, nor did she express religious delusions herself though she was already ill. By the time *B* fleetingly believed she was the Virgin Mary, *A* was a Scottish lady being experimented upon by evolutionary scientists; later she believed she was God. Though the twins differed in personality, there is no reflection of this in their illnesses.

DZ 8, male, born 1908

A: 1st hosp. 44 (Sc 44), 2 spells: 42 weeks; FU 55, working.
B: ——; FU 55.

	A	*B*
Consensus Diag.	Sc (6)	Normal (6)
Slater Diag.	Paranoid Sc	Normal
Meehl Diag.	Acute paranoid reaction	Normal
Global rating	4.5	1.0
MMPI	2″170586′934-	2503-69184/7:

Zygosity Evidence. Seen, quite different.

Background. Social class 4, *Fa* died 59; *Mo* 77; 1 *F* 65, 2 *M* 63, 3 *F* died 2, 4 *F* 61, 5 *M* 59, 6 *B* and *A*, 7 *M* 53, 8 *M* 51, 9 *M* 49, 10 *F* 47, 11 *F* 44 (abn).

Fa, semiskilled, good provider; no show of affection, strict but fair. *Mo* "a good mother." 11 *F* puerperal psychosis, good recovery. 8 *M* (MMPI 429-1837/506:)

B the weaker baby. Both of avg. intelligence although by *A*'s account, *B* was a couple of classes ahead in school. Family noted to be "very normal." *A* and *B* always had a close, warm relationship and as adults saw each other every week. *A* farm laborer in Canada (20–26), later unskilled laborer; single. *B* tradesman, married; twins, daughters (20) and son (14).

Psychiatric History of A. At 37 suspicious of subversive activities at work (*onset*) and again (at 40) when called up for jury duty. Later disturbed by significance of his dreams and ideas of murder being "put into his head." Hospitalized (at 43) for physical observation (bleeding from rectum), became obsessed with belief that he had cancer with but a few months to live. Referred to *Maudsley* OPD (at 43) in anxious, hypochondriacal state. Paranoid ideas soon evident: doctors withholding information from him, accusations of rape and of murders, mind filled with obscenities put there by others, body cut in two with the top half dead (diagnosis: paranoid Sc). Admitted to OW (at 44) after found in gas-filled room having taken overdose of aspirin; to MH for 35 weeks (diagnosis: Sc). Remained well for 8 years after spectacular change in mental state with 12 ECT. At 52, after bad cold became fearful of diabetes and developed visual and auditory hallucinations of devils chasing him. Admitted to OW then MH (diagnosis: recurrent Sc, catatonic element). Treated with trifluoperazine, within 7 weeks in total remission.

When Interviewed. At 55, well clinically and working, lived alone, had girl-friend. *How are you feeling in general, your mental health?* "Well quite good, but, eh, having the heart trouble probably has quite bearin' on . . . on . . . the outlook that . . . it wouldn't do to have another breakdown on top of the heart business. . . I'm tryin' to keep mentally stable anyway." Mentioned that the father had died of heart trouble (twins 25). *Have you figured out what led up to your illness, what caused it?* "Well no, it came in the first, through some fear, and yet I couldn't tell you what fear it was. . I had so much fear . . and it drove me so much that . . I tried to gas myself . . . I'd never do it again (laugh) but I did it, something drove me to do it." *What led up to it do you think, what caused it?* ". . .right from childhood I had nervous tendencies. . ." Later ". . .do you remember . . . that Jones man, he murdered that girl . . . readin' it, it played on me mind, and I more or less took the guilt. . ." *Did you ever hear any voices or see anything like that?* "I did the first time. . . I used to yell back at them . . . (laugh)." *And you just laugh at that now?* "Well, I can, yes, but it wasn't really laughable."

Later History of B. Air-raid warden and Home Guard during World War II. Fair health, "good nerves," great courage, and no fears according to *A*.

When Interviewed. B made normal impression. *What kind of person would you say you were, how would you describe yourself?* "Oh. . . I talk a bit too much sometimes . . . a bit moody . . . . there are several people I can't get on with, but on the whole I get on with the majority of people." Later, ". . .I don't think I'm any different from the average person. . ." *How is your ability to concentrate?* "For me age, fair. . ." *What do you think other people think about you?* "Well, I suppose quite the average person. . . ."

Comment. This normal working class family produced twelve children of whom only *A* was schizophrenic, while one sister had a postpartum psychosis with a good recovery. *B* is quite unremarkable from a psychological point of view. *A*'s illness had a late onset (at 37) as might have been expected from his good premorbid history; his total remission aided by phenothiazines after two attacks gives reason for optimism in such cases.

DZ 9, male, born 1937

A: 1st hosp. 19 (Sc 19), 6 spells: 74 weeks; FU 26, home invalid.
B: ——; FU 26.

	A	B
Consensus Diag.	Sc (6)	Normal (6)
Slater Diag.	Hebephrenic Sc	Normal
Meehl Diag.	True Sc	Normal
Global rating	6.0	1.25
MMPI	827*5 4 3 '' 0 6 '-19/	9'42-38/05 1 76:

Zygosity Evidence. Blood, fingerprints, seen.

Background. Social class 1. *Fa* 57, *Mo* 49; 1 *F* 31, 2 *A* and *B*.

Fa (MMPI 2'0-15768/34:9#), an academic, unemotional, introverted. *Mo* (MMPI 4-236071/89:5#) excitable, voluble. *Fa* Jewish, *Mo* not. *A* right-handed, *B* left-handed. *A* "sweet disposition," obedient, solitary child with neurotic traits (thumbsucking, swaying habit, night fears); *B* "cried more as baby," naughty, aggressive, defiant child. Both in boarding school from age 6. *B* learned to read first, a "little brighter" although *A* excelled in math and twins were in the same classes. *B* popular with classmates and teachers, *A* began to stammer in elementary school and was bullied in secondary school. *A* and *B* did not get on well with each other and never had a close relationship. *A*'s education was interrupted at 16 when he contracted t.b. and again with his psychiatric illness and recurrence of t.b. *B* did advanced level physics at a technical college in preparation for study of medicine, but failed several attempts at Medical Board entrance exams. As young adults, *A* was taller but lighter weight than *B*; *A* was a perfectionist and easily worried, *B* ambitious, inconsiderate and extroverted.

Psychiatric History of A. Stammer since age 8. At 16, developed fears of attack in bedroom, took compulsive precautions (*onset*). Suicide attempt (at 18) by gassing while depressed, stammer worse. Seen by Jungian therapist with initial improvement, then increased anxiety, terrifying dreams, refused to go out, remained in bed. Hospitalized at *Maudsley* (at 19) for 1 month (diagnosis: anxiety state and stutter in schizoid personality); discharged with favorable prognosis. Attempts made at occupational training but after a minor set back he refused to return to work and was soon readmitted with florid symptoms of thought disorder, hallucinations and delusions. Treated with 70 insulin comas and discharged to OP in less than a year. At 21, he spent 3 months at night hostel as he found it difficult to live at home while undergoing rehabilitation; then shallow affect, manneristic, auditorily hallucinated, yet quite cheerful. A few months later (at 22) he spent 2 days at MH, then left; then a short time later entered Maudsley (diagnosis: chronic Sc); deluded that the nurses got into bed with him when at the MH where he recently spent 2 days. Walked out of Maudsley after a few weeks but back again within a month; apathetic, withdrawn. After 5 weeks at Maudsley, he was maintained at home by his *Fa*, but unemployed, unimproved and on trifluoperazine, 30 mg per day.

When Interviewed. Four years later (at 26) single, still maintained at home. *And you find that the tablets work for you...?* "I find I don't hear any voices or things like that, that's the great advantage there. None of the old tablets I took had any effects on the voices at all ... these ones really stop them altogether." *How do you feel about the kind of life you're leading?* "Oh dear, it's most useless, isn't it?" *Do you have a circle of friends?* "I don't go out at all... I don't want to meet anybody... I was at school with ... want ... I don't want them to know that I've been ill. I loathe meeting them, I hate meeting them. Always tell them I work at home . . . job . . . which I'm not . . always tell them I work."

Later History of B. At 22, doing National Service as commissioned officer on Continent. At 26, salesman for large company (for 2½ years). Girlfriend, but no marriage plans.

When Interviewed. At 26, inclined to be concrete in his responses or to miss the point: *Are you a driving individual—do you drive yourself?* "Yes, I have to drive all day long." *How would you say your mind works in general?* "Well, I'm pretty open-minded—people who normally see me, see me in a certain way—I don't think I have any front or anything like that." *What do you think led to his (A's) nervous breakdown?* ". . .Well, I wouldn't like to say really . . he was always sort of quite a little bit reserved. . ." Later ". . .I've accepted it, also I accept any other illness, really. . ." *Do you have any feelings that a similar thing could happen to you. .?* "No, no, it never worried me, anyway." *What kind of things do you do for recreation?* ". .I just bought a sports car. . . I always flash off here or go away for the weekends. . ." Later, ". . .I do quite a lot of fishing and shooting. . . I'm very interested in Zoology, funny enough. . ."

Later Information. A was readmitted to MH 1 month after interview.

Comment. Interesting parallels exist between this family and DZ 6. *A* was maintained as a home invalid on an inpatient dose of drugs; father is a well-compensated schizoid person, and *B* is quite normal and extroverted. The constitutional weakening by tuberculosis at age 16 may have led to more chronic and severe schizophrenia than might have been expected; even early psychotherapy for *A* did not serve a preventative role.

DZ 10, male, born 1907

A: 1st hosp. 49 (Sc 55), 2 spells: 13 weeks; FU 56, recent hospital discharge.
B: 1st hosp. 25 (Sc 25), 3 spells: 1383 weeks; FU 56, hospital.

	A	B
Consensus Diag.	Sc (4)	Sc (6)
Slater Diag.	Paranoid Sc	Hebephrenic Sc
Meehl Diag.	Acute paranoid reaction	True Sc
Global rating	4.5	6.0
MMPI	138"25'4760-9/	2*"98'750-46/13:

Zygosity Evidence. Seen, no doubts.

Background. Social class 3. *Fa* died 85, *Mo* died 52 (abn); 1 *M* 64, 2 *M* died 43, 3 *A* and *B*, 4 *F* died inf., 5 *F* died inf.

Fa and *Mo* second cousins. *Fa* clerk, *Mo* died of chronic nephritis 17 days after admission to MH; agitated, depressed, hallucinated, history of puerperal (at 38) and menopausal (at 50) depressions. Maternal grandfather died in MH, as did maternal uncle (feebleminded); another maternal uncle and maternal aunt committed suicide.

A right-handed, *B* left-handed. *A* more delicate as a child; *B* overactive child, excitable and enthusiastic. Both good in school although *A* considered a "little brighter." *B* did not get along well with his parents and was "kicked out of the house" by step-*Mo* at age 14. No evidence of marked attachment or dominance in the twin relationship. *A* bookkeeper, employed in Central America at one time. Married (at 26) (wife's MMPI 2'301-74589/6:), one daughter at age 28 (MMPI 97160-38254/); solitary, "rather fussy", worried about premature ejaculation. *B* various types of employment, including several clerical jobs, after leaving school, 3 years in U.S. where lived "fast life" and reportedly where knocked unconscious during a brawl. Single, no sexual interest until 19.

Psychiatric History of A. Between the ages of 23 and 42 years he had five short attacks of depression during which he was unable to work. At 49 his energy output slackened, he developed the delusions that his boss was watching him, that police of a foreign power were spying on him and that his residence was "bugged" (*onset*). At *Maudsley* OPD (at 49) schizophrenic illness suspected, admitted as IP. In hospital he was convinced that the patient in the next bed was spying on him. He learned to live with his paranoid ideas and was discharged after 7 weeks (diagnosis: paranoid psychosis), good social recovery. Three years later he was home from work for 3 months in an excited state during which he believed he had won a stocks and bonds newspaper contest for which he demanded the money. In another 3 years he again became excited and delusional, convinced that the money from a train robbery was hidden in the seams of his sitting room curtains; admitted to MH (diagnosis: paranoid Sc). Delusions soon remitted on chlorpromazine; discharged after 6 weeks.

When Interviewed. A few months later along with wife, still quite well but displayed a slight undercurrent of excited affect. "If I had money I would invest . . ." later. *You'd rather have stocks than bonds?* "I think so, yes." *It's a little bit more exciting?* "Yes, yes, I think so." *Do you have a circle of friends. . .?* ". . .Well, not a lot . . . I like to be alone a lot you see and quiet, I like quietude a lot, peaceful I suppose." *How does your liking excitement fit in with your liking quietude?* "Well, it depends upon the excitement. . . I like meeting people really, I don't meet enough but it's my fault because I don't go out to look for them. . ." *You're not afraid of people?* "Oh no, no, not a bit, no. I'm not a bit, no, if King Kong walked in here it wouldn't worry me."

Psychiatric History of B. At 24 fired from 2-year employment at a London tavern, perhaps because of breakage. Later (at 25) restless, unable to concentrate, vague and evasive (*onset*). Diagnosis: Sc at Maudsley OPD (at 25) (before the current twin register was begun) where he sought help because he felt "all queer and clashing." Admitted to MH; many mannerisms, unable to give detailed history, inappropriate affect. Oriented for time and place, no hallucinations or delusions but required close supervision because of impulsive behavior. He absconded after more than a year, still very schizoid and lacking initiative. A month later at Maudsley OPD, unable to work, complained "inside of head not as other people." A few months later admitted (at 27) voluntarily to MH for 38 weeks. Reattended OPD (at 29) while living with eccentric uncle and doing odd jobs; thought-blocking observed (diagnosis: hebephrenic Sc). At MH admission some insight, e.g., "Once you've had this trouble it's very hard to get it under again. . . You see, the mind, it overlaps day after day, I suppose, and it's not overlapping properly, you see"; 6 months after admission, attempted to hang himself (diagnosis: Sc). He remained deluded, depressed, untidy, occasionally absconding. At 41 leucotomy was considered. He developed frequent maniacal outbursts from age 46 which were treated with ECT (discontinued after a fractured vertebra) and repeated courses of chlorpromazine with no improvement. At 53 there were periods of acute agitation and self-mutilation; morphine was prescribed and there were signs of Parkinsonism which were not evident 2 years later.

When Interviewed. At 56, continuously hospitalized for past 24½ years, gave the impression of possible superadded brain damage and was able to hold only an extremely limited conversation. Very poor performance on copying simple geometric figures. He asked how long he was to remain hospitalized and perseverated on his age. *How is your mind working?* "Pretty fair." *You feel pretty good in the morning when you wake up?* "Yeh . . . yeh. . ." *Is there anything in particular that worries you?* "No, no, no." *Is there anything that you're afraid of?* "No, no, I don't think so." *How do you spend your time?* ". . . in the morning laundry . . . . and occupational therapy . . . . make the mats. . ." *Do you enjoy reading?* "Not much . . not a lot, not a lot. . ." MMPI was read to him.

Later Information. A rehospitalized (at 57) excited, paranoid.

Comment. Although one of our three concordant DZ pairs, they are quite different in ages at onset and course and (so far) outcome. The severity of *B*'s schizophrenia sets a record in our series but is contributed to by presumptive brain damage and the effects of state hospitalization per se. *A*'s late onset led two judges to withhold a diagnosis of schizophrenia. This was our only pair where there was parental consanguinity and where there was such conspicuous genetic loading on the maternal side for mental illness. The array of psychopathology is compatible with a broad polygenic theory.

DZ 11, female, born 1930

A: 1st hosp. 18 (Sc 18), 4 spells: 141 weeks; FU 33, working.
B: ——; FU 33.

	A	*B*
Consensus Diag.	Sc (4)	Other (5)
Slater Diag.	Catatonic Sc; psychopathic	Psn, anx state
Meehl Diag.	Acute undifferentiated Sc	Psn, anx reaction
Global rating	4.75	1.75
MMPI	4″6-51873/290:	30-627/15489:

Zygosity Evidence. Seen.

Background. Social class 2. *Fa* NK; 1 *B* and *A*, 2 *M*, maternal half-sib 4; Adoptive *Fa* died 73, Adoptive *Mo* 69.

Twins adopted at 15 months. Biological *Fa* salesman, *Mo* unmarried domestic. Well-to-do adoptive parents refused to adopt one other child born illegitimately to twins' biological *Mo*. Adoptive *Fa*, executive, shrewd, rigid, unsympathetic, generous but lacking in understanding; adoptive *Mo* (MMPI 50-92781/3:46#) cooperative and warm on interview, good perspective on twins, not "schizophrenogenic". Easygoing, overgenerous (*A*), unable to show affection, neither very maternal nor very domesticated (*B*). But both seen as essentially good parents by both *A* and *B* now.

B thumb-sucker until 4; *A* measles at 8, enuretic until 14, a nail biter until 16. Twins had a private vocabulary until age 3. Governesses until 8, to Canada at 11 with adoptive *Mo* during World War II, then back to London at 13. Always attended private boarding schools where *B* was a little better than *A* at both lessons and games. Some rivalry between *A* and *B* in that *B* was better looking; *A* nonetheless popular, plenty of initiative. *A* attended secretarial college and *B* worked as a receptionist after completion of secondary school.

Psychiatric History of A. At 18, while attending secretarial college, her manners deteriorated, she kept undesirable company, put on weight and menstruation became irregular (*onset*). She became depressed, weepy, confessed to hectic sex life, fears of pregnancy and VD; then quiet, inaccessible, and would not eat. Admitted to *Maudsley* as IP (at 18) where mute, inaccessible, dishevelled, smelt of acetone, remained in semistupor; total of 26 ECT, after 6 months still apathetic and listless, showed marked improvement on thyroid 6 gr daily. IQ 126. Still somewhat listless at discharge after 36 weeks (diagnosis: probably catatonic Sc). After discharge was told for first time that she was adopted. At OPD about 4 months later alert, lively, extravagant, cheaply and flashily dressed, impulsive, obstinate, not hypomanic. She became increasingly wayward, was sentenced to 6 months imprisonment for stealing, sentence withheld on psychiatric grounds, to MH as VP. At about this time (at 20) had induced abortion. After 52 insulin comas and 5 ECT no insight, impulsive, at times violent. Leucotomy considered but after 6 months improved with no further treatment and after 43 weeks discharged as recovered (diagnosis: chronic Sc). She remained quite well until *B* announced her marriage plans; *A* became boisterous, careless, and, after she returned home inebriated, her adoptive *Fa* returned her to the MH where she remained for 20 weeks. No psychotic features were observed and she was regarded as a psychopath. Married at 23, divorced at 26, no children. Worked as secretary, lived with aunt and was well until age 27 when she became restless, couldn't sleep, lived life of an amateur prostitute. When family heard she was pregnant (had second induced abortion), she was admitted to MH under an urgency order where diagnosed as psychopathic personality with (?) underlying schizophrenic illness. She left the hospital 4 weeks later and travelled around sleeping with truck drivers. When brought back 2 days later, saw nothing wrong in her escapade; depression and thought disorder had vanished. After 42 weeks discharged considering marriage to ex-patient.

When Interviewed. Five years after discharge, had not remarried, living on own, working, no further treatment, stable and symptom free. *You have no trouble in your concentration?* "No, I don't think so. . . I've got a very good memory." She felt confident about her work as a secretary. Explained that she didn't recall how she felt when she found out she was adopted as her mind was "blank" for that period of her life (the time of onset of psychosis and first hospitalization) but that presently "I just think that I've been very lucky. Two very good parents. I don't think now I'd want it any other way really, not now." *What kinds of things give you the most pleasure?* ". . .I like tennis and bowling and I enjoy dancing . . . television . . . reading. . ." *Do you have any habits which seem a little bit silly to you?* "I bet, not heavily. I have a little bet on the horses now and again and perhaps that's a waste of money."

Psychiatric History of B. Married (at 22) a successful lawyer with whom she was happy. Had three children, two boys (10, 8) and a girl (5). She worked as a receptionist until her first pregnancy; then busy with children, home, golf, bridge, community activities—but described herself as "a home bird." At 28, her adoptive *Fa* died the same week her third child was born; then she was incapacitated with thrombosis. During the following year her husband had major surgery, the children were ill with childhood ailments and *A* was very ill; "I stood it for a year before this actually happened . . . and everything just reached a pitch with me. . . I was out one day and just seemed to panic for no apparent reason. . ." After panicking a second time, *B* consulted GP who treated her with chlordiazepoxide HCl and amphetamine. She lost her appetite, became fearful of panicking again then became concerned that she might become ill like *A*. She was never seriously indisposed by her symptoms but it was a year before she regained her confidence.

When Interviewed. At 33, there had been no recurrence of anxiety and she was feeling "very well, very fit," and looked and acted that way. *Do you consider yourself to be an emotional person, do you show your feelings?* "No I don't. . . I hadn't shown any feelings . . . about my sister. . . I had heard about my own adoption, sort of between the ages of seventeen and. . . over a period of ten years I suppose I had taken a lot more than the average person would without showing any emotion to it. . . This had built in to this sort of feeling . . . sudden panics for no apparent reason." The feelings of panic didn't affect her thinking at all. After the first time, "I got back to the car and I knew I was perfectly safe to drive that car home. . ."

Comment. *A*'s first breakdown defies easy explanation on environmental grounds; she had led a comfortable middle class life with good adoptive parents, and, like her sister, had been evacuated to Canada during the war with the adoptive mother. A number of judges commented on the psychopathic nature of *A*'s subsequent breakdowns and they may have been exacerbated by the punitive reaction of her adoptive father and inappropriate treatment with ECT and insulin coma. Although *B* had a neurotic diagnosis from our judges, she had no schizophrenic features and made a perfect recovery; at follow-up, she was one of the healthiest people in the entire sample. *A* at follow-up showed no schizophrenic features, had mended her psychopathic ways, and was only suspect because of her marked elevation on the MMPI scale indicating psychopathic deviancy.

DZ 12, male, born 1937

A: 1st hosp. 23 (Sc 23), 5 spells: 45 weeks; FU 26, recent hosp. discharge.
B: ——; FU 26.

	A	*B*
Consensus Diag.	Sc (6)	Normal (6)
Slater Diag.	Hebephrenic Sc	Normal
Meehl Diag.	True Sc	Normal
Global rating	6.0	1.75
MMPI	21'63584-709/	9-504/21 7836:

Zygosity Evidence. Blood, seen.

Background. Social class 3. *Fa* died 56, *Mo* 56; *A* and *B*.

Fa storekeeper, a quiet, knowledgeable, conscientious man with few diversions. *Mo* dull, anxious, suspicious when first seen (1952), treated for thyrotoxicosis, was over-protective of *A*, MMPI(6'831247-90/5:).

Toxemic pregnancy, 6-week premature breech birth, *A* 5 pounds, *B* 3 pounds, 11 oz. By 2 years, *B* heavier than *A*, thereafter, taller and heavier and by age 15,

5½ inches taller. *A* and *B* never got along well with each other (*B* described as "bossy") and at age 12 attended different schools where each was below average academically; *A* attended a secondary modern school where he graduated 22nd of 24, *B* attended a technical school; differences in PMA scores consistent with this. Family seen when twins almost 15 as part of study of normal twins: *Mo* of dull intelligence, anxious, suspicious; *Fa* stable, knowledgeable craftsman. *A* looked perhaps 4 years younger, still occasionally enuretic, also friendly and compliant. *B* shy, awkward, unhelpful, had badly bitten nails. *A* became a signwriter, a policeman, then clerk but remained single. *B* a toolmaker, married (at 26) and had plans to have a family.

Psychiatric History of A. At 20, believed he was accused of homosexuality and that people were talking about him (*onset*). He left his job and later, when he met the man he believed accused him, he went into a "state of semiconsciousness" and contemplated killing the man. Three years later and about 1 year after *Fa*'s death, seen at *Maudsley* OPD (at 23); GP reported that he had had bizarre ideas and hypochondriasis for some time. At Maudsley he was vague and ruminated on methods of killing his accusers. Diagnosis: "uncertain (? early Sc with obsessions)." Within 2 weeks he was admitted to MH unable to concentrate, deluded regarding sexual organs and auditorily hallucinated (diagnosis: hebephrenic Sc). He responded well to trifluoperazine, chlorpromazine and ECT and was discharged 11 weeks later, free of delusions. Six months later he was readmitted with relapse of florid ideas of reference and delusions of sex change. He refused medication, was compulsorily detained, then responded well to trifluoperazine and was discharged after 9 weeks. During a third hospitalization (for 5 weeks) he expressed suicidal intentions. Diagnosis was again that of Sc with improvement on chlorpromazine. When seen during this hospitalization for the Genetics Unit he was friendly, at ease and with no psychotic symptoms. A few months after release he appeared at Maudsley OPD with a letter addressed to another hospital asking advice about entering a leper colony. There were two hospital admissions (at 25) within the next year following suicidal attempts and complaints of bizarre hypochondriacal delusions and incongruous affect. He made a good recovery on phenothiazines and ECT.

When Interviewed. At 26, a month before discharge from fifth hospitalization; 2 months later was still well. *How do you usually handle your emotions?* "Well, I'm not neurotic, I don't think. . . No, I'm not neurotic. . . I take it calmly." He stated that he had no plans for the future. . . "I wait 'til I see what tomorrow brings. . ." Expressed opinion that 6 years in the art world and the "some not so decent" people he met there were the cause of his illness. He described his most recent suicide attempt as "euthanasia, mercy killing . . . on myself actually, I'm glad to say." *And you call that euthanasia?* "Well, you'd call it suicide if you want to." *And did you think that you knew what you were doing when you. . .?* "Oh, yes." *And how do you feel about it now?* (long pause)

"I don't think I would have done it now." *Was it just something that passed through your mind and now it's over with?* "Eh .. well, my brother's getting married next Saturday week so . . . I shall have to go to the wedding." Later, . . . *any things that you were afraid of in the past?* "Yes, fear of homosexuality . . . I've still got that fear . . . fear of becoming one." *. . .reasons for this fear. . .?* "You've got two types of homosexuality, one is the mental type and one is the body type. . ." *And which type do you fear?* "Eh . . . both (long pause)."

Later History of B. Seen for Twin Investigation at 25: reluctant to be interviewed, ill at ease, facial twitch, very tense, would not complete questionnaire. He had been satisfactorily employed by one firm except for the time in Army. Enjoyed dancing and football but had no close friends.

When Interviewed. A year later (at 26), had been recently married. Described self as not very ambitious, reserved, "a bit suspicious," and commented that he liked to have his own way. He didn't think he was much like either parent, didn't recall admiring any one person as he grew up and said that he and *A* "used to get on each others nerves." He expressed opinion that *A* could overcome illness if he "set his mind to it . . . he should grow up actually. . . ." *Is there anything in particular that worries you?* "The only thing that I've thought about actually is if I was out of work you know . . . we don't get paid for sickness or anything like that. . ." *Do you have any questions that I might be able to answer?* "Only one. . . I was going to ask if anyone would know how I am up top" (laugh).

Comment. *A*'s schizophrenia is marked by severe thought disorder but he makes a good recovery from his attacks. *B* is seen as normal, though he is somewhat restricted and inflexible, and has difficulties in social relationships. Indeed, *B* seems to have been the more schizoid premorbidly. It is of interest that the mother was noted prospectively as being anxious, suspicious, illogical and overprotective toward *A*. On follow-up she proved to have an abnormal object-sorting score and an abnormal MMPI, highest on the Schizophrenia and Paranoia scales.

DZ 13, male, born 1916

A: 1st hosp. 33 (Sc 31), 1 spell: 303 weeks; FU 48, working, symptoms +.
B: ——; FU 48.

	A	*B*
Consensus Diag.	Sc (6)	Normal (6)
Slater Diag.	Paranoid Sc	Normal
Meehl Diag.	True Sc	Normal
Global rating	5.75	1.5
MMPI	8*4"29'063-571/	042-539/7186:

Zygosity Evidence. Seen, no doubts.

Background. Social class 4. *Fa* died ? 39, *Mo* died 78; 1 *M* 50 (Sc), 2 *A* and *B*.

Fa semiskilled laborer, died when *A* and *B* 18-months old. *Mo* had married at 35 hoping to avoid having children; described variously as "an extremely possessive and very bitter woman" and as a good *Mo*, the latter description by 1 *M* whom the *Mo* favored 1 *M* (MMPI 98'5260-14/37), MH twice (28, 30)—God speaks to him, voices guide him, has a mission to study the origin of the pyramids, has discovered the secret of Yoga (diagnosis: Sc). At 50 still deluded (e.g., bus-route map contains the secret of the pyramids), but regularly employed (as capstan operator), author of privately published "Find the Integrator-Operator (or the Spiritual Neutron), Parts I, II, and III," "Open Thine Eyes that Thou Mayest See, Volume I," and other works.

A more childhood illnesses, scarlet fever followed by rheumatic fever; *B* unaffected by either. There was no close relationship, *A* and *B* both dominated by 1 *M* when children, "went separate ways when older." Both left school at age 14; *A* usually a class behind *B* throughout school years.

Psychiatric History of A. Ideas of reference perhaps began at 14 after homosexual episode with older man. When in the army (at 29) felt victimized (*onset*). Later (at 31) believed he had insight into other people's minds, felt he must study the Chinese language, was upset by vibrations at work and believed his brain moved about according to attractions surrounding it; referred to a clinic but did not see psychiatrist until nearly 2 years later (at 32) at *Maudsley* OPD (diagnosis: schizophrenic defect state). He continued work as skilled laborer until excited behavior and bizarre letters to employers necessitated transfer to MH (at 33). He believed a letter to his elder brother concerning the secret of the pyramids had been intercepted: "All gates are open and that must link up with Aldgate and Aldersgate"; "there are three primary movements, Right, Left, and Center . . . and all human movements are controlled by the Universe." Six months later, after 33 insulin comas, no improvement, found thinking "all a mess" and proclaimed that "there is stuff at the base of the spine which rises when you're in a certain state and sets the nerves a-tingle." Later hallucinated, disturbed by "vibrations coming from brother," 9 ECT. Four years later, noisy, abusive, made homosexual assult on another patient, further ECT. Following year fairly quiet, discharged after nearly 6 years (diagnosis: paranoid Sc, liable to recurrence). Subsequently lived with elder brother and returned to work in factory.

When Interviewed. At 48, at home jointly with elder brother, still single, believed he had kept well by dieting, evidence of thought disorder, did not consider the MMPI had been drawn up by "pure intelligence, so I have attempted to counter impure intelligence with pure intelligence. . . I look upon this chart more as a political and religious dynamite chart as far as this country is concerned and not a personality chart." *Do you think there is such a thing as mental illness or as physical illness, is there such a thing or do we make that up?* "You've got to bring the physical, the mental and the spiritual together as one. . Now there's an argument on the society plane . . also it does exist, and if a searching person is in a ward . . in ward 4 of the same hospital, it can be proved. When the nurses . . when the nurses used to come in on the morning, occasionally, I've not told me brother. Occasionally she'd walk up to his bed and say, come on get out of your bed. This bloke would get up out of the bed, he'd point to the left, the bloke in the left bed and the bloke to the right of him and he'd say he wants me up and the nurse would say go on get out into the wash house, and don't talk such stupid rubbish, and then he'd wave his hands around; it took me ages to puzzle it out and I came to this conclusion, he must have been in what I would call a conscious trance, he's on his bed he can't move . . . the only way I could puzzle it out was that through the sexual organs . . . eh . . . psychic cloud sent up from all those people in the ward must descend and must draw the air out of the body through the sexual organs and then he got this sense that he was in the seminal fluid and he called it . . . on top, that means to say there must be a current of air going through his body as I see it."

Later History of B. Worked for same company from age 14, interrupted only by army service overseas where prisoner of war for 3 years. Age 23 married a woman who had been nearly blind since childhood, no children.

When Interviewed. At 48, appeared perfectly normal in all respects; home owner, avid gardener, keen cyclist. Proud of achieving clerical position after beginning as a manual laborer. *You've worked with them, how long?* "Thirty-four years." *What will you do after you retire?* "Well, it's a way off yet, isn't it? It remains to be seen, just depends on whether I decide to go on to 60 or 65. 'Course I've been fortunate in health, I've not 'ad a day off sick, since 1938." *How did you get on during the war when you were POW? Where were you captured?* "I was captured in North Africa, held prisoner in North Africa for a while, then we went across into Italy. I spent I should think . . . about two and a half years in Italy. Then we thought the day of freedom had come with the invasion, unfortunately the Germans dropped parachutes down. Parachutists recaptured us and took us across into Germany; eventually we were released with the Russians." *Did you feel you were pretty well adjusted at the time you were released? Feel you could put it all behind you and didn't have any nightmares or things like that afterwards?* "Oh no, I had no nightmares or anything like that. No, I think I can safely say I had really adjusted myself."

Comment. A, in the absence of drugs or psychotherapy, can still work for a living without attracting attention despite extreme thought disorder. If lack of a father, a possessive, embittered mother, and subjection to the dominating influence of a prepsychotic or already decompensated brother are considered to be sufficient to explain *A*'s schizophrenia, it is surprising that *B* is so completely normal. Not only was he exposed to the same factors, but he has had to cope with the stresses of captivity as a POW and taking care of a disabled wife.

DZ 14, male, born 1939

A: 1st hosp. 22 (Sc 22), 1 spell: 46 weeks; FU 24, working.
B: ——; FU 24.

	A	*B*
Consensus Diag.	Sc (6)	Other (2)
Slater Diag.	Paranoid Sc	Normal
Meehl Diag.	Acute paranoid reaction	? Simple Sc
Global rating	5.0	2.5
MMPI	4'358-1672/90:	7"2'80 341-56/9:

Zygosity Evidence. Seen.

Background. Social class 2. *Fa* 59, *Mo* 58, 1 *F* 26, 2 *A* and *B*.

Fa (MMPI 629-5410/378:) butcher, *Mo* (MMPI 54-89/132760) hard of hearing. Conventional, undemonstrative, unperceptive, well-meaning family. Birth weight: *A* 4½ pounds, *B* 3½ pounds. *A* smaller as child, right-handed; *B* left-handed. *A* better educated in that he attended commercial college until 17; at 22, Wechsler IQ 83; later, Progressive Matrices and Mill Hill IQ 110. Vocabulary IQ based on Mill Hill: *A* 103; *B* 89. Neither twin married.

Psychiatric History of A. At 18, after misinterpreting TV program on the Wolfenden Report, feared he would become a homosexual because he was a twin; tried to suffocate himself on a cushion. During army service abroad (at 20) felt "run down" mentally, heard a voice, which might have been God's, telling him to seek treatment in England (*onset*); popular songs and TV commercials had special message for him. Unable to cope with former routine work on return to previous job, was increasingly "far away." Not long after return to work, found wandering by the police, admitted to *Maudsley* IP (22) via OW where expressed delusions of guilt—"I think I was meant to jump into the river, I would have been cleansed"—ideas of reference and auditory hallucinations. At Maudsley vague, retarded, consistently depressed, ruminated about homosexuality. Even with intravenous sodium amytal revealed little about himself. Imipramine relieved depression but did not influence delusion that the Wolfenden Report had caused the left side of his brain to decay, he had seen Sir John Wolfenden in a pub, laughing at him (incongruous laughter noted). No improvement on chlorpromazine 400 mg. After 8 months at Maudsley, behavior was normal but delusions of a plot against him were unshakable. Within a few weeks he no longer mentioned his delusions and was discharged (diagnosis: Sc), remained remarkably well during 6 months OPD FU.

When Interviewed. At 24, still working as clerk, single, nothing schizophrenic observed. Stated that he had been "very well, actually" since he left the hospital; spirits were "Oh, very good," . . .*sleep, O.K.?* "Yes," *And you eat O.K.?* "Yes." *Is there anything that is bothering you at all?* "Nothing at all." Defensive in attitude but cooperative. *Do you spend any money or time with girls?* "Occasionally yes, not very often . . . I . . . you know I like going out with girls." *Is there anything you're afraid of?* ". . .well I'm obviously afraid of some things. I mean I'm not necessarily afraid of anything I shouldn't be afraid of."

Later History of B. National service at 20, and year abroad. Observed by family to be easy-going, practical, determined with short-lived, fiery temper; frequented conservative political club like *Fa*; skilled worker.

When Interviewed. Anxious, noncommittal, lifeless. Says he's "a bit depressed". . . . "I suppose normal things depress me. . ." Homesick when in army, wept when reprimanded. "I'm just a bit quiet, that's all." Characteristic responses: *What's in the future for you?* "I don't really know." *Look forward or afraid?* "A bit of both I think." *What do you look forward to?* long pause—"I don't know—I'm supposed to be getting married in the near future, well, I suppose I look forward to that." *Just suppose?* "Well, I do, but it's—I do, I mean it's—on the financial side—it all depends on—financially." Expressed anxiety about girlfriend. *About her leaving you, or about her health? . . . Is she not well?* "Yes, she's perfectly alright that's the funny part about it." *How do you sleep at night?* "Not too good. . . I worry about . . . silly things. . . ."

Comment. The chief interest here is the psychiatric state of Twin *B* as revealed by the interview and the MMPI. Meehl regards him as a borderline schizophrenic (? simple schizophrenia) no doubt on account of his anhedonia and possible pan-anxiety. Slater originally saw him (like MZ 4*B*) as an extremely lifeless, colorless personality, within normal limits, but as very liable to succumb to a reactive depression in the future. Later on, reviewing Meehl's diagnosis, he agreed there was some suggestion of schizotypy, though to his mind it was not specific enough for him to make such a diagnosis. Our second U.S. judge (LM) saw him as a schizoid-depressive character with a high risk for psychosis, while two United Kingdom judges, like Slater, diagnosed Normal, ? neurotic personality. He is the only nonschizophrenic *B*-twin whose MMPI is more abnormal than that of the *A*-twin. Retrospectively, it is tempting to see the diagnoses of both twins as lying on the same dimension. Perhaps Kety *et al.* (1968) would diagnose *B* as an inadequate personality of a kind which they would include within a schizophrenic spectrum.

The parents appeared normal if somewhat restricted and defensive and the older sister was not noteworthy. Despite many inquiries at the Observation Ward, at the Maudsley and subsequently, nothing emerged from the premorbid history, up to the time when *A*, at the age of 18, first expressed unusual ideas about homosexuality, to suggest that either twin would become a schizophrenic. Nor is there anything from the differential twin history which would have led one to predict that, given one would be schizophrenic, it would be *A*.

The pair is still young. A follow-up in 30 years' time might be instructive. Will *A*'s recovery last? Will *B* fall ill, and if so will he show unmistakable schizophrenic symptoms? Or will the pair turn out to be one in which a paranoid schizophrenia is matched with a depression (cf. DZ 17, 21; MZ 24)?

DZ 15, male, born 1927

A: 1st hosp. 33 (Sc 34), 5 spells: 43 weeks; FU 36, recent hosp. discharge.
B: ——; FU 36.

	A	*B*
Consensus Diag.	Sc (6)	Normal (5)
Slater Diag.	Chronic paranoid Sc	Normal
Meehl Diag.	True Sc	Normal
Global rating	5.5	1.25
MMPI	Not done	2-908/473 15:6#

Zygosity Evidence. Blood, seen.

Background. Social class 5. *Fa* 31 (abn), *Mo* 63, 1 *M* died 35, 2 *B* and *A*, maternal half-sibs, 4 older (1 *F* ? Sc), 1 younger, NK.

Fa unemployed on 100% pension from army for shell-shock, deserted family soon after birth of *A* and *B*. *Mo* inconsistent, unreliable, affectionate, of dull intelligence. Cohabited with three men in turn; by the 1st had daughter who was said to have been a recluse for 12 years (? Sc) and other children, including a pair of twins who died in infancy; by the second had a son and *A* and *B*; by the third a son.

Difficult birth (birth weight: *B* 5½ pounds, *A* 6 pounds). *B* was the more robust twin, *A* taller, later in sitting up. On desertion of the *Fa* the family was placed in an institution for the indigent. At 7, *A* and *B* were placed in a special residential school for the retarded, perhaps only for social reasons in the case of *B*; as young boys *A* tried to bully *B*. After age 10, *A* and *B* were separated (perhaps only to different parts of orphanage). *B* placed in foster home at 15, where he "felt a bit inferior." He adjusted well in a second foster home and was still living with the foster parents, unmarried at age 36. He was employed in essential factory work during World War II (took his turn as a civilian fire-watcher); later became a porter. *B*, although placed in a special school, did learn to read and write. *A*, at 15, was sent to a hospital for the mentally deficient. On admission his Stanford Binet IQ was 51, he was illiterate, difficult to control, and in excellent physical health. Five years later he absconded while working from the hospital as a gardener and was subsequently discharged. Although rather unsociable, moody and sullen in nature and despite many job changes and heavy drinking, he was able to provide for a wife whom he married at 30, and who was already pregnant, and a son, now 6.

Psychiatric History of A. At 31, while employed as a fireman on a merchant ship he imagined he was accused of being a homosexual and that a lumbar disc lesion, for which he was hospitalized, was caused by his food having been tampered with (*onset*). He also made references to an imaginary large sum of money which he had won but not received. A psychiatrist at a GH diagnosed a paranoid psychosis due to alcohol, prescribed trifluoperazine and referred him to *Maudsley* OPD (at 33) from which he was admitted as IP. He appeared unintelligent and rather depressed. A Wechsler IQ was 83. There was a specific reading disability, "possibly based on poor auditory verbal learning." He responded well to perphenazine and was discharged after 8 weeks (diagnosis: paranoid reaction in mildly defective personality, ? related to alcoholism, ? to intellectual impairment). Within 2 months of discharge, paranoid symptoms returned—food was drugged, police were watching him, co-workers talked about him. Eventually he was afraid to leave the house, suffered from insomnia, loss of appetite and lack of libido and was unable to work. At about this time his wife gave birth to twin daughters, now 2. Diagnosis at readmission to Maudsley (at 34) was changed to paranoid Sc. He was treated with chlorpromazine at this and at two subsequent admissions to Maudsley (final diagnosis: chronic paranoid Sc intellectual impairment). Within 7 weeks after discharge from Maudsley admitted to MH having appeared at a police station in response to auditory hallucinations stating that he had committed many unrecorded crimes (later at MH insisted he had committed many sexual offenses over Christmas holiday). He made considerable improvement with 11 ECT and trifluoperazine, discharged home after 10 weeks (diagnosis: paranoid delusional system due to exacerbation of chronic Sc) and worked as a gardener at the hospital while attending OPD.

When Interviewed. In MH during fourth hospitalization (at 36). Described self as confused—"That's right, confused. I don't know what's goin' on 'alf the time. Because when . . . once you've been on a lot of ships . . . (laugh) and you get hit hard by the State and you come out, work ashore, and then someone finds you're . . . you don't know where you're going to turn, and you don't know what to do, not in this country, but in this country it's very 'ard. . . I found out it all depends on how you do it in the first beginning, if you make a muck of it at first, then you've had it. That's the way I look at it now like. Now if anything happens . . say if I was framed again I would know what to do. . . I'd go straight to the law." Attributed his illness to "the stuff put down on men on the ship . . . they put something in the salt . . only for me, only for me cause they wouldn't use salt, they had it all planned that they wouldn't use salt, they had it all planned . . . that salt." MMPI not completed because of illiteracy.

Later History of B. Good natured, enjoyed work as porter because "you see people and talk to people," never bored as "too much to do." He had few friends but had a dog and a motor bike. Still lived with foster parents.

When Interviewed. At 36, expressed no worries except regret for his inadequate education. Fears marriage because of lack of background and family life, although "couldn't wish for a better family than foster parents." *What do you do that gives you the most pleasure, what is your idea of a really good time?* "I wouldn't say I was ... I'm not good in company I'm afraid ... I like company ... I feel a bit lost ... I've a good time with a crowd of fellows, actually .. a coast ride or something like that or to a show .. or theater." *Have you ever come close to getting married yourself?* "I have once, I would like to .. but there's this fear all the time .. not having the background or the family life, the proper family life. ... I've been around them that have got married, I see some of the predicaments they're in ... they would worry me and perhaps I'd be in the same boat as (A). ..." Seemed normal, no evidence of subnormal intelligence.

Comment. The twins had very little contact with their abnormal family or, from the age of 10, with one another. If they had been reared by their father or in the home which produced the schizoid or possibly schizophrenic half-sister, *A*'s illness might by some have been attributed to the faulty family environment. As it is, their institutional background has no doubt had an influence, perhaps more on *B*, single and still living with his foster parents, than on *A*'s paranoid illness.

DZ 16, male, born 1921

A: 1st hosp. 27 (Sc 27), 2 spells: 478 weeks; FU 42, working, symptoms +.
B: ——; FU 42.

	A	B
Consensus Diag.	Sc (6)	Normal (6)
Slater Diag.	Hebephrenic Sc	Normal
Meehl Diag.	True Sc	Normal
Global rating	5.75	2.75
MMPI	Not done	Not done

Zygosity Evidence. A seen, also photographs and *B*'s wife's description, no doubts.

Background. Social class 5. *Fa* died 63, *Mo* 69; 1 NK, 2 NK, 3 *M* 52 (abn), 4 NK, 5 NK, 6 NK, 7 *A* and *B* (? order), 8 NK, 9 NK, 10 NK, 11 NK.

Fa builder's laborer; bad-tempered, a drinker. *Mo* cheerful. Early home environment somewhat disturbed by *Fa*'s quarrelsomeness. 3 *M* hospitalized (49) for 7 months for psychoneurosis; MH (at 50) for 11 months with chronic psychogenic abdominal pain and secondary depression, resistant to treatment with 7 ECT and chlorpromazine; "intense hypochondriasis" dominated the clinical picture; later improved.

B was always much larger and more robust than A as well as more intelligent. Although A and B left school at the same age, A was three grades behind B; A became a common laborer and B a skilled laborer. A served in the Pioneer Corps for 4½ years during World War II; B served 4 years in the Army as a private. A, although "bad tempered," had numerous friends, married and had one daughter; B married twice, had no children by second wife and it was unknown whether or not he had children by the first marriage.

Psychiatric History of A. At age 26 (*onset*) he was found wandering and expressed various paranoid accusations about wife. He threatened his mother, attacked his wife, who then left him, and soon lost his job as a laborer. Within 2 months admitted as *Maudsley* IP for 16 weeks (diagnosis: paranoid Sc). At admission he expressed vague persecutory ideas, was auditorily hallucinated (voice told him to steal) and he smelled a suspicious smell of gas. After 38 insulin comas and no improvement, he was discharged. At home he was unemployed, laughed inappropriately, was jealous, violent (his wife left him again) and within 6 months was readmitted to MH (diagnosis: hebephrenic Sc). There was no improvement until he was put on chlorpromazine 7 years later. In two more years (at 37) he went home on a trial visit where he improved even more in that he became neat and clean and worked around the home and garden. His mother cared for him and he was on generally good terms with his wife.

When Interviewed. At 42, he was still living with his elderly, deaf mother, working as a sod cutter. Formal interview or testing was not possible as he had deteriorated; he stared out the window and did not reply to questions. He had attempted to strangle his wife who then moved to another village. He returned the Briggs questionnaire uncompleted except for frequent scrawling of the word "volt" and the phrase "C of E" (Church of England). He was not taking medication.

Later History of B. He refused to be seen for the Twin Investigation (at 42), remaining upstairs out of sight, but his wife was seen at the door and an MMPI form left for B. He did not consider completing the MMPI and his wife reported "...once his mind is made up, nothing will make him change." He was regarded as a healthy, level-headed, solid, happy person who got on well with others, had a few close friends and was somewhat of a "home bird."

Comment. It is a pity that such a large sibship should be so uninformative in regard to dominant and polygenic expectations, but entrée to the family was frustrated by A's deteriorated condition, his mother's deafness, and B's refusal to cooperate. As in DZ 2, we have a nontwin sib of a schizophrenic presenting with severe hypochondriasis and depression. The MZ co-twin in MZ 14 was also a severe hypochondriac. Such instances give weight to the clinical impressions of, e.g., Meehl and H. S. Sullivan that hypochondriasis may be a schizophrenic "equivalent" deserving some status in considerations about what goes into the schizophrenic spectrum.

DZ 17, female, born 1899

A: 1st hosp. 52 (Sc 59), 2 spells: 29 weeks; FU 65, died at 65.
B: 1st hosp. 59, 1 spell: 7 weeks; FU 65, working.

	A	*B*
Consensus Diag.	Sc (5)	Other (6)
Slater Diag.	Paranoid Sc	Reactive depression
Meehl Diag.	Acute paranoid reaction	Depression, kind unknown
Global rating	4.5	3.25
MMPI	46'' ' 308271-5/9:	43''1826'79-05/

Zygosity Evidence. Blood, fingerprints, seen.

Background. Social class 3. *Fa* died 71, *Mo* died 71; 1 *M* died 66, 2 *F* died 17, 3 *A* and *B*, and 3 *M* died 42.

Fa skilled laborer; alcoholic, abusive when drunk. Insecure home atmosphere.

Twins were two of triplets: *A* born first, weakest of three; *B* second born, delivered by instruments, strongest of 3; 3 *M*, a boy, instrumentally delivered, who was killed in London air raids in 1941 after normal life, skilled worker, married, five children. Early physical differences were not noteworthy. *A*: menarche at 17, at which time she fainted, then was acutely self-conscious at time of periods; *B*: menarche at 14 with no trauma. *A* attended secondary school, left to care for grandmother. *B* a poor scholar but liked needlework and later did needlework as piecework for various firms. She was sociable but anxious (always a nail-biter), married at 28, and had a son at 32. *A* was sensitive, reserved, never married, lived with parents until age 37. Twins were very attached to each other and spent later lives living together; *A* lived with *B*, even after *B* married. Both were members of Pentecostal church.

Psychiatric History of A. At menopause (aged 46), believed people accused her of being wicked (*onset*); remained paranoid but continued working until 52 when she complained of headaches, became increasingly suspicious, and resigned her job. Admitted to *Maudsley* IP (diagnosis: paranoid state and involutional depression). Depressive features cleared with modified insulin and 13 ECT, but remained somewhat suspicious. On discharge, after 19 weeks, resumed work (dressmaker), in a new job. She remained fairly well for 5 years when the "accusations" began again (at 57); suspected people of saying she was immoral.

Recurrence of illness coincided with what she thought was return of menstrual periods. Treatment at DH (10 weeks) with 9 ECT provided temporary improvement (diagnosis: paranoid personality). Later in the year, although at home, she was no better. Blind diagnosis (Slater) in Twin Study, paranoid Sc. At 62 still secretive, persecuted and much preoccupied with some "crime" in her past but managed to work; saw accepting GP weekly for "talks." Completed MMPI yet refused to be seen for Twin Investigation. Died later in year (at 65) of cerebral hemorrhage.

Psychiatric History of B. At 43, when husband temporarily absent, had illness with weight loss in which she temporarily lost use of the left eye. Menopause at 46, uneventful but at about this time developed throat illness "in sympathy with" her sister. At 51 "on verge of nervous breakdown"; husband ill with what she feared to be cancer although this was not diagnosed. A year later seen for Twin Investigation; warm friendly, talkative, preoccupied with illnesses in the family, complained of heaviness and pressure on the head. Five years later, when seen again, still worrying about family affairs, needed medication for sleep, observed that although she used to be cheerful and happy, she now felt tired, lacked energy and then became much more religious. Two years later (at 59) hospitalized at Neurosis Center for 7 weeks (diagnosis: reactive depression); husband deceased 7 months previously, adult son releft home, complained of sleeplessness, loss of appetite, headaches, felt that "her mind stops at times." Improved on sodium amytal and chlorpromazine. At 62, insisted she had had no treatment at MH, denied symptoms other than insomnia; had many friends, church activities, attended classes.

When Interviewed. At 65, still complained of headaches and took "Sparine" ... "only at nighttime"; talked at great length about health and domestic problems. When asked *And yourself?* (i.e., her health) answered "Yes, oh but my knees are shockin'. So of course when I was at the hospital, I told you, they took X-rays and that, and they examined me there and they said, what's this 'cause I've got a lump 'ere and that's fatty tumor really, but I 'ave got arthritis in the spine, so I told them, then they said, well it looks to me as if your knees are like your neck." ..."I went to the doctor and eh I told 'im about my legs, and I said, I told 'im about my nerves, I said—I said I think I'm worrying about my legs, so he gave me tablets, and every time 'e gives me tablets, they seem to knock me over, and I can't take them. . . .I look to the Lord for healing, and I do, I really do think, if it wasn't—I've been going to church ten years now—if it wasn't for that church, I don't know where I would be . . . it is only the church that keeps me going."

Comment. A suffered from a chronic paranoid illness of late onset with prominent affective features, which developed in a sensitive, seclusive personality. B, who was a poor student, had a sociable, anxious, dependent personality and developed anxiety, insomnia, and depression when her husband or son was ill or absent. Her rambling, hypochondriacal talk in the interview was not inappropriate in a woman of her age and cultural setting. She was seen by most judges as having a neurotic depression. She has little in common with *A* insofar as schizophrenic or paranoid elements are concerned; however, both show affective features.

DZ 18, male, born 1937

A: 1st hosp. 18 (Sc 18), 4 spells: 88 weeks; FU 26, recent hosp. discharge.
B: ——; FU 26.

	A	*B*
Consensus Diag.	Sc (6)	Normal (6)
Slater Diag.	Paranoid Sc	Normal
Meehl Diag.	True Sc	Normal
Global rating	5.5	1.0
MMPI	25'473-810/69:	05-238/674:19#

Zygosity Evidence. Seen, not similar at all.

Background. Social class 1. *Fa* 55, *Mo* 59; 1 *M* 28, 2 *B* and *A*, 3 *M* 24.

Fa (52-136078/49:) executive (department manager) in professional concern, a "good father albeit a little strict." *Mo* (MMPI 9-48361/720:5#) intelligent, vivacious, excitable, did not favor any particular son. 1 *M* (MMPI 52-0371684/9:); 3 *M* (MMPI 278-5104369/).

B ahead in milestones and earlier in appearance of secondary sex characteristics. *B* more dominant than *A* and protective of him. *A* identified with the *Mo*, *B* with the *Fa*. Twins were very close as they grew up. Raised by governesses, attended private schools, passed exams. *A* regular church-goer with strong religious feelings.

Psychiatric History of A. From 15 (*onset*) gradually more indecisive, preoccupied with feelings of being evil and of possible sex change, estranged from *Mo* and thought she was going to die. At 18, while on reserve training, was suddenly unable to account for himself and was admitted to a military hospital, hallucinated and deluded. He was transferred to *Maudsley*, treated with 40 insulin comas and, though depressed during treatment (made one suicide attempt) he then improved and after 30 weeks was discharged (diagnosis: Sc). Wechsler IQ at Maudsley 114. He worked in a solicitor's office and was relatively well for several months; then told *Fa* that the body of a woman had been burned in his office, thought he was being chased by Egyptians and Communists; readmitted to Maudsley (at 19) where attacked a nurse in response to aural hallucinations, exhibited thought disorder in association of ideas and neologisms. Treated with reserpine, improved, discharged after 32 weeks. At home vague, listless, unable to work on farm or in advertizing agency. After 3 years developed insomnia, heard voices: trees or wind saying "you're going to be ill," became overactive, readmitted to Maudsley (at 22). There was some improvement on trifluoperazine in that his elation subsided, then became hallucinated, withdrawn, believed *Fa* died; and deteriorated further until reserpine added, then improved to state before admission; discharged after 12 weeks (diagnosis: Sc). Managed at home with OPD supervision for next three years: 3 short-lived breakdowns, continual employment difficulties because of unrealistic plans. Soon after *B*'s marriage, *A* (at 26) heard voices accusing him of sin, thought he was the devil, admitted to MH. Improved after 5 ECT and trifluoperazine, discharged after 4 weeks. Began training for associate manager's post.

When Interviewed. Four months later, at home living with parents, still working, remained fairly well in protected environment (sympathetic employer). Still single, indecisive and unrealistic in his plans for the future. "I think it's quite possible that I can get into something .. something like television . . . producing and that sort of thing. . . ." *How is your ability to concentrate now?* "It's alright on superficial things .. things like television . . . I think on anything which requires a lot of concentration, it's not at all good." *What would you say was the matter with you . . . how did it (illness) look to you. . .?* ". . .Well, I think mainly it's losing reality, losing touch with what is really wrong and having one's mind full of unreal things . . . eh . . . not so much that as one's response to them being unreal . . . and therefore, living half in touch with . . . what one is usually in contact with normally. . . ." *There was a time when you were seeing things that other people didn't see, how do those seem to you now?* ". . . they don't have any meaning or connection with anything else, just completely isolated, just an illness, really."

Later History of B. Had wanted to be a farmer from the time he was a small boy and at 26 had a farm of his own (financed by parents); had been married for 3 years and had two young sons.

When Interviewed. No psychiatric symptoms, said he was very happily married although didn't think he and his wife had "all the problems of sex . . . sorted out, that sort of thing. . . ." He expressed an irrational feeling that something bad was about to occur; had rational worries about farm and livestock and felt "somewhat apprehensive about the future". *Have you worked out some kind of a philosophy of life for yourself that you can put into words?* ". . . I think mainly I want . . . to remain . . . loving my job . . . my family, of course . . . yes, I want to be happy working hard farming .. I love nature very much. I prefer animals to people . . . really . . . collectively anyway . . . I couldn't live in a city I don't think." Had no particular intimate friends; interested in music and politics. He was concerned that he was still a nail-biter after concerted attempts to break the habit.

Later Information. A readmitted to MH (at 28).

Comment. Family solidarity and normal parents did not prevent the initial or subsequent breakdowns of *A*. No unique environmental stress can be identified. All three brothers were within normal limits when tested with MMPI but *B* highlights the difficulties in distinguishing between normal introversion and schizotypy. Unless it is known that such a person is related to a schizophrenic, he will most likely be seen as normal without qualifications.

DZ 19, male, born 1936

A: 1st hosp. 23 (Sc 23), 1 spell: 232 weeks; FU 27, hosp.
B: ——; FU 27.

	A	*B*
Consensus Diag.	Sc (6)	Normal (6)
Slater Diag.	Paranoid Sc	Normal
Meehl Diag.	True Sc	Normal
Global rating	5.75	1.5
MMPI	8726*54"30'19-	53468/12907:

Zygosity Evidence. Seen, no doubts; different eye color.

Background. Social class 4. *Fa* died 48, *Mo* 68; 1 *F* 34, 2 *M* 28, 3 *A* and *B*.

Fa warehouseman, died of tuberculosis when twins aged 11 after 2 years of invalidism. *Mo* (MMPI 50-238614/98:) overpossessive, deaf, "warm and motherly." Difficult maternal aunt lived in overcrowded home, adding much tension and quarreling.

Birth weight: *A* 6¾ pounds, *B* 6¼ pounds. Both had pneumonia, only *A* had scarlet fever. As a child *A* was a nail-biter and afraid of the dark, as he grew older he was more quiet than *B* but not shy. Both were good students (*A* at the top of his class), had many friends, were active scouts and superior sportsmen; *A* an almost professional caliber cricket player, *B* an avid football club member.

Psychiatric History of A. At 20 on return from 2 years National Service, socially withdrawn, studied books on psychopathology, had unrealistic ideas of obtaining university education while employed in college laboratory. He lost this job (at 22) for irregular attendance, lived secluded life in own room (*onset*). Four months later (at 23) became violent in public with vague paranoid ideas of a political nature and was admitted to *Maudsley* IP where he was friendly and cooperative to a degree, but vague, unrealistic, pseudointellectual, and at times aggressive. Mill Hill Vocabulary IQ 90. Treated with perphenazine, trifluoperazine, and nialamide with no benefit; became depressed, 4 ECT, then fatuously happy and amorous; within 1 month depressed again with suicide attempts. Later, emotionally flattened and incongruous, preoccupied, wandered about; at times hostile, suspicious. Transferred to MH, unimproved (diagnosis: paranoid Sc).

When Interviewed. At 27, 3½ years later, still at MH on chlorpromazine 100 mg t.i.d. plus 6 months individual psychotherapy; at time of interview, group therapy once a week. Worked on hospital farm and went home on weekends; "superb cricket-player" for MH team. Had believed the Communists were after his brain and that he was Jesus Christ. On interview, troubled by thoughts of his sexual make-up ("I've got a split mind, sometimes I think I am masculine and feminine at the same time"). Suspects a plot is going on to harm him—something to do with "latent homosexuality," thereby believes he is safer in the hospital where he will not commit sexual crimes against smaller or younger people. *Do you ever see or feel or hear anything that other people don't?* "See, feel or hear... I've had delusions, what you call grandiose I think, alleged thoughts of ... I'd like to have been a scientist ... thoughts of getting up to the moon ... and setting up some (mumble) where I could force people to abandon all their ... this present war, cold war situation ... I've had these thoughts." Believed his illness might be "due to too much religion pumped into me at school"; explained that thoughts came into his head over which he had no control; related fears of not being able to "stand on my own two feet in a job of work..." Said "...I've always wanted to work on a farm, now the nearest I've got to it ... is in this hospital ... but at the same time I suffer constantly with fatigue ... I probably couldn't keep up the pace ... I often look ... I would take my coat off and say to you look I'm not ... I'm schizophrenic in as much as I haven't got a ... man's own strength ... size of a man's arm ... I consider my growth is stunted...."

Later History of B. Left school at 15½, National Service aged 18–20 which he enjoyed so much he "nearly signed on." Married at 24, one son (18 months), wife pregnant.

When Interviewed. At 27, living with *Mo* but planning to set up own household, employed as route salesman, work which he enjoyed. Described himself as "happy-go-lucky" and able to "shrug off" worries, found it "very hard to dislike anybody." *Have you cried in the last five years?* "No." *Have you ever been really depressed?* "Really depressed, no. Not depressed, no." Explained that at his football club "...we've got our own bar, and what ... every Saturday night we're out there, there's always dozens and dozens of people out there you can talk to." No psychiatric symptoms or treatment. Very cordial and cooperative. Mill Hill Vocabulary IQ 95.

Comment. Both twins had unusually good premorbid histories, notably friendly, athletic, and scholarly. Given a choice, one would have guessed that *B* would become schizophrenic if either did but it was *A*, heavier by one-half pound at birth and the top of his class who broke down with *B* remaining completely normal, happily married with two children. The paternal deprivation before adolescence, a deaf mother with whom communication was difficult, a quarrelsome aunt in the house, together with exposure to *A*'s illness were not sufficient to cause a breakdown in *B* with his different genotype.

DZ 20, male, born 1931

A: 1st hosp. 28 (Sc 28), 2 spells: 14 weeks; FU 33, working.
B: —— OPD 18; FU 33, working.

	A	*B*
Consensus Diag.	Sc (6)	Other (3)
Slater Diag.	Hebephrenic Sc	Reactive depression
Meehl Diag.	Acute undifferentiated Sc	Normal
Global rating	4.75	2.0
MMPI	2'47 0-839156/	53 924/78061:

Zygosity Evidence. Fingerprints, seen.

Background. Social class 1. *Fa* 62, *Mo* 63; 1 *M* 35 (abn), 2 *B* and *A*.

Fa architect, psychopathic personality (persistently unfaithful, periods of excessive drinking, egotistic and inconsiderate, no interest or affection for his children); absent for 5½ years during World War II (MMPI 5214-380 67/9:). *Mo* (MMPI 243 1985/076:), Frenchwoman, "self-contained," one of her brothers ? homosexual, suicided. 1 *M* difficult youth, unsuccessful in school, fell in love with "unsuitable" girl, discharged from army on psychiatric grounds, emigrated to Canada where had many job changes and was arraigned for embezzlement before eventually settling down to a University education and marriage.

B 2 pounds heavier at birth; *A* had to be revived after birth. *A* longer at birth and considerably taller than *B* as an adult. In infancy *A* banged his head and had convulsions when teething; throughout childhood he had temper tantrums. *A* was very weak after a severe bout of scarlet fever at age 3 and had to "relearn walking" (*Mo*). *B* was sent to Switzerland at 13 with suspected t.b., tests negative. Only *A* is colorblind. *A* left-handed, *B* right-handed; *A* first to shave. *B* was generally a class ahead of *A* at school, but *A* who was more hard-working, gradually overtook *B*. The twins resented their twinship. Neither twin dominated the other as the most dominating influence in the family was the elder brother; *B* fell in more with the elder brother who tried to bully *A*. The twins did not differ in preference for one parent over the other, although there was a great deal of hostility directed toward the *Fa* by both twins because of *Fa*'s attitude.

Psychiatric History of A. At 13 saw consultant psychiatrist because of rages, bad dreams, and restlessness. At school he showed some reading disability and vestiges of mirror writing. *Fa* procrastinated about treatment and *A* improved. When seen for Twin Investigation at 19 he had been accepted in a medical school and was not regarded as a problem although minor obsessional traits were observed (collected tin boxes, could not bear to throw things away). After 5 years (at 24) without qualifying for practice, he was dismissed for forgery (*onset*). While a student he developed a duodenal ulcer. He married at 24, did not want children, wife did, *A* not interested in sex. At 27 seen again for Twin Investigation; looked worried, seemed rather naive and was extremely indecisive and given to ruminating and self-criticism. His wife left him about this time (they were later divorced). One year later he appeared at *Maudsley* emergency clinic (at 28) complaining that his thinking was wrong and that experts in the mind could put it right in some unspecified way. Intangible nature of his talk, curious withdrawn affect and bewilderment suggested diagnosis: ? Sc. The following day he called the clinic to say that he had quit his job as one of his counselors didn't "believe in the Kingdom of God . . . I had to get away from the place so I left"; affect was inappropriate and he failed to keep a later appointment with a consultant. Same month admitted to MH (at 28) for a period of 10 weeks (diagnosis: paranoid Sc); symptoms included perplexity, thought blocking, affective blunting, and hallucinations; acute episodes of catatonia; it was thought that the schizophrenic process might have been going on for as long as 12 years; improved on large doses of chlorpromazine and ECT. A year later readmitted; improved on 6 ECT and thioridazine. Discharged OPD after 4 weeks and after 6 months seemed well.

When Interviewed. At 32, two years later, no further hospitalization, still on thioridazine. Self-employed, lived alone, *Mo* and others lived separately in same house. Attributed onset of his illness to overwork and relapse to neglect to take medication. At the beginning of the interview he commented that he had "never felt so well'" and had no difficulty sleeping. He later said that he often had difficulty sleeping, that he could not concentrate as he used to and could no longer memorize. *Do you show happiness or depression?* "I don't suppose I'm very happy . . . I feel as if I'm in a cage to a certain extent . . . before I was more full of energy but I'm not the same now . . I've lost a fair bit of confidence from what I had before I was ill." The *Fa* was also seen and spoke of *A*'s life as filled with "chaos and pandemonium" at the onset of the illness, reported that ? one episode was precipitated by *Mo* leaving for visit in Canada. The *Fa* had continued to help support *A* financially. At 33 *A* wrote briefly, "I am much better in health now and can cope with business so much better."

Psychiatric History of B. During later childhood he worked less and less at school, began to rebel against his *Fa* and at 16 was expelled from school. He tried various types of work and study but gave them all up. Had consultation with psychiatrist (at 16). At 17 began stealing from home, became increasingly off-hand and depressed; at 18 took chloroform in a suicidal demonstration and was seen at Maudsley OPD. On examination he was bored, sulky, said he hated his *Fa*, could not trust his *Mo* and had much preoccupation with sex. An emotional problem of adolescence diagnosed and after 3 months of psycho-therapy was completely well. He soon emigrated to Canada where he began working for a mining company. Married at 26, at 33 he was still with the same firm, happily married having never regretted emigrating (letter from *B*, information from parents).

Later Information. A (from newspaper report): Age 35, no permanent address, charged with forgery, forfeited his bail, remanded for medical report.

Comment. The family background was disturbed and all three boys showed behavior disorders. Only one became schizophrenic. The pair was first investigated at 19 before there was any sign of schizophrenia in *A*. At that time *B* might have been assessed as more at risk for schizophrenia on account of his deteriorating performance, his greater suspiciousness, his sexual preoccupations and his more disturbed relationship with both parents. However, events—to which brief psychotherapy, emigration with escape from the family environment and maturation may have contributed—showed his adolescent depressive reaction to be merely a transient affair and within normal limits according to three of our judges. The role of organic factors in *A*'s personality development and eventual somewhat atypical schizophrenia is unclear.

DZ 21, female, born 1917

A: 1st hosp. 32 (Sc 32), 3 spells: 118 weeks; FU 47, working, symptoms +.
B: 1st hosp. 23, 3 spells: 31 weeks; FU 47, working.

	A	*B*
Consensus Diag.	Sc (6)	Other (6)
Slater Diag.	Paranoid Sc	Endogenous depression
Meehl Diag.	True Sc	Psychoneurosis conversion and depressive
Global rating	5.0	3.5
MMPI	Not done	48-37026/519:

Zygosity Evidence. Fingerprints, seen; no doubts.

Background. Social class 3. *Fa* died 75, *Mo* died 76; 1 *M* 51, 2 *A* and *B*.

Fa in civic government, bad-tempered, unpredictable, drank to excess. *Mo* anxious, dependent, devoted to family, susceptible to headaches, early in *A*'s illness *Mo* sought OP medical advice for her own depression and sleeplessness.

A heavier than *B* at birth by about 5 pounds. *A* dark, *B* fair. *A* more difficult to manage during early development; *A* always required more attention but *B* demanded attention. *A* "St. Vitus' Dance" at age 8. Twins of equal ability at a school where most pupils of above average intelligence; both took commercial training. There was no close twin relationship; both twins preferred their *Mo* to the difficult *Fa. B* more dominating in personality. *A* reserved, stubborn, oversensitive, labile, fastidious, worried excessively about her work where she did a "good job" as a secretary, few friends or interests outside *Mo* and home. *B* sociable, hypersensitive, anxiety prone and liable to hysterical reactions. *A* never married; *B* married (at 29) wealthy man 13 years her senior with a 9-year-old son by a previous marriage; later had a son (16). In early 20's *B* told she had otosclerosis; fenestration of one ear at 30, other at 46.

Psychiatric History of A. Appendectomy at age 32; while convalescing, developed first paranoid ideas (*onset*): thought a doctor's remark reflected on her morals and that the nurses and patients in the ward ridiculed her. On return to work became suspicious of "peculiar chatter," tearful and sleepless; then everywhere saw talk, nudging, meaningful glances as signals about her. Six months after appendectomy, attended psychiatric OPD (diagnosis: acute paranoid Sc); paranoid symptoms were severe but affective response, although cool and suspicious, was appropriate. Admitted to *Maudsley* at 32, 5 ECT with little change; 14 insulin comas, then left hospital after 19 weeks (diagnosis: paranoid Sc). Remained at home unimproved and unemployed; was seen for Twin Investigation—admitted she was sensitive, critical, independent, and reluctant to accept help but asked for referral to woman psychiatrist. Continued at home, unemployed, interpreting everything said to her as having some deprecatory meaning. Woman psychiatrist recommended readmission to Maudsley at 34. At admission many ideas of reference, thought blocking, feelings of inadequacy and failure; soon noted to be depressed. IQ: Mill Hill Vocabulary 109, Progressive Matrices 126. Treated with insulin comas and ECT with little change. About this time *B* was hospitalized for second time with a depression; *A* was quite upset but responded to methedrine and discharged to home after 55 weeks (revised diagnosis: paranoid personality).

At 38 saw woman psychiatrist privately but "completely uncooperative." Admitted to MH (at 39) with ideas of reference and hypochondriacal symptoms of "typically schizophrenic variety" (bowels and speech have been changed), also depressed. Thought not to be psychotic after psychotherapy begun. After 44 weeks to work while attending OPD (diagnosis: Sc). Information from psychiatrist in contact since last MH discharge reported relapse of paranoid symptoms (at 47) which caused her to leave her job as she thought everyone was talking about her. She responded well to small doses of trifluoperazine. Although chronically paranoid, resumed working full-time, living with and caring for 75-year-old *Fa* until his death (*Mo* died the same year, twins aged 46). The psychiatrist noted: "Even in her most psychotic stage she has a surprising amount of insight and ability to relate." Refused cooperation in twin study.

Psychiatric History of B. At 16, on first date, vomited in smart tea garden; thereafter fear of vomiting in public places; holiday at 17 spoiled on this account. Hospitalized at neurological hospital for investigation of sciatica (at 23), probably genuine on admission. Proved to be a very difficult patient with ambivalent attitude toward treatment, was focus of hysterical patients (diagnosis: neurosis, presenting as sciatic pain). Transferred to teaching hospital, improved with physical therapy. Discharged within a month (diagnosis: chronic back pain, no psychogenesis elicited); no mention of phobias. At 33 seen for Twin Investigation, no complaints except dislike of domesticity and some anxiety. Admitted to neurosis center (at 34) with history of several months' depression with recurrence of phobia of vomiting after recommendation by dentist that she have two wisdom teeth extracted. She complained at length of unsympathetic husband and reported 14 pound weight loss, sleeplessness and indecision. Regarded as reactive depression which cleared after 8 ECT although sleeplessness remained a problem. Discharged on medication after 14 weeks (diagnosis: anxiety hysteria). Another psychiatric opinion at this time: endogenous depression. At 38 attended psychiatric OPD at teaching hospital, recovered after 7 ECT (diagnosis: endogenous depression with obsessional features). Attended same OPD at 45, when, while learning to drive a car, she became exceedingly nervous (opinion: depressive picture), treated with phenelzine sulphate, recovered, passed driver's test at first attempt.

When Interviewed. At 47 well-integrated but dissatisfied with marriage, very intelligent, coherent and articulate, unusual self-confidence, very well dressed, only medication an occasional sleeping pill when on edge; said felt fine, "I'm not a person to go up or down too much. I stay rather the same—I don't enthuse too much . . . I can get pretty low. . . Normally I'm fairly all right apart from the odd breakdowns I've had . . . but I got over them very well . . . I have really. . ." *How does it feel at the times you've had to end up having some electrical treatment?* ". . . .a little bit deadly . . . the thing that's bothering you sort of keeps coming up into your mind . . . you don't seem to be capable to sort it out with yourself. . ." *How do you make use of your day?* "Well, I play golf, I paint a little. . . I belong to a political organization. . . I generally go up to town one day a week . . . see around the shops and look around . . and . . . a garden which I do."

Comment. The paranoid schizophrenia in *A* is paired with a recurrent depression in *B*, with hysteroid coloring. It has been 2 years since *B*'s last attack and at follow-up she was quite normal. We were tactful in not pressing for a confrontation with the paranoid *A* twin, as in DZ 17 and 29, so as to maintain rapport with the *B* twins.

DZ 22, female, born 1943

A: 1st hosp. 20 (Sc 20), 1 spell: 54 weeks; FU 22, recent DH discharge.
B: ——; FU 22.

	A	*B*
Consensus Diag.	Sc (6)	Normal (5)
Slater Diag.	Paranoid Sc	Normal
Meehl Diag.	True Sc	Normal
Global rating	5.25	1.5
MMPI	8<u>26</u>*<u>79</u>″04′3-15/	<u>2307</u>-89<u>41</u>6/5:

Zygosity Evidence. Blood, seen.

Background. Social class 2. *Fa* 59 (abn), *Mo* 59; 1 *F* 32, 2 *B* and *A*.

Fa (MMPI 9′5-6<u>34</u> <u>87</u>/21:0#), successful private business; psychopathic life history, including bootlegging, compulsive gambling with absences from home of 2–3 days at a time; psychiatric OP past 7 years (diagnosis: hypomanic psychopath with obsessional drive), fears *A*'s illness may result from his long-cured VD; now "the only one to get through to *A* is *Fa*." His sister IP at 22 and 61 with paranoid, depressive psychoses. *Mo* (MMPI 5″27′<u>31</u> 80-<u>69</u> 4/), anxious, not loving, links *A*'s condition to own failure as a mother, contrives too much and unsuccessfully to encourage self-reliance in *A*, has recently considered vegetarian health camp to cure her. 1 *F* (MMPI 2′0-6<u>89</u> <u>734</u>/5:1#) "highly strung," ?anxiety reaction, resents *Mo* and *A*. Jewish family.

Toxemic pregnancy, premature birth (birth weights: *A* 4 pounds, *B* 4¼ pounds). *A* always looked more robust. Residential wartime nursery till 2, daily contact with *Mo*. *B*, no significant problems, pined for *A* when briefly separated, "a little mother to her." *A* developmentally backward, clinging, difficult about food, wanting to be made a fuss of, relied on *B* and jealous of her. *Mo* favored *B* because it was easier to make contact with her. *B* ahead in school, and only *B* went to grammar school at 13. *A* average achievement but retiring, always a dreamer, "difficult to reach," excelled in dancing and drama; upset from about 7 by family rows over *Fa*'s gambling.

Psychiatric History of A. CGC at 10—resentful of *Mo*, jealous of *B*; sensed atmosphere of secrecy at home; stole from school. From teens on, unpredictable aggressive attacks, e.g., at 16 smashed china because she thought family was making fun of her. At 16-19 art school and drama course, then casual jobs, such as telephonist. At 19 returned from drama tour looking wild and unkempt (*onset*); soon after this, attempted to choke *Mo*, threatened her with knife, wrote note "To Darling Mummy" on one side, "Drop dead" on the other. Three months later private psychiatric consultations (recent paranoid developments, not considered schizophrenic) and in another 2 months referred to *Maudsley* OP (at 20). Further development of symptoms and loss of job resulted from paranoid pursuit of colleague whom she unrealistically thought was spreading gossip about her led to serious consideration of a schizophrenic illness and admitted to DH 5 months after first referral to Maudsley. About this time filled diary with remarks such as "She used you to cancel out all *her* mistakes. . . You were influenced by two sides." In hospital, withdrawn and guarded, some compulsive treading and rocking, believed *Mo*'s making sandwiches signaled some sexually significant joke to *B* behind her back, preoccupied with and ambivalent about sex, romantic fantasies about older men verged on the delusional, heard voices saying, "Yes, No, good girl," persistent ideas of reference. IQ Mill Hill 100, Matrices 117. Treated with trifluoperazine and fluphenazine. Became less aggressive at home, concealed her ideas of reference. After 54 weeks discharged from DH (diagnosis: paranoid Sc) to attend Industrial Rehabilitation Unit.

When Interviewed. After 3 months in DH, docile, cooperative. *What kind of person would you say you were?* Long pause. "I don't know. I've been told lots of times I'm . . . selfish, but it isn't really that . . . I'm just not complete, if you know what I mean, I'm just sort of . . . um . . . a bit unsteady." She felt lacking in self-confidence. "If something goes wrong, there's got to be a reason for it, it's got to be like a test and I just—I've got this thing about God just testing us—and it's just a thing to cover up the fact if something does go wrong." Spoke of the intensity of her mistrust and fear of her mother; saw the father as warm-hearted but overpowering at times, while at other times "he can be wonderful, you know."

Later History of B. After leaving grammar school, to secretarial training and employment. Enjoys her work, leads active social life, has boy friends. Like *A*, is single but looking for a husband.

When Interviewed. At 21, admitted to spells of depression for a few days, sees doctor occasionally for pep pills, at one time saw him every week—"there was always something I was complaining about," e.g., headaches. Not a consistently good sleeper or eater. She seemed able to maintain an accurate perspective of the chaos in the family and to have coped satisfactorily with a difficult situation. Traced difference between twins to early difficulty of *A* in getting on with people.

Later Information. A found "unemployable" by rehabilitation workers because too slow, and 1 year after leaving DH, no change.

Comment. This case on the face of it offers grist to the mill of several purely psychodynamic views about the etiology of schizophrenia—in particular, the "chaotic" family, the nature of the abnormality in the parents and the singling out of *A* for special attention. Attempts to explain the case on the basis of a single "schizogene" run against the difficulty that the parents are diagnosable along an affective dimension rather than a schizophrenic one. However, the fact that Twin *B* is the least disturbed member of the family, while the older sister, who was as split off and rejected as Twin A, is anxious and depressed but not schizophrenic, indicates that the environmental stressors are not etiologically specific. Twin A's illness is compatible with her being critically different genotypically from both her sisters. The case fits the view that in this instance the effects of the genetic and environmental factors involved augment one another. Slater regarded the illness as developing progressively on the basis of a personality deviation of a schizoid, obsessional, and aggressive kind to which the disturbed relationship with the mother had contributed. On the genetic side the illness of the paternal aunt permits the inference that the father is the carrier of a more than average number of polygenes predisposing toward a psychosis of a schizophrenic or at least a paranoid kind and that he has transmitted them disproportionately to Twin A. When combined with the mother's contribution, a genotypic loading was attained which made Twin A likely to break down under the regime described above with a specifically paranoid schizophrenic illness.

DZ 23*, female, born 1911

A: 1st hosp. 36 (Sc 28), 11 spells: 108 weeks; FU 53, home invalid.
B: ——; FU 53.

	A	B
Consensus Diag.	Sc (6)	Normal (4)
Slater Diag.	Chronic paranoid Sc	Normal
Meehl Diag.	True Sc	Normal
Global rating	5.75	1.75
MMPI	Not done	201'6-7834 59/

Zygosity Evidence. Fingerprints, seen (ES, 1939).

Background. Social class 3. *Fa* died 61, *Mo* 77; 1 *F* suicide 21 (abn), 2 *F* 57 (Sc), 3 *M* 54, 4 *A* and *B*, 5 *F* 44.

Fa common laborer after failure of small business; violent when intoxicated, which was often in later life, and attacked *Mo*. *Mo* possibly overprotective, did not get along with *Fa*. 1 *F* committed suicide when pregnant by a married man; 2 *F* (MMPI 480-125763/9:) certified paranoid Sc, suicide attempts, illness had fluctuating course, affective coloring, later satisfactory long-stay patient at halfway house.

A larger than *B* at birth; *B* "very tiny." Both measles and whooping cough, only *A* diphtheria, only *B* scarlet fever. *A* very backward in walking. Both quit school at 14, *A* left top class, *B* left 2 classes behind *A*. There was no exceptionally close twin relationship. *A* was quiet, sensitive, shy and "a real

---

*A is proband No. 255 in Slater (1953).

home girl," who always relied on the *Mo* and always confided in her, later, a "good steady worker." *B* was sociable, cheerful and energetic. *A* worked in factory from 14 until marriage at 24; one child (son, 26). *B* worked as packer in dog biscuit factory from 14, married to co-worker at 20; no children (*Mo* reported *B* not fond of children). *Fa* died when twins 34.

Psychiatric History of A. Three years after marriage, husband's work as a baker necessitated moving household (at 27). She shortly became depressed (*onset*), was unable to cope with her child, and believed neighbors were talking about her, saying she was dirty. There was general improvement after staying with *Mo* for a week but she still lacked initiative and was unable to care for child. Six months after onset, seen at Maudsley OPD (at 23) before present Twin Investigation was begun—very depressed and tremulous, symptoms unchanged (diagnosis: "well-marked depressive state with ideas of reference and big loss of weight.") Attended OPD weekly and within a month no longer hallucinated. Seen by Slater for Twin Investigation at this time: total impression a schizophrenic one, although equally compatible with a brief paranoid episode in a mental defective, but no later evidence of such defectiveness. A few weeks later "perfectly well." At age 36, after a few weeks depression and a suicide attempt (drank lye), admitted to GH for 2 weeks. Five months later (at 37) 11 weeks *Maudsley* IP (diagnosis: recurrent depression). Mill Hill IQ 97. Treated with 7 ECT. Lived in same house with parents after discharged but reserved, often depressed, still hallucinated. Impression when seen again for Twin Research (at 37): paranoid schizophrenic, with excellent preservation of personality and large affective elements. Three months after seen, jumped into a river because she felt hypnotized, but on immersion remembered she had not cooked her husband's supper so she swam ashore and went home. Subsequently Maudsley IP (22 weeks) where treated with insulin comas and ECT (diagnosis: hebephrenic defect state). In another hospital bilateral prefrontal leucotomy performed (at 38), favorable results; husband reported "best she'd been in years." Within 3 years symptoms recurred; neighbors not only talked about her but controlled her by hypnosis and could see her in the bath; she refused husband intercourse. Between ages 42 and 49 there were seven hospitalizations, generally diagnosed as schizophrenic illness. There were episodes of agitation and sleeplessness but little severe depression in the later episodes. On chlorpromazine until final hospitalization (at 49), then looked and felt better on chloral hydrate. Refused to be seen (at 53) for follow-up (would not answer door), reportedly not at all well, no longer attended OPD, a recluse with no contact with neighbors and little contact with family.

Later History of B. Seen for Twin Investigation at 28; rather hard-faced young woman, said never worried or depressed, not nervous or oversensitive, mentality such as would be expected by her poor school record. Widowed at 44.

When Interviewed. At 53, lived alone in almost excessively clean and tidy apartment, looked ten years younger than her age. Thought the family had had more than their share of troubles but "I think I'm hard. I can take it"; devoted to family and believes family relies on her—"I'm always there." Affect rather flat but spirits "all right." When asked about her work replied "I'm a biscuit packer, dogfood biscuit packer. Been there 25 years. . ." *And you've enjoyed that kind of work?* "Oh yes, oh yes I liked it very much." *Do you live a happy life by yourself?* "Yes, I'm quite contented . . . rather a busy life really, all day work see, 'til a quarter to five and I pop around to Mum's and come home and do me own work so I'm busy all the time . . . I have dinner at work. . ." *How has your appetite been?* "Alright yes, I eat alright." *And how do you sleep at night?* "Alright." "I think there's more people in the world today on the mental side than there used to be. Since the war you hear more of it don't you? And I've noticed that there always seems two in the family. There is two people up at my firm and they've both got two in the family that is on the mental side."

Comment. Without invoking genetics, it is difficult to say why *A* and not *B* developed a recurrent and eventually chronic paranoid illness with affective features similar to that of another sister. *B* had a low birth weight and poor school record. She had the stresses of having an alcoholic father and having to deal with the development of psychiatric disorder in two sibs. Later she was widowed and lived for many years on her own. According to the MMPI her personality may be more than averagely depressive, introverted, and hypochondriacal. Yet, at the age of 53 she is coping well and appears to be within normal limits.

DZ 24, female, born 1927

A: 1st hosp. 27 (Sc 29), 7 spells: 202 weeks; FU 36, recent hosp. discharge.
B: OPD 36; FU 37, working, symptoms +.

	A	*B*
Consensus Diag.	Sc (5)	Other (5)
Slater Diag.	Recurrent Sc, atypical	Anxiety state
Meehl Diag.	Schizo-affective psychosis	Pseudoneurotic Sc
Global rating	5.5	3.75
MMPI	Not done	Not done

Zygosity Evidence. Both twins stated they never looked alike.

Background. Social class 3. *Fa* died 63, *Mo* died 53; 1 *M* 42, 2 *A* and *B*.

Fa shoemaker—strange, silent, anxious, strict, undemonstrative; leg amputated in World War I. *Mo* Roman Catholic; preferred 1 *M* to twins, died when twins 14 years of age. 1 *M* jealous of and cruel to both twins.

A difficult infant, thin, irritable, temper-tantrums, much thumb-sucking. *B* good-tempered, placid baby, no developmental problems. Only *B* hospitalized at age 5 for diphtheria. Menarche: *A* 15½, *B* 12½. *Mo* insisted they dress alike; they both objected. Both had average intelligence, in same class at school and same placement in class, *A* was a shy, self-conscious child who thought teachers and others were against her; the strict, punitive, educational methods frightened *B* who was also frightened by 1 *M*'s cruelty and of *Fa* and 1 *M* after death of *Mo*. *A* unhappy with *Fa* after *Mo*'s death and, after violent quarrel with *Fa*, left home at 19, came to London from Scotland and married soldier of foreign origin. *B* kept home for *Fa* and 1 *M* from 15 after short period of factory work, then left home to live with girlfriend until marriage to grocer at 18; son born when she was 20.

Psychiatric History of A. Depressed over domestic difficulties for a time (*onset*) during first pregnancy (at 24) which ended in miscarriage at 4 months. 10 days after birth of first daughter (at 27) was confused, afraid of husband, called him "daddy," ran out of house half-dressed, picked up by police, told them husband was dead, kept asking for 1 *M*; admitted to OW, transferred to *Maudsley* where became increasingly unpredictable, restless, noisy, impulsive, and obscene; mood incongruous. Symptoms virtually ceased after series of 12 cardiazol treatments. Discharged after 25 weeks (diagnosis: puerperal psychosis Sc). Affect normal 1 month later at Maudsley OPD. Admitted to MH (at 29) 14 months later after birth of second daughter (29); depressed, agitated, crying and smiling simultaneously (diagnosis: puerperal psychosis). Repeatedly in and out of MH; aggressive, anxious, ideas of reference; sterilization agreed to, then refused. Left hospital after 36 weeks (final diagnosis schizo-affective psychosis, initiated by puerperal psychosis). Always depressed and often violent at home, e.g., smashing panes of glass, returned to MH for 4 weeks (at 30) where treated with chlorpromazine and 6 ECT. *Fa* died at about this time. Readmitted to MH (at 31) for violent behavior during seventh month of third pregnancy; sterilized after birth of third daughter. The four subsequent hospitalizations, over a period of 4 years, followed episodes of violence with almost continual depression with paranoid features—neighbor was "taking the rise out of her" and "my husband doesn't like me being experimented upon." Treatment consisted of phenothiazines, then ECT. Perphenazine, reserpine, and imipramine hydrochloride used successfully during fourth admission after seizures developed on chlorpromazine. (Diagnoses of Sc and of affective psychosis were made.)

When Interviewed. At 36, she was quiet and ready to leave the hospital that day, looked 45. She tried more than once to end the interview—"Can I go now, doctor? I get awful embarrassed . . . I'm very shy, I want to go home . . ." but was persuaded without difficulty to continue. She commented that she felt more like 80 than 35; her idea of a good time was "a little drink, doctor, that's all."

She felt disliked by the staff of the hospital and compared herself unfavorably with the rest of her family.

Psychiatric History of B. At 27, very upset and worried by psychiatric illness of twin. At 30, similar, though milder, symptoms (*onset*) to those for which she later attended psychiatric OPD at 36, i.e., weight increase, crying, fear of roads, she imagined pain in back was cancer and could not be reassured by GP. Also, slept well but did not feel rested, lost all interest in keeping house or dressing well, both of which had been important to her previously. Treated with tranylcypromine sulfate (diagnosis: reactive depression). Depression attributed to domestic life (husband gambler, heavy drinker, different religious faith; 1 *M* lived next door demanding meals and housekeeping). Seen for Twin Investigation (at 37) by local PSW on our behalf; complained of depression but gave impression of a very anxious, unhappy, self-centered, irritable woman. She was drinking markedly more (began drinking at age of 34), six whiskies plus three pints beer per evening, often felt more depressed and regretted thus squandering her earnings. Had recently become sensitive to noise, developed lapses in concentration and on one occasion tore tiles from fireplace in a rage when husband home late for dinner. Also, feared people might be looking at her and talking about her; afraid of dying suddenly; insisted husband paint entire ceiling because of tiny spot on it.

Later Information. Since discharge, *A* at 41 had been admitted to a different MH.

Comment. *A*'s illness is remarkable for the association of the first three attacks, i.e., until she was sterilized, with the puerperium or with pregnancy. Affective features remained prominent throughout, and there appears to have been little personality deterioration. The epileptic seizures were presumably due to drugs or drug withdrawal. If her schizophrenia is atypical clinically, it is also conceivable that it may have developed on a different genetic basis from most. Certainly, in the opinion of most judges, the resemblance between the twins is on the affective side, *B*'s illness being termed depressive by three judges and by the psychiatrist who treated her, and *A*'s illness seen as schizoaffective. Meehl regards *B* as a pseudoneurotic schizophrenic, possibly on account of features such as chronicity, rage and sensitivity to noise, which Slater sees rather as the possible effects of alcohol. Premorbidly *A* was the less sociable and more suspicious, and it is in keeping with this personality difference that *A* should be the schizophrenic.

DZ 25, male, born 1930

A: 1st hosp. 18 (Sc 18), 2 spells: 33 weeks; suicide 20.
B: ——; FU 33.

	A	B
Consensus Diag.	Sc (4)	Other (3)
Slater Diag.	Catatonic Sc	Juvenile delinquent
Meehl Diag.	True Sc	Normal
Global rating	6.0	2.25
MMPI	Not done	1'8029-54736/

Zygosity Evidence. Blood, fingerprints (favor MZ), seen.

Background. Social class 4. *Fa* 63, *Mo* 77; 1 *F* 33, 2 *M* 30, 3 *F* 28, 4 *M* 24, 5 *F* died 10, 6 *A* and *B*.

Fa house painter, still working at 63. By various reports strict and punishing, easy-going with little warmth or understanding, a small, voluble, defensive man, more intelligent than *Mo*. *Mo* employed as domestic until age 60, described variously as gentle, motherly, a "good mother" but little in touch with children, and as a dull, emotional woman who suppressed information. She had two convictions for shoplifting, serving 1 month in prison where she spent most of the time in the prison hospital as the sentence "nearly broke her heart." 5 *F* choreic, later died of heart disease. Deprived childhood (no overcoats in winter); most of family of low intelligence.

Birth weights: *A* 6½ pounds; *B* 7 pounds. No notable differences in early development, both extremely thin children (*B*'s thinness of special concern to *Mo*). Quite different physically as they got older, *A* 5 feet 11 inches, darker hair, broader face; *B* 5 feet 9 inches. Both twins of low intelligence, *A* a particularly poor reader although slightly better than *B* at other subjects. Twin relationship no closer than with other sibs. *Mo* regarded *A* as a leader and readily admitted *A* to be her favorite son, *Fa* had no preference. *A* lively child, affectionate, lots of friends; *B* quiet, normal, bookworm as child (*Mo*). *A* sustained stomach puncture wound at age 8 when he fell on iron spike. Age 10 *A* and *B* convicted, with others, of stealing cigarettes from bombed-out store, on probation for 1 year. At 14, *A* put on 1 year probation for larceny, a year later a further 2 years for "breaking in." *B* no further offenses during school age, joined regular army at 17.

Psychiatric History of A. Caught stealing from railway at 16; ten other offenses considered, sent to Approved School (was there 14 months). Rejected from military service because of stomach injury. When *B* sentenced to Borstal (at 18), *A* became anxious about his own delinquencies, dressed in his best suit, left the house and later in the evening presented himself at a GH having taken an overdose of aspirin. Three days later to OW, then MH because of alternating periods of excitement (smashed windows, once attacked nurse with dagger) and catatonic stupor (would only shake or nod head in answer to questions). First few days after admission to MH kept in padded room; 1 month later emerged from stupor. Four weeks later made unprovoked attack on another patient with sharpened metal rod while in hostile, excited state. Self-inflicted scratches on wrists also reported. Treated with 9 ECT, improved, discharged to *Fa* after 21 weeks. Appeared unaccompanied at *Maudsley* OPD (at 19) 2 months after discharge from MH, complained of "funny feeling . . . smash six windows . . . or take some tablets." Impression: aggressive psychopath with EEG abnormality; refused to accept treatment. Admitted to Maudsley (at 20) after suicide attempt by inhaling chloroform. Behavior at Maudsley normal, then destructive when not allowed to remain in bed (repeatedly smashed windows, unable to explain why). There was never any stupor or overt schizophrenic behavior and his insight remained good; however, he did create an impression of strangeness and one consultant psychiatrist (Slater) considered him a catatonic schizophrenic and potentially homicidal (hospital diagnosis: pathological personality of schizoid type). He absconded, was returned, then admitted to MH where regarded as psychopath with aggressive outbursts. After 12 weeks *Fa* removed him from MH. Three months later, after disturbance at work, he left a note "Do not disturb," took overdose of aspirin and died 4 days later (at 20). Coroner's verdict: suicide.

Psychiatric History of B. He was "never any good at soldiering" and was often absent without leave from the regular army (age 17–20). Twice at 18 and at 20 he broke in and rifled the gas meter at home; on the first occasion sentenced to 2 months in prison, the second time sentenced to Borstal training after *Mo* reported *B* to police in order to clear herself of suspicion. He learned painting and decorating at Borstal where described as obstinate and evasive and did not mix freely. Another observer noted that he "matured a great deal at Borstal" and stuck to a job once he made up his mind to do so. When seen for Twin Investigation at 20, his Commanding Officer reportedly said he had been courting trouble in the army by remaining away and getting caught. He himself said he preferred Borstal to the army where there were too many people giving orders . . . tough, aggressive, a "hard Nut!" After serving overseas there were no further offenses. Redecorated parents home when on leave at 22, Borstal supervision discontinued. He was still doing well at 33, reportedly married, with one child. Refused to be seen but completed MMPI in return for payment.

Comment. It is hardly surprising that both twins from this background were juvenile delinquents. The question arose whether *A* was truly schizophrenic or was an aggressive psychopath with mood swings. The consensus diagnosis seems justified, and there are in addition good grounds for regarding *B* as schizoid. Indeed, he seems to have been the more evasive, withdrawn and awkward personality of the two. If he eluded us personally on the follow-up, at least he could be persuaded to complete the MMPI which substantiates the schizoid impression gained earlier. The profile was abnormal and highest on the Hypochondriasis, Schizophrenia, and Social introversion scales.

DZ 26, female, born 1935

A: 1st hosp. 21 (Sc 21), 3 spells: 38 weeks; FU 29, not working, symptoms +.
B: ——; FU 29.

	A	*B*
Consensus Diag.	Sc (5)	Normal (6)
Slater Diag.	Hebephrenic Sc	Normal
Mcchl Diag.	True Sc	Normal
Global rating	5.0	1.25
MMPI	Not done	4'38-27 1̲6̲/095:

Zygosity Evidence. Blood, fingerprints, seen.

Background. Social class 3. *Fa* 54, *Mo* 54; 1 *B* and *A*, 2 *M* 26.

Fa, dissatisfied printer; difficult, obstinate, quick-tempered. *Mo* looked stern and unsympathetic yet was "extremely kind." Parents favored *B* and rejected *A*, the more difficult child, in every way the ugly duckling of the family.

Premature birth by 1 week (birth weight: *A* 3½ pounds, *B* 6¾ pounds), *A* breech delivery. *B* easily ahead in development. *A* bilateral ptosis upper eyelids and "twisted right leg," both of which repaired age 13, snub nose, about which she was often teased, repaired later, only *A* rheumatic fever in childhood. *A* left-handed, *B* right-handed; menarche at 15½ and 14½, respectively. *B* night-mares in infancy. *B* brighter at school although a "pale, dreamy child" who disliked school. Twins were not close, did not go around together and *A* was jealous of *B*'s all-around superiority. *A* trained as domestic, later employed as children's nurse and as cook; single. *B* commercial course when *Fa* disapproved of occupational therapy training. Later awarded grant for nurses training, completed course and worked for 1 year before marriage to dentist (at 23).

Psychiatric History of A. CGC at 12 for obstinacy. Admitted to MH OW from GH after overdose of seconal (at 21). Suicide attempt precipitated by depression and recent emotional attachment to vicar (*onset*); tried "to frighten parents, to get some sleep because couldn't get reasonable job." At admission, lack of affect, said she felt alright. Later, vague, circumstantial—saw Christ, believed others could read her thoughts, believed (without evidence) that she had cancer or was pregnant; 60 insulin comas, 6 ECT, then reserpine; discharged after 21 weeks (diagnosis: Sc). Within a month, three suicide attempts, to MH for 12 weeks then to GH for second ptosis repair. Referred to *Maudsley* OPD (at 24) by GP as "a cause of gross anxiety to parents and myself for many years, intelligence pretty low." Slow and awkward in movements, hesitant in speech, rude, insulting, unreasonable; considered character disorder, but way of talking "compatible with schizophrenic defect", (diagnosis: ?schizophrenic residual state). Letter from GP (at 25) to Maudsley OPD said *A* had "gone into a hypochondriacal depressive phase . . . mainly . . . self pity". Later abusive and threatening to clergymen who befriended her and their families (300 letters to one in a short time) culminating in an acid-throwing threat. Now often dirty and untidy. *Fa* thought *A* deliberately antisocial. To MH (at 28); described as ". . . a most irritating creature." Discharged after 5 weeks (diagnosis: paranoid state— possibly high grade subnormality due to birth injury). Later, expresses suicidal ideas, sought admission to a therapeutic community, considered unsuitable because "far too disturbed."

When Interviewed. At 29, had just been dismissed from a job after usual story of superiors being "beastly" and "unfair." Used psychiatric jargon in interview; considered she was "not ill, just slightly unbalanced because wrong environment. . ." Referred to wrong environment or not being in right job 12 times in course of 30-minute interview . . . *are you an emotional kind of person, do you show your feelings?* "No, not really . . ." . . . *are you especially sensitive to other people's feelings?* "Oh, I can read people like books. I'm very sensitive to atmosphere. Immediately I go into a room I can sense a certain atmosphere." . . . *Did you ever feel as if you had communication from God?* "Well, I don't want to discuss it . . . that is private . . . But I've had most horrid experiences, most uncanny . . . as I was talking to him (friend) . . . I had a horrible uncanny experience that he was going to end up in a motor crash, you know, he was going to die . . . and it happened. . . . I get warnings like that . . . preconceived warnings. So I've definitely got proof that there's a Creator."

Later History of B. Seen for Twin Investigation (at 24), impression entirely normal, a little shy, affect warm, plenty of personal resources.

When Interviewed. At 29, very normal impression, happily married, with three children (sons 6 and 4; daughter 1). Described period of mild depression in early marriage when husband away a lot, baby going through difficult phase coinciding with other environmental stresses. *Can you tell me how you've been feeling?* "Oh well . . . I feel perfectly alright . . . em . . . as far as . . I mean I think . . I like being a housewife actually . . I mean I was a nurse before we married, but I think one thing about being a housewife . . . is . . . you are much more in charge of your own day aren't you? . . . . . the repetition is boring. I enjoy making a home . . . em . . . . we've really only just started I suppose. . . ." *What kind of things do you do that gives you the most pleasure?* "Oh I don't know, all sorts of things really, I enjoy doing all sorts of things . . . em, well I can't think of any one particular thing. I enjoy going out with my husband without the children . . . (laugh). . . ."

Comment. A started life with a 3¼ pound weight disadvantage and a breech delivery followed by a string of physical insults. She was in all senses the ugly duckling. Perhaps nothing could prevent the vicious circles in *A*'s life which finally led to her schizophrenia; without a genetic predisposition for schizophrenia we would have predicted some other kind of psychiatric disability. *B* was different on every count, attractive, successful and, at follow-up, very sure of herself.

DZ 27, female, born 1927

A: 1st hosp. 17 (Sc 18), 12 spells: 656 weeks; died 36.
B: ——; FU 37.

	A	*B*
Consensus Diag.	? Sc (3)	Other (5)
Slater Diag.	Epileptic psychosis	Psychopathic personality
Meehl Diag.	True Sc	Borderline ambulatory Sc
Global rating	5.75	3.75
MMPI	Not done	Not done

Zygosity Evidence. Never mistaken for each other, even as infants.

Background. Social class 4. *Fa* died 57, *Mo* died 47 (abn); 1 *F* 50, 2 *M* 45, 3 *M* 40, 4 *B* and *A*, numerous half- and step-sibs.

Fa furnace tender, described variously as a good, placid, steady man and as "woman-mad," unfortunate in his choice of four wives; violent tempered, imprisoned for manslaughter of first wife. *Mo*, married twice (*Fa*'s second wife), MH for 6 weeks following suicide attempt; alcoholic; kept home meticulously clean but made children, who were terrified of her, do the work. *Mo* died, unmourned, when twins 9 years of age. Chaotic family with much violent temper, many suicide attempts, some seizures and prostitution among members which included three siblings, at least 9 half-sibs and numerous step-sibs. Maternal half-sister (MMPI 168'49-3250/7:).

A heavier birth weight at 2 pounds, 10 ounces; *B* a little less. *A* always larger, taller and developmentally ahead of *B* who was a smaller, more delicate baby and was "none too bright;" difficult pregnancy and instrument delivery. Twins reportedly thought alike and one cried if the other were away or unhappy. *A* taken off by a man at age 9, court case. Life of wandering and neglect after death of *Mo*; *Fa* in pub nightly. First step-*Mo* died soon after marriage to *Fa*, second step-*Mo* very strict. Evacuated during World War II (10–14). Lived unharmoniously with *Fa* and step-*Mo* until 17th birthday, then to live with older sister who urged twins (*A* refused) into prostitution. At about this time *A* described by probation officer as "lovable and affectionate"; *B* as "of very simple make-up, very unstable, stays anywhere." *B* described (by relatives) as "mischief-making and shocking liar." Had some domineering ways with, and fascination for men; "none of them can stand up to her"; *A* didn't like men . . . animals yes, not men . . ." *Fa* died when twins 18.

Psychiatric History of A. Suicide attempt at 14 (took iodine), chronic psychiatric invalid from 17 (*onset*) starting at time of much domestic stress with suicide impulses, hysterical behavior, and wandering. Hysterical astasia-abasia from 18, amnesia and seizures from 19 at which age ? acute pyogenic meningitis following aspirin poisoning. Clearly epileptic at 20, epileptic psychosis diagnosed (at 21) but no seizures after 23. ECT, worse after 2nd course, rarely on anticonvulsants. From 21 mostly lay curled up in bed, developed many contractures; paraplegic, DS often considered but finally excluded. At 29, still hospitalized, marked thought disorder, poverty of affect, disjointed and meaningless conversation ("I betted you an Aspro," name is "Girl," age is 6¾ but 9 if she "wants to be"), destructive and noisy. Chlorpromazine begun with remarkable mental response and physical rehab; 1 year later cheerful, well-oriented, persevering in walking exercises. Remained out of hospital for 3½ years from age 30 after chronic hospitalization for over 10 years. *Maudsley* OPD (at 34), chronically deluded (diagnosis: paranoid psychosis). Developed cancer of tongue during unhospitalized period, which caused death at age 36. She had been in at least 21 different hospitals. Diagnosed schizophrenic stupor (at 18); schizophrenic reaction (at 23); and hebephrenia, as well as paranoid illness (at 34). Earlier diagnoses included manic-depressive, psychopath, moral insanity, and high-grade defective.

Psychiatric History of B. Prostitute sister soon regretted having *B* (at 17) with her as *B* much preferred this life and truanted from work to help her sister. Married (18) when already pregnant, to unemployed, deaf man. Appeared emotionally immature and shallow to PSW; thought she might be admitted to MH where sib was (and from which *B* had twice engineered *A*'s escape) for confinement. In-laws doubted paternity of baby after its birth. Second child born 1 year later, father foreign army officer (this son attempted suicide and was hospitalized at 17). Third child, fathered by sister's husband, born later following year (this daughter attempted suicide at 12 years). From 31 used name of man with whom she was living, while her husband lived with her older sister for 7 years before returning to home of *B* and her co-habitee. Referred to Maudsley OPD (at 35) after suicidal threats but refused to attend. Borrowed money for *A*'s funeral which she spent in decorating home and was still paying for funeral a year later. Unwilling to be seen (at 37) for Twin Investigation; was said at this time to keep hat and gloves on for days, to run out of house screaming and to be very elusive to callers, very dirty, and to keep to her bed.

Interview with Older Sister and Maternal Half-sister. Corroborated chaotic family life and undirected childhood and adolescence of the twins; also provided much of *B*'s history.

Comment. *A* is diagnostically the most doubtful DZ proband. Though Meehl calls her "true Sc," he raises the possibility of an organic component in his comments, as do the other two judges who admit the possibility of schizophrenia; and it is highly likely that some such organic factor accounts for the difference between the twins. Much of their similarity in respect to psychopathic behavior can be attributed to the chaotic family environment.

DZ 28, female, born 1930

A: 1st hosp. 23 (Sc 24), 3 spells: 37 weeks; FU 33, working.
B: ——; FU 33.

	A	*B*
Consensus Diag.	Sc (5)	Normal (5)
Slater Diag.	Manic-depressive psychosis, ? Sc	Normal
Meehl Diag.	Acute undifferentiated Sc	Normal
Global rating	4.5	1.75
MMPI	4'1378 - 62/90S:	4-893571/062:

Zygosity Evidence. *Mo*'s and twins' description, no doubt; green-blue eyes in one twin, other twin brown-eyed.

Background. Social class 1. *Fa* died 72 (abn), *Mo* 75. 1a *M* died 25 and 1b *M* 34, 2 *M* 29, 3 *A* and *B*.

Fa, lawyer, Canada, a chronic, agitated melancholic with delusions of financial ruin; alternately hospitalized where treated with ECT, and nursed at home by rather histrionic wife. *Fa* died when twins 22. *Mo* "the one sound, solid sensible member of the entourage" (MMPI 54386/12709:).

A for some years thought to be retarded but developed reasonably well and equalled her contemporaries. Always shy, "peculiar" called herself "serious, introverted"; age 12, a brother attempted sexual intercourse. Age 19, nurses training interrupted by infectious hepatitis, never married. *B* poor scholar although later secretary to an accountant for 10 years, then married professional man.

Psychiatric History of A. Depressive periods, neurasthenic attacks, jaundice (infectious hepatitis at 19), made her unable to work regularly from age 18 (*onset* ?) until 23 when arrived in London where depression and lassitude led to admission to nursing home for 3 weeks (at 23). Treated with 6 ECT (diagnosis: endogenous depression). Recurrence of depression a few months later (at 24) led to second nursing home admission for 3 weeks with modified insulin treatment. Casual affairs with strangers (intercourse denied) preceded third nursing home admission for depression. More assertive than previously; spoke freely about her sexual difficulties, interpreted psychiatric interviews as designed to test her sexual reactions. Treated with chlorpromazine (only 75 mg/day), then with dextroamphetamine sulfate, 10 mg, which produced a sharp behavioral change; became hilarious and mildly destructive, then almost immediately morose. At this time struck doctor on the face saying "It's all your fault." (Diagnosis:

inadequate personality poorly adjusted in the sexual and social spheres and preoccupied with these difficulties in a schizophrenic way). At private psychiatric clinic where transferred; complained of confused thoughts, physical exhaustion and headaches. Psychotherapy attempted but abandoned after 2 months as *A* more frankly paranoid and aggressive (believing other patients were talking about her); deluded and hallucinated. (Diagnosis: paranoid Sc), transferred to *Maudsley*. Wechsler IQ 77, hostility and resentment subsided after 40 insulin comas. Discharged (24) after 31 weeks (diagnosis: paranoid Sc). OPD (25) headaches, tiredness, no thought disorder. Returned to Canada where, at 33, still well and successfully employed as secretary for 6½ years. GP attendance for "lack of pigmentation in my skin, causing white patches . . . and a burning sensation and numbness in some parts," treated with an antihistaminic. Very friendly and clearly written letter giving recent history.

Later History of B. At 33, two children; described as "gay, attractive, energetic, mildly excitable, sociable, never incapacitated by nerves." Briggs questionnaire states that for short periods she used sleeping pills and has sought medical assistance for her "nerves." Lives one block away from *Mo* and *A*.

Comment. In the absence of first-hand information, it is difficult to evaluate this pair, apart from underlining that the twins are clearly DZ and discordant for psychosis. Though *A* seems to have made a satisfactory recovery, there is fairly general agreement about her having been schizophrenic. Reviewing his diagnosis, Slater now gives schizophrenia as his first choice, though one notes the "melancholia" of the father. *A* was the more introverted.

DZ 29, female, born 1905

A: 1st hosp. 52 (Sc 52), 2 spells: 38 weeks; FU 59, working, symptoms +.
B: ——; FU 59.

	A	B
Consensus Diag.	Sc (6)	Normal (5)
Slater Diag.	Paranoid Sc	Normal
Meehl Diag.	Acute paranoid reaction	Normal
Global rating	4.5	1.75
MMPI	Not done	430'2618 79-5:

Zygosity Evidence. Photos, no doubt.

Background. Social class 2. *Fa* died 72, *Mo* died 32; 1 *F* died 36, 2 *B* and *A*.

Fa businessman, devoted to family; to London from Central Europe with twins after losing property, died when twins 40, Jewish family. *Mo* died at twins' birth.

Twins in Children's Home until 2, when home again, happy despite jealous, nagging step-*Mo*. *A*, the better student, trained in dressmaking, then clothes model. Successful chef and housekeeper after arrival in England at 34. *B* handicraft worker, then chef in England and manageress of restaurant.

Psychiatric History of A. Premorbidly timid, cheerful, pleasant, conscientious, introspective, hypochondriacal; no close friends, but acquaintances and a 10-year affair with married man, ending age 32. At 51 (*onset*) still single, suspected people of trying to prove her a thief and men of following her; moved home and job several times, but found everyone involved in a plot against her; symptoms increased until unable to work. To OW, then *Maudsley* IP (at 52); fairly well systematized paranoid delusions with many ideas of reference, no hallucinations. Unable to accept treatment, her delusional system expanding to include patients and staff (diagnosis: paranoid Sc). To MH where one of her delusions was that Maudsley staff had poisoned her food. 4 ECT, became increasingly tense; still evasive on chlorpromazine, discharged (at 53) to sister after 24 weeks. Five months later readmitted to MH after sudden relapse; gave up job because food poisoned, bizarre ideas of reference, auditory hallucinations of a frankly sexual nature. Improved on trifluoperazine (diagnosis: paranoid Sc). Not seen for Twin Investigation (at 59) as *B* feared, and GP confirmed, that interview might disturb her precarious mental balance. Always slightly paranoid; on trifluoperazine. Held domestic position with difficulty and with much help from *B*.

Later History of B. Described (at 52) by *A* as capable, friendly with everyone, devoted to her work. Social workers saw her as a pleasant, voluble, demonstrative person; rather excitable and quite tearful about twin's illness. Twins observed to ,"live for each other." Follow-up for Twin Investigation (at 59) described by GP as "incredibly devoted to twin, does a full day's work, and afterward the chores her sister is supposed to do."

When Interviewed. Confirmed GP's observations. *You've been holding down a good job ever since you came to this country?* "Oh yes, oh yes. I work very very hard here, very hard, because certainly I wasn't used to it to do cooking and things like that you see and for nineteen years I've (mumble) . . . . . here and certainly worked very very hard and sometimes I was standing fourteen and fifteen hours on my feet. You see I've operations on my legs, varicose veins, because I wasn't used to it, to this kind of work and so the doctor told me no you can't, you must take a sitting job. So I found it you see and now I'm seven years . . I'm working as a cashier." *How do you spend your weekends?* "With my sister, with my sister, mostly I help her . . . otherwise I'm really ashamed because I didn't ask you before to come, you see, but I've no time whatever, perhaps you wouldn't believe me but if you speak to my friends . . you will . . because I lost nearly all my friends about it." (i.e., twin's illness and demands). Reiterated several times her satisfaction with England. Stated that *A* and *B* were both engaged during World War II but "so many other worries hindered marrying."

Comment. Regardless of the very close relationship between *A* and *B*, with its symbiotic overtones, *B* remains free of mental illness with no hint of *folie à deux*. Despite an assortment of environmental stresses such as no mother, the two earliest years in an institution, and being displaced immigrants with its attendant cultural isolation, both twins stayed healthy until *A*'s paraphrenia at age 51.

DZ 30, male, born 1928

A: 1st hosp. 19 (Sc 20), 2 spells: 21 weeks; FU 36, working.
B: ——; FU 36.

	A	*B*
Consensus Diag.	Sc (5)	Normal (6)
Slater Diag.	Atypical Sc	Normal
Meehl Diag.	Acute undifferentiated Sc	Normal
Global rating	4.75	1.25
MMPI	82'304-15679/	35867-1942/0:

Zygosity Evidence. Seen.

Background. Social class 2. *Fa* 71, *Mo* 69; 1 *F* 38 and *M* (died inf), 2 *B* and *A*.
Fa (MMPI 6'45-9270138/) business executive; *Mo* (MMPI 4''2' 716038-9/5:) highly strung. Parents separated but not divorced after twins grown up. Sister was pushed into background in family relationships after birth of *A* and *B*. Trained as mental nurse, employed as medical secretary but is "not really balanced" (*Fa*). At 35 still single and "difficult"; was once dismissed from post for striking another girl.

Gestation 8 months (birth weight: *A* 6½ pounds, *B* 6¼ pounds). *A* sustained minor head injury age 2½ when he fell on his head 17 feet from window onto snow beneath. *A* shy, reticent, studious, inclined to lean on *B* who was idolized by *Fa* and favored by *Mo*. Both attended boarding school; *B* possibly more intelligent than *A* in that he had a university degree in business administration; overshadowed *A* in almost all areas. *B* described by *Fa* as "very adventurous, a complete extrovert" and as very successful in his work, both abroad and for the government at home; *A* obtained some qualifications in pharmacology, later, after his psychiatric illness, successful career as assistant plant manager. *A* married at 28, 1 daughter (3); *B* married with 2 daughters (5, 2).

Psychiatric History of A. During army service (19) while at research station (classified work) complained of mild headaches, "feeling a bit vague," went to doctor complaining of "gland trouble" (*onset*) and spent about 2 weeks at army psychiatric hospital (diagnosis: unknown). At 20, difficulty in concentrating on studies, unreality feelings, thought blocking, grimacing and social withdrawal with referral to *Maudsley* OPD (diagnosis: Sc), next day to private MH. After 4

ECT and 40 insulin comas continued to show blocking and flattened affect; however, grimacing, inappropriate laughter and manneristic mouth-twitching no longer present. Discharged after 6 months (diagnosis: Sc).

When Interviewed. At 36, stated parents had found him unimproved at army discharge and arranged for private psychotherapy. "I started off twice a week and then once a week. I don't think it did a lot of good. I think over that period of time I could have stopped it, just a passage of time perhaps. It must have been some help I think. My father seemed very happy and very good of him to foot the bill (laugh) .. but in the end I decided that ... I had made as much progress as I was going to make and that was that. I think after that I steadily improved .. and I haven't considered myself, you know, in any way ill for some time. I feel quite certain that, you know, I won't have a relapse at all, feel very confident that I shall be O.K." *What kind of things do you do that give you the most pleasure? Do you have hobbies?* "Well, I haven't got many hobbies at the moment, I used to have a few..." *What's your idea of a really good time?* "... I suppose just like relaxing perhaps by the coast somewhere, do a bit of swimming..." *What kind of advice would you have for somebody who had similar problems for getting themselves back with it?* "... the first thing one has got to keep with people, don't wander off by myself, keep in contact with people .. I find it a bad thing to be alone you know, although one felt one wanted to be, it was certainly not the right thing to be at the time."

Later History of B. Had never sought help for psychological problems although psychotherapist who interviewed him at 20 recommended therapy. Very successful in a rewarding job. Almost physical fitness cultist (member of cycling and rowing clubs).

When Interviewed. Found difficulty in describing himself, eventually said that he was sociable and disliked being alone. Appeared somewhat sentimental, but caught self and became suspicious about breach of confidence. By end of evening spent with *B* and his wife, interviewer had the clinical impression that many of *B*'s traits of extroversion and bravado were in the nature of reaction formations to feelings of inadequacy. *Well how has your life gone, how would you describe what has happened to you?* "Well I've certainly never met my own expectations ... em ... of course every year or two there's some ... completely unexpected twist as far as my life is concerned and I've long since come to ... accept the fact that I can never really be at all certain where I shall be or what I shall be doing in a year's time, so expectations don't really .. come into it very much as far as I'm concerned. I don't know that I ever had any ... very clear expectations of anything. Obviously one expects in a general sense to ... to get involved in an interesting career and gradually bring in a bit more money (laugh) so as to be able to give one's own children the same sort of chance or better than you had yourself, but he ... this is very general ... it's difficult to say how I ... how have I showed up against those expectations I should say pretty average. I wouldn't say I've been eh . . such a running success that I'm very pleased with myself, on the other hand I'm not a howling failure, ups and downs naturally and em . . . the final result to date has been I should say about average for a person of my opportunities . . ." *That's not much of an answer.* "It wasn't much of a question . . . (laugh)"

Later Information. A rehospitalized (at 37) after suicide attempt (carbon monoxide), wife left him, got new job like old, on chlorpromazine, "surprisingly socially effective" (diagnosis: depression in mild ambulant schizophrenic).

Comment. Interpersonal relations in this family were probably poisonous while the twins were growing up; emotional divorce between the parents was followed by physical separation after the twins had left home. With a paranoid father and a hostile, agitated mother (inferred from MMPI and long interview with father) it is a wonder that *B* and his older sister are not seriously incapacitated like *A*. Again we must invoke degree of genetic predisposition to account for the difference. At follow-up *A*'s MMPI was indicative of decompensation although he was clinically fine and had been free of serious symptoms for 15 years; one year later he attempted suicide and was rehospitalized.

DZ 31, male, born 1926

A: 1st hosp. 25 (Sc 25), 5 spells: 346 weeks; FU 37, working, symptoms +.
B: ——; FU 37.

	A	*B*
Consensus Diag.	Sc (6)	Normal (4)
Slater Diag.	Catatonic Sc	Normal
Meehl Diag.	True Sc	Normal
Global rating	5.5	2.5
MMPI	Not done	249-860751/3:

Zygosity Evidence. Blood, fingerprints, seen.

Background. Social class 5. *Fa* died 83; *Mo* died 54; 1 *F* 48, 2 *F* 45, 3 *M* 43, 4 *M* 41, 5 *M* 39, 6 *A* and *B*.

Fa janitor, three times married; inconsistent and vague. *Mo* nervous and excitable but a "good mother."

Both weighed 6¾ pounds at birth. *B* was a class behind *A* at school. There was no really close twin relationship; *A* always a little less sociable than *B* and other siblings. Reportedly *B* had Sydenham's Chorea at 15 for which he was hospitalized for about 6 months (left side "all in movement"); no recurrence and no personality change identified. *Mo* died when twins 16. *A* a skilled laborer, army private, overseas 19–21; *B* domestic staff in RAF 18–20. *B* a general laborer with history of frequent job changes. All changes voluntary, attributed to a variety of excuses, i.e., "couldn't stand noise" of pneumatic drill; heights "bothered him" while fitting window frames.

Psychiatric History of A. Hospitalized twice while in the army, once with grippe; again (at 21) because he "pulled an officer up" for not saluting him (*onset*). Considered unfit for further service. Unable to make decisions, concentrate or hold job after discharge. Later (at 24) "doubled-up" on left side, depressed, deluded (thought still in the army and while in the street would suddenly stop and walk as if on sentry duty), irritable and short-tempered with brother-in-law who "persecuted him." *Maudsley* OPD (at 25) very suggestive of schizophrenic thought disorder; walked with head fixed slightly turned to left. Maudsley IP within 2 months (at 25) with periods of bizarre activity followed by periods of "normality." Blamed brother-in-law for loss of jobs. Discharged home after 31 weeks (diagnosis: schizophrenic dementia, paranoid). A month later believed he was King, intended to marry Princess Margaret—to MH (at 26), walked out after 6 weeks and 8 ECT; a short time later working satisfactorily. Admitted to MH for 80 weeks (at 29) after locking sister out of house; exhibited inappropriate and bizarre gestures, motiveless grin and grimaces, gave inappropriate answers to attempts at conversation, at times mute, walked with trunk bent and twisted to the side. Paranoid and catatonic features both present during fourth and fifth hospitalizations (at 30 and 32); treated with ECT. Believed he was doctor at MH, appeared deaf but as catatonic symptoms became more apparent his deafness became less so. On trifluoperazine (at 34) in chronic rehabilitation unit with successful week-ends at home; later worked as a road-sweeper (one of the best, according to foreman). Discharged MH (at 36) after spell of 204 weeks (diagnosis: catatonic Sc). *Fa* died about this time. Attempt at interview failed (at 37), uncooperative, and landlady feared interview might upset him. Still single, employed.

Later History of B. Described by relatives as nervous, slow, unmarried, no girl friends and few men friends. First seen for Twin Investigation (at 25) shy, lacked warmth and energy, inarticulate with a speech mannerism—spoke slowly with rounded, almost closed lips. No hobbies or interests.

When Interviewed. At 37, slow and ponderous in interview. Still single, lived at home of married sister, very schizoid impression. In good health, in good spirits, no "substantial" worries, concentration "pretty good", *What kind of reputation do you think you have, what do people think about you?* "Oh eh . . . they think that I'm a good worker, honest, and em . . . I think the firm thinks well of me . . . yes." *What gives you the most pleasure?* ". . . I get a good deal of pleasure out of me work and em . . also I get a good deal of pleasure when I go out and have a drink and have a game of cards and darts." *Close friends?* "Men . . . oh em . . . oh yes, oh yes. Yes, eh . . . people who I have a game of cards with and darts and so forth." *Do you read the newspapers or watch TV?* "Oh yes, yes." *Do you have any opinion about the present world situation?* "Do I have any opinion about the wor . . . the present world situation? Well em . . . I suppose I have, yes, all I've got to say is I hope peace is maintained throughout the world and good will . . (sigh)." . . . . *concentration?* "Pretty good." . . . *mind work O.K.?* "Yes, yeh, yeh. . . ."

Comment. After nearly 7 years in hospital, the last four continuously, *A*'s rehabilitation is remarkable. Premorbidly he had a better work record than *B* and he appears to have retained a certain persistence. *B* is among the most schizoid of the co-twins, Meehl going so far as to consider him possibly to be a simple schizophrenic. Slater would agree to calling him a schizotype: he has no girlfriends and few men friends; has had "hundreds of jobs," though never sacked; is lacking in warmth, energy and spontaneity; he has a speech mannerism and is very taciturn. It is perhaps surprising that more such personalities were not found amongst the co-twins of schizophrenics (cf. MZ 3, 6, 16; DZ 11, 25).

DZ 32, female, born 1942

A: 1st hosp. 12 (Sc 18), 7 spells: 143 weeks; FU 22, recent hosp. discharge.
B: ——; FU 22.

	A	B
Consensus Diag.	?Sc (3)	Normal (4)
Slater Diag.	Severe psychopathic pers	Normal
Meehl Diag.	Acute undifferentiated Sc	Schizoid pers
Global rating	5.0	2.5
MMPI	1'32 48 - 9670/5:	0'824-5967/3:1#

Zygosity Evidence. Never mistaken, different hair and eye color, *A* 3 inches taller.

Background. Social class 4. *Fa* 56 (abn), *Mo* 52; 1 *B* and *A*, 2 *F* 19, 3 *M* 12 (abn); Maternal half-brother 29.

Fa self-employed in flea market, chronic neurotic, sensitive to noise, migrainous, psychiatric OPD attendance in RAF and later ("inadequate, dependent character"); much medication for years, also one-quarter bottle of liquor nightly. Unhappily married from start, constant rows; very jealous of young son who ousted him from marital bed; disliked by family, long-term sexual interference with daughters. *Mo* total deafness after birth of unwanted son, later deaf only when worried; at 43 doctored for "nerves," 5 years later hospitalized for "blackout," later depressed, self-blaming, anxious and talkative. *Mo* married to give illegitimate son a home; liked by her children but they did not bring their problems to her. 3 *M* attended CGC (at 10), behavior problem (anxious, aggressive).

Difficult breech birth (*A* 7 pounds, 5 ounces, *B* 5 pounds +). *A* walked and talked first, *B*'s development enough later to be of concern to *Mo*. *A* right-handed, *B* left-handed. *A* hospitalized for 3 months at age 4 with enteritis. Menarche at 16 for both twins. *B* considered a little brighter at school although neither more than average. At 18 *A* received below average scores on group IQ tests; these scores thought to be underestimates but 3 years later tested with similar results. *A* "nerves bad" at 5 when face burnt by striking box of matches. Later sensitive, lacked confidence; also sociable, obedient, many friends. *B* temper tantrums, shy, moody child; later sensitive, very nervous (especially 11–12). From 13–15 sexual interference by *Fa*, often shouted, screamed and fought to prevent him.

Psychiatric History of A. At 10 *Fa* beat her bare buttocks severely when she set fire to a drawer of papers; at 11 stammered, cried easily, unwanted sib born, discovered that she had an illegitimate half-brother. 10 15 *Fa* often sexually interfered short of penetration, about which she told no one. From 11 (*onset*) plucked out her lashes and eyebrows, developed bald patches, excoriated face and burned herself with matches. Years later said "the things I done when I was ... in the Maudsley I done in a small way when I was young ..." GH OPD (at 12) for alopecia (diagnosis: hysterical reaction–artefacta). To special residential school (at 12) for 26 weeks; returned "a different child" (did not mix well) but alopecia had cleared up. From 12 on continued minor self-injuries but led normal school, occupational, and social life. Increasingly depressed, self-destructive, difficulty in concentration (at 17) with subsequent GH hospitalization for 4 weeks (diagnosis: anorexia nervosa). Swallowed crucifix (at 18), GH crucifix extracted at esophagascopy. Two months later appendectomy (18); resentful, withdrawn, tearful. Readmitted GH, 4 weeks after discharge as wound broke down (interference) several times. Transferred to *Maudsley* IP deeply depressed, refused food, repeated attempts at self-mutilation (scratched until raw, swallowed needles, and thermometers, rubbed pins into wound which healed only after immobilization in cast). Episodes of depersonalization and derealization "nothing seems real-like, a nightmare." About this time consultant's opinion: "flat, almost split affect; very strange, odd affective illness." Treated with trifluoperazine with transient improvement and marked Parkinsonism, abreaction with LSD failed, 6 ECT ineffective. Continued destructive, ambivalent, unable to concentrate, auditorily hallucinated (one voice *Fa*'s), no ideas of influence but felt people were against her; eventually (at 19) seen by treating psychiatrist as presenting full-blown schizophrenic picture, though regarded by others as long standing personality disorder; thioridazine to 900 mg with gradual improvement; discharged after 53 weeks (final diagnosis: Sc). Admitted as emergency to neurological hospital (at 20), deliberately withholding history of previous Maudsley admission, EEG suggested organic pathology. Later admitted psychiatric history, transferred to psychiatric ward where denied stress within family, with fiance (engaged to be married 2 months before admission to neurological hospital) or at work. Planned to marry though unsure of loving fiance and upset by his slightest attempt to fondle her (diagnosis: hysteria). After discharge seen weekly, arm and leg in casts to control self-mutilation (diagnosis: severe adolescent behavior, hysterical traits most prominent). At seventh MH admission (at 21) unable to stand, scissors gait (but played table tennis same evening), no CNS changes. During next 49 weeks continued self-lacerations, broke off engagement (he was "too good" for her). Absconded at least twice, made an unsuccessful attempt at nurses training. Treated with ether abreactions, imipramine hydrochloride and chlorpromazine (diagnosis: severe hysteria).

When Interviewed. On day of discharge (at 22) defensive, took 2 days to complete MMPI with assistance from solicitous nurses. *What are your plans?* "To take up nursing . . . the psychologist said I'm intelligent enough to get into nursing, but . . . I'd find it very hard after being in a mental hospital." Preferred not to talk about illness as "you get yourself in trouble . . . when you're ill it's not so bad to think about it, but when you're better, I think it does more harm than good." Had thought of reasons for her illness . . "each reason I seemed to come up against was entirely different from the first one . . . and the one before . . . in the end I just gave up." *How is your ability to concentrate?* "Now very good . . . when I was in the Maudsley, I was pretty bad . . ." *What's your idea of a good time?* "Dancing, parties, social . . . the same as anybody else."

Later History of B. Described by *A* as "temperamental, never shows any feeling, likes her own way a lot"; at Maudsley OPD (at 18) as informant for *A* found to be nervous. Of family life said "It's been terrible all the time . . . parents rowing . . . I don't dream nice things," remained at home because of twin's illness and considered that she had some influence in stopping family rows. Then (at 18) had had a number of boy friends, little sexual activity: always thought of *Fa* and was frightened. Refused to be interviewed (at 22) but corresponded as a substitute, gave clear information and completed MMPI. Still living at home, engaged to be married, working as sales model.

Comment. The dynamic interpersonal factors in this family provide a field day for clinicians and illustrate the difficulties of disentangling genetic and environmental contributors to pathology. *A* presented diagnostic problems for most judges because of the gross hysterical features and emotional instability. *B* was the only consensus Normal with two votes of ?schizophrenia and her MMPI and age suggest she may have a poor prognosis for staying well. The father looks like a schizotype, i.e., a spectrum case while the mother has few noted assets. In some sense *B is* concordant with *A*, but this could reflect environmental communalities as well as genetic ones.

DZ 33, female, born 1911

A: 1st hosp. 36 (Sc 47), 3 spells: 19 weeks; FU 53, working.
B: ——; FU 53.

	A	*B*
Consensus Diag.	Sc (5)	Normal (6)
Slater Diag.	Paranoid Sc	Normal
Meehl Diag.	Chronic undifferentiated Sc	Normal
Global rating	4.5	1.5
MMPI	34'81-9267/50:	239846/7051:

Zygosity Evidence. Fingerprints, seen.

Background. Social class 1. *Fa* died 43, *Mo* died 43; 1 *M* died 51, 2 *A* and *B*. *Fa* in medicine, died when twins 9 years of age. *Mo* "gay and laughing" but "always something wrong with her" (physical complaints). Left *Fa* when twins 7 for naval officer whom she later married; died when twins 11 years of age.

Birth weights not known but *B* heavier and stronger baby and larger, more athletic as adult. Twins had little contact with parents; nurse until 7, then brought up in boarding schools by committee of parishioners; home life either unhappy or nonexistent. *A* higher class at school, *B* sensitive about it but "blossomed out later." Twins always had fights, *A* tended to dominate *B*. *A* attended university, did teaching and secretarial work. Married at 26, 3 daughters (23, 22, 16) husband architect, very quiet; second daughter psychiatric OPD treatment at 17 for headaches and behavior disorder. *B* left school at 18, nursing student until 21, then trained as occupational therapist.

Psychiatric History of A. Possible breakdown at 17 (no details). Considered intelligent, attractive, mildly hypochondriacal, domineering and with mild mood swings. At 35 became fatigued, unable to do housework; 5 ECT privately with improvement. At 38, after appendectomy symptoms as before, upset by lack of sympathy from husband, admitted to *Maudsley* IP for 9 weeks. Mild depression and hypochondriasis observed. Psychotherapy: desire to dominate and play masculine role stressed. Treatment for further 2 years as OP (diagnosis: variously as anxiety state or as psychogenic fatigue). Well until 47 when began to bombard Maudlsey consultant with letters, newspaper clippings, cryptic messages about the psyche, the pyramids, etc., but refused to see him (*onset*); thought to have severe affective illness. Suicide attempt by swallowing glass and rings and attack on 17-year old daughter led to MH admission (at 47) in state of catatonic excitement. Said to have believed previously that she had been treated through radio connection with Maudsley doctor. After chlorpromazine and 9 ECT excitement subsided leaving schizophrenic talk and delusions: "I claim to be Jesus Christ and the Virgin Mary, I'm not out of my mind" (diagnosis: paranoid Sc). Discharged, much improved after 2 months; remained well.

When Interviewed. At 53, smartly dressed and composed, satisfied with busy domestic and social life—"Well, I think competent housewives really, normally as worth their weight . . . so that I couldn't ever think that I would want for service or for occupation." *Can you recall the content of your insanity? Can you look at it objectively now without it bothering you?* "Yes, I can discuss it, I can view it I think so . . . em . . . eh . . . So many things happened and I think any experience that you em . . . grow from or learn from, can enrich your own living and, I imagine, your own understanding, I would hope so. I find I'm just as rigid and just as likely to em . . . reach an impasse, but I do know that I . . . that there it is, and that it isn't an issue now as to whether I'm right or wrong, that isn't the end of it at all. It's really what is best, and I may need to be much more . . . em . . . fl . . . fluid . . . not fluid . . . em . . . I'm . . . I'm much more yielding if you like—resilient—in a way." Very normal impression.

Later History of B. Enjoyed work, liked to travel, generally shared flat with woman friend. First seen for Twin Investigation (at 39); independent, uncompromising, contemptuous of *A*'s illness and treatment. Single, many women friends, men friends older. Hysterectomy (at 47). Refused to be interviewed (at 53) but agreed reluctantly to complete questionnaire—her "final participation" in twin research. Single, lived in spare room in twin's home because of difficulty of finding own flat.

, *Further Information. A* readmitted to MH age 55.

Comment. The course of *A*'s illness is of interest in that her first hospitalization at 37 did not suggest schizophrenia so much as some affective disorder. By 47 she was clearly schizophrenic but made a quick and apparently complete recovery on chlorpromazine. On follow-up at age 53 she was quite well and insightful but 2 years later decompensated again. The unhappy or nonexistent home life has had no great effect on *B* although she remained unmarried.

5
MULTIPLE BLIND
DIAGNOSES OF THE TWINS AND
CONSENSUS CONCORDANCES

Even a neo-Kraepelinian would be forced to infer the presence of unconscious conflicts from the polemics and emotion surrounding the debates about psychiatric nomenclature. Axiological issues are smuggled into the kinds of classification preferred for ordering severe mental illnesses because the terms used have etiological implications; when the etiology is unsettled, obscure, or unknown, partisan debate is bound to ensue. As we have noted in Chapter 1, research on the etiology of schizophrenia has been beset by such difficulties. The split between "organic" psychoses and "functional" psychoses is a case in point as is the word "reaction." Although Adolf Meyer was reasonably concerned lest the label on a psychiatric patient lead to a counterproductive reification of a "disease" with a denial of the patient's psychological functioning or "dynamics," nothing productive was accomplished in the United States by the routine addition of the word "reaction" to the term "schizophrenic." Everything is a reaction to something. When appended to "schizophrenic," the word "reaction" had the seductive effect of focusing attention on the environmental circumstances and slighting the intrinsic nature of the person; it should be obvious that whatever the circumstances, they do not have a uniform effect on all persons exposed to, or apperceiving them. The latest revision of the Diagnostic and Statistical Manual (DSM-II) used in the United States has deleted the word reaction and is thus much more in line with the ICD; we do not feel as some (Cancro & Pruyser, 1970) that it was a "serious loss" for American psychiatry. With Gruenberg (1969, p. 371), we would "hope to destroy the stereotypical thinking which believes that he who says 'schizophrenia' is a reactionary, mechanistic, organistic, narrow-minded obstructionist and that he who says 'schizophrenic reaction' is the last best hope on earth for humanistic, scientific psychiatry." Obviously both stereotypes are equally incorrect.

Our point of view, or bias, about schizophrenia is that it is a relatively specific syndrome and neither a disease entity nor an arbitrary figment nor an epithet (cf. Chapter 1). The validity of the concept will depend ultimately on bio-physical data from patients; in the interim, data from genetic-family studies, together with its clinical utility, should suffice (cf. Chapter 2) as justification for the continued use of the concept. When the identical twin of a diagnosed schizophrenic becomes psychotic, the odds are overwhelmingly in favor of that psychosis being schizophrenia rather than something else. This kind of predictive value lends credence to the scientific respectability of the concept. We feel safe in asserting that it is more specific than mental disorder, psychosis, or ego weakness, and less specific than, say, Huntington's chorea.

We deplore mindless labeling as much as any competent therapist does. Diagnosing a patient presupposes a clinically useful degree of validity in a system of classification, the purpose of the diagnosis being to facilitate the management of the case by suggesting differentiated treatment and by approximating the prognosis, or, in our situation, to permitting research on etiology. David Shakow's crystal-clear and sagacious overview of the issues involved in classification for a science of psychopathology concludes in part that ". . . we cannot achieve our results cheaply. We need to be everlastingly concerned about standards in observation, standards in description and communication as well as standards in syndromization and theorizing [1968, p. 140]."

Dissatisfaction with the diagnosis of mental patients has pervaded the field of psychopathology at least from the time of Tuke (1892) until the present day (Kreitman, 1961; Zigler & Phillips, 1961; Shepherd et al. 1968, Menninger, 1970). Both nihilistic disavowal of diagnosis and cautious optimism about the possibilities of improvement for the process by analytical research were generated by the dissatisfaction. Contrary to prevailing misinformation, a reliable diagnosis of schizophrenia can be made. The lack of unanimity in the literature stems from a mixture of diagnostician variables, patient variables, and design variables (cf. Kreitman, 1961). The poor reputation of diagnostic processes seems to have originated with the results of efforts in the United States during the first three decades of this century to compare the percentage agreement between diagnoses made of patients seen first at the Boston Psychopathic Hospital and then at a Massachusetts state hospital. Wilson and Deming (1927), for example, found a 34% disagreement between the two settings with dementia praecox being especially unreliable. Much of the unreliability could be traced to the "professional attitude" of the hospital superintendents (Shakow, 1968) toward the concept of schizophrenia. Wilson and Deming could take comfort in the fact that their results for psychiatric diagnostic reliability were of the same order as those reported by Cabot (1912) comparing the physical diagnoses of 3000 living patients with those made for them after autopsy.

Studies frequently cited as proving the unreliability of psychiatric diagnosis (e.g., Ash, 1949; Hunt *et al.*, 1953; and Schmidt & Fonda, 1956) have been judiciously criticized by Kreitman (1961). Not so well known are studies in which we take some comfort. Norris (1959) in a series of over 6000 patients, seen first in an observation ward and subsequently in London mental hospitals, found 89% agreement for the category "functional psychosis," even though the interval between interviews was from 2 to 4 weeks. For the specific diagnosis of schizophrenia the percentage agreement was 68%. In the process of building their "Present Psychiatric State" interview schedule, Wing *et al.* (1967) obtained complete two-judge agreement on a diagnosis of schizophrenia for 69 of 75 (92%) patients called schizophrenic by the first clinician. An important outcome of the work by Wing and his team was the demonstration that of the 400 symptoms that could be rated from their interview schedule, those yielding a psychotic section score had very high reliabilities compared to those related to neurosis; the reliabilities for "delusions and hallucinations" (4 items) and "behavioral and speech abnormalities" (5 items) were .933 and .945, respectively. The further fruits of such efforts at refining the reliability of diagnoses are reflected in the cross-national study of mental disorders in the United States and the United Kingdom (e.g., Cooper *et al.*, 1969).

The intention of these introductory remarks is both to counter the widespread sentiment that "schizophrenia" is a mere semantic convenience with no existential properties and to introduce our experiment with multiple blind diagnoses of the 114 twins that led to the consensus diagnoses set out in the headings of the case histories in Chapter 4.

PURPOSE OF THE EXPERIMENT

There are two main points that can be made about how methods of diagnosis of schizophrenia might affect twin concordance rates and hence conclusions about genetics. The first point is the criticism of author-made diagnoses. Experimenter bias, it has been claimed (Jackson, 1960; Rosenthal, 1962a), may lead the investigator to make diagnoses contaminated by his knowledge of the zygosity of the pair and biased in the direction of his expectations. Some workers might unconsciously use a looser standard for concordance in MZ than in DZ pairs. Others might lean over backwards in their efforts to avoid any such bias. It is generally agreed that if the investigator, in the interests of consistency and good psychiatry, has to change old hospital diagnoses, he should at least indicate the fact and/or provide the case histories. Better still, he can invite another clinician to make a blind diagnosis from pertinent information. The latter would not know either the zygosity of the pair or the psychiatric history of the other twin. The researcher could then abide by this decision, or at least compare the result with the diagnosis reached after comparing the twins' histories side by side.

The second point relates to different standards for the diagnosis of schizophrenia. Quite distinct from the possibility of an investigator using a looser standard for schizophrenia for the co-twin than for the proband, is the question of whether concordance rates might vary according to the strictness or looseness of diagnostic standards adopted in general. Kallmann has been criticized for his wide concept of schizophrenia, implying that this is one reason why his concordance rates are high. But it is not clear whether wide criteria, applied equally to *probands* and co-twins, would lead to higher concordance rates than would strict ones. It might be reasoned that the adoption of loose criteria for schizophrenia in general would dilute genetic factors by the inclusion among *probands* of cases that were not "true" schizophrenia; lower concordance rates might then be obtained than if strict criteria had been used because the co-twins could have even less specific forms of impairment or be normal. The effect of diagnosis on concordance rates can best be looked at by having the twins in one series diagnosed by different methods and comparing the results. This approach has not been tried before and is one of the unique features of our schizophrenic twin study.

PROCEDURES

For the experiment a special psychiatric summary was prepared for each twin. The summaries did not refer to the zygosity of the pair or to the diagnosis, if any, of the co-twin. They included a brief account of the background, including any environmental or personal factors that might be relevant, and a fuller account of the morbid history or personality, based on all hospital records and on our own investigations and follow-up. All spells of treatment were mentioned and the diagnosis and main symptoms given. Needless to say, information was better in some cases than in others. Of course, when we prepared the summaries, *we* knew about both twins, and we tried to do justice to both similarities and differences. But by systematically including all the objective data, such as length of hospitalization and details of symptoms, in this (on the whole) long and well-observed series, it is unlikely that the results were seriously biased by what we chose to include or exclude when preparing the summaries. If any of the hospital diagnoses themselves were biased through the psychiatrist's knowledge of the other twin, the information, we hope, was sufficiently detailed and our judges astute enough for them not to be taken in. The summaries were fairly different from those in Chapter 4 in that more detail was included and they were, of course, separate for each twin.

The Judges and Their Tasks

All 114 summaries were diagnosed by each of 6 judges of different backgrounds from the United Kingdom, United States, and Japan. They were K. Abe (Japan), J. L. T. Birley (U.K.), P. E. Meehl (U.S.), L. R. Mosher (U.S.), J. S. Price (U.K.), and E. T. O. Slater (U.K.). After a consensus diagnosis was obtained from these six judges, as described below, the cases were each diagnosed by E. Essen-Möller (Sweden) as a representative of Scandinavian psychiatry. We are much obliged to all these colleagues for so generously giving their skill and time to our project.[1]

Our task was different from that of a reliability study. We did not try to define in advance what we meant by schizophrenia—or anything else—and then ask the judges to rate the cases accordingly. The idea was for each to apply his personal clinical criteria in his own way so that we could see how concordance varied depending on diagnostic standards. We hoped to obtain a consensus diagnosis of each case that could then be used as an alternative to hospital diagnosis in the calculation of MZ and DZ concordance rates.

Each judge was therefore given a free hand as to how he approached the task. But whatever else he offered to undertake, he was asked to make a diagnosis that could be classified as: S, or schizophrenia; ?S, or uncertain schizophrenia, a class preferably to be kept to a minimum; O, or other psychiatric abnormality (to be specified); and N, standing for normal or within normal limits. The summaries were presented in representative batches with the twins of a pair separated by at least one batch of 20 cases.

Consensus diagnosis was reached on a unanimous vote of 6 judges in 58 of 114 cases, with only 1 minority vote in 27 cases, and 2 minority votes in a further 16 cases. The remaining 13 for whom there was no majority decision were allocated to a consensus diagnosis by adopting conventions which we believe make clinical sense and which were, of course, applied regardless of zygosity. In practice, all twins with three or more S votes and none with less than three were classified as S. Twins with an accumulation of at least 3 S or ?S votes (e.g., S S ?S O O N) were called ?S. All with four or more N votes were consensus N, and the remaining 18 twins were consensus O.

---

[1] Equally we thank the Danish psychiatrists Margit Fischer and C. Flach who between them diagnosed one of each pair. For this reason their opinions could not be incorporated in our consensus. However, the "Scandinavian" diagnosis of Table 6.1 was obtained by pairing their diagnoses with those of Essen-Möller for the other twin.

Judges' Diagnostic Orientation and Usage of S/?S

We will briefly characterize the orientation of each judge, describe how he approached the task, and show the frequency with which he diagnosed S or ?S. The judges' 684 (6 × 114) decisions are set out in Appendix C. Table 5.1 gives the frequency with which each of our judges used each category, together with the number of consensus diagnoses. From the Grade I concordance rates reported in Table 3.2 of 10/24 MZ pairs and 3/33 DZ, it could be seen that 70 twins had previously been diagnosed as schizophrenic. The consensus diagnosis rejected two of them (MZ 10A and 12A) as being neither S nor ?S and moved a further five to qualified schizophrenia, ?S (MZ 6B, 9A, and 13A; DZ 27A and 32A). The consensus left 63 of our original 70 schizophrenic cases as S. Since one twin (MZ 23B) never previously diagnosed schizophrenic is now ?S by consensus, the number of S plus ?S twins is 69 of the 114.

Slater's (ES) approach to psychiatry is well known (Slater & Roth, 1969). All his first-choice diagnoses of schizophrenia were classified as S, even though some were atypical or uncertain. Three cases in which another diagnosis was preferred on the evidence provided, but where he specifically stated that the possibility of schizophrenia should not be ruled out, were classified as ?S. The single case in which he diagnosed symptomatic schizophrenia, i.e., an organic psychosis, was grouped with the "Other" diagnoses in accordance with Slater's views. He comes about midway in order of frequency with which schizophrenia was diagnosed.

Meehl (PM) is a clinical psychologist and past president of the American Psychological Association, an analyzed psychotherapist influenced by the psychoanalytic school of Rado, and at the same time the proponent of a monogenic theory of schizophrenia in some respects like Slater's (Meehl 1962, 1970). He is a strong believer in the concept of pseudoneurotic schizophrenia. Among other tasks, he agreed to classify all cases which he regarded as schizo-

TABLE 5.1

Judges' Usage of Diagnostic Categories for 114 Twins

Judge	Diagnoses				
	S	?S	O	N	S+?S
ES	59	3	24	28	62
PM	77	2	8	27	79
KA	55	1	27	31	56
JB	43	7	37	27	50
LM	52	11	29	22	63
JP	64	6	18	26	70
Consensus:	63	6	18	27	69

phrenia according to a schema used in the NIMH provided by David Rosenthal. This indicates whether they were chronic schizophrenia (or one of a number of approximately equivalent terms), acute schizophrenia (or its equivalent) or borderline schizophrenia (or its equivalent). In keeping with Meehl's personal concept of schizophrenia, "borderline" cases were counted as S. The two cases diagnosed by him as "schizotypes" were counted as ?S. The result is a high rate of schizophrenia in the sample.

The 4 other judges were all younger clinicians with 5 to 7 years' experience of psychiatry at the time and with individual research interests of a very varied kind.

First, in alphabetical order, is Kazuhiko Abe (1965, 1966), a Japanese psychiatrist who was a guest worker in our Unit. His psychiatric training, like that of most Japanese psychiatrists, was basically Kraepelinian, but his diagnostic practice has been modified by his own experience of research on familial psychoses and patterns of relapse. He first diagnosed the cases in his own preferred manner using terms such as "remitting paranoid-hallucinatory psychosis." Although he personally prefers not to use the term schizophrenia, he classified as S any paranoid-hallucinatory psychosis, whether chronic or remitting, that was not atypical or symptomatic. He used ?S once only, and is among the more conservative in his diagnosis of schizophrenia.

Most conservative was James Birley, a Maudsley-trained consultant, now Dean of the Institute of Psychiatry. He has worked in the MRC Social Psychiatry Unit on precipitating factors in schizophrenia (Birley & Brown, 1970) and with John Wing on the development of a diagnostic schedule for schizophrenia. He is of the opinion that schizophrenia is often too loosely diagnosed. In most cases he checked the presence of specified clinical symptoms such as thought disorder, using items on the Wing schedule, to assess the likelihood of schizophrenia. His own 6-point diagnostic scheme was collapsed[2] with his consent to S, ?S, O, or N yielding the results in Table 5.1. Only 50 cases were classified as S or ?S by Birley as compared to 79 by Meehl at the other extreme.

Loren Mosher is a psychodynamically and research-oriented American psychiatrist with previous experience in W. Pollin's NIH Twin and Sibling Unit, working on the families of MZ twins discordant for schizophrenia. He is interested in patterns of communication in families, has worked with T. Lidz at Yale University, and is now Chief at the Center for Studies of Schizophrenia at the NIMH. He preferred to diagnose schizophrenia on a rating scale from 0 to 7 that had been developed at the NIMH, rather than to record it as present or absent. Hence he has rather more ?S cases than the other judges. With his

---

[2] Thirty-one first-choice schizophrenia and 12 first-choice ?schizophrenia; 7 first-choice other diagnosis but ?S; 14 other diagnosis but just possibly schizophrenia (??S), and 23 other diagnosis with no possibility of schizophrenia; and normal.

agreement the most suitable cutting points were at numbers 4 and 5 on his scale, counting 5, 6, and 7 as S, and counting 4, which approximated to borderline or doubtful schizophrenia, as ?S.

The last of the judges was John Price, who has special interests in genetics, ethology, and the affective disorders. He was engaged in research in our Unit on the parents and sibs of psychiatric patients (Price, 1969). Though personally sceptical of the value of a diagnosis of schizophrenia, preferring, like Abe, a more detailed formulation of clinical and etiological factors in each case, he found no difficulty in classifying the summaries applying the criteria taught him at the Maudsley by Sir Aubrey Lewis and others. He anticipated and secured good agreement with Slater, except that possibly organic psychoses with a schizophrenic picture were counted as S. In cases classed as ?S his probabilities were fairly evenly balanced between schizophrenia and affective disorder.

British social psychiatry, American clinical psychology and psychodynamic psychiatry, and neo-Kraepelinian psychiatry are all to some extent represented in our judges. Later we shall be able to see how Scandinavian psychiatry fits into the picture.

Diagnostic Agreement

So much for the backgrounds of the judges and the frequency with which they diagnosed schizophrenia in the material as a whole. Now we come to the extent of their agreement in diagnosis.

Here Slater has contributed an analysis of variance, showing the contribution to the diagnostic disagreement found of variation between judges, on the one hand, and of variation between classes (probands vs. co-twins, MZ vs. DZ individuals) of subject, on the other. Differences between judges were significant at the .001 level, with $F = 4.41$ (5 degrees of freedom). Differences in agreement between zygosities and between probands versus co-twins were nonsignificant. Since there was no difference between classes of subject, the analysis suggested that all the judges were attempting to measure the same things but had different cutoff points for doing so. The analysis depended on the extent of agreement, case by case, of the diagnosis of each judge with that of the other 5 judges on the panel. As a convention, 2 points were scored for each agreement as to S, ?S, O, or N. ?S was regarded as poised halfway between schizophrenia and some other diagnosis, so that ?S by one judge matched with S or O by another counted as 1 point. If all the judges had agreed all the time, each would have scored 1,140. The observed range of points per judge was from 343 to 396. However, the average agreement by the six judges across all 114 twins was 79.4%. Slater was most often in agreement with the others. It is not surprising that for opposite reasons Meehl and Birley were in least agreement with the others. Agreement would have been remarkably closer if Meehl's borderlines and schizotypes had been grouped with O diagnoses and all Birley's possible schizo-

TABLE 5.2

Analysis of Variance for Interjudge Diagnostic Agreement[a]

Source of variance	Sum of squares	df	Mean square	F
Differences between				
Proband and co-twins	41.26	1	41.26	1.27
MZ and DZ	2.79	1	2.79	0.09
Interaction: Pro : Co X MZ : DZ	74.14	1	74.14	2.28
Total, between groups	118.19	3	39.40	1.21
Remaining between patients	3578.14	110	32.53	
Total variance (patients)	3696.33	113		
Between judges	60.24	5	12.05	4.41
Variance accounted for	3756.57	118		
Within groups	1541.09	565	2.73	
Total variance	5297.66	683		

[a]Courtesy of E. Slater.

phrenics, including those with two question marks, had been counted as S–103 out of 114 cases.

Two quantitative measures of agreement between pairs of judges can also be reported. In effect Birley and Mosher diagnosed schizophrenia on their own 4-point scales of certainty or severity. In the circumstances the observed correlation between them of .62 is quite good. More impressive is the agreement between Meehl and Mosher when they rated each subject on a 7-point scale of global psychopathology, which has been used by Kringlen and at the NIMH. Their ratings correlated .88 (cf. Chapter 6). All six judges agreed with the consensus diagnosis in 51% of cases, five judges in 24%, and four judges in 14%. There was no clear majority in the remaining 11%.

Further evidence about the degree of consensus or homogeneity among the elements entering into a diagnosis of schizophrenia is generated by treating the six judges as a 6-item scale; each item can be "passed" or "failed" by each twin. We are indebted to our colleague Professor Auke Tellegen for the idea, its evaluation, and the interpretation of the results. The homogeneity of the 6-judge scale can be viewed as an indicator of the agreement among all judges as well as an indicator of the reliability of their combined judgments. Diagnoses were converted into a one, zero dichotomy with S and ?S equal to a score of one, a "pass," and O and N equal to a score of zero, a "fail." Scores on each twin could range from zero to six. All twins with a consensus diagnosis of Normal, 27 of them, were removed from the analysis to avoid an inflated view of the homogeneity obtainable in psychiatric diagnosis.

The statistical estimate of the scale's homogeneity was obtained by applying the Kuder-Richardson Formula 20 (Guilford, 1956, pp. 454-455) to the proportion passing each of the 6 items, and from the variance of the scores on the total 6-item scale. A KR20 coefficient of .91 was obtained. This means that the consensus diagnoses of our 6 judges would be expected to correlate .91 with the consensus diagnoses of a subsequent panel of 6 comparable judges. Further, the consensus diagnoses of our judges would be expected to correlate .95 (the square root of .91) with the "true" classification of schizophrenia obtained from an infinitely large panel of comparable judges. From a *reliability* point of view, our consensus diagnoses contained little error due to variations between judges. The data above provide only indirect support to the *validity* of the concept of schizophrenia. The average between judge correlation was estimated to be .62. In terms of percentage agreement in diagnosis between judges, this correlation corresponded empirically to an average of 83% in our sample, with a range of 69-92%. Obviously more than one diagnostician should be consulted, whether for clinical or research purposes.

RESULTS

Concordance for Consensus Diagnosis of Schizophrenia

We now come to the effect of the blind diagnoses on concordance rates for schizophrenia in MZ and DZ pairs. Any pair in which a proband did not satisfy the revised diagnostic requirement was omitted. Thus the number of MZ pairs was reduced from 24 to 22, since two former Maudsley probands were rejected as being neither S nor ?S. One was a young man with a short-lived adolescent anxiety reaction (MZ 12A), at one time thought to have been schizophrenic. The other rejected proband was a psychopath (MZ 10A) who was once diagnosed in a mental hospital as having a schizophrenic episode. Neither of their co-twins was schizophrenic, though the twin of the psychopath was similarly affected. In 11 pairs both twins were S or ?S according to consensus diagnosis, raising the MZ concordance rate to 50% from the figure of 42% previously reported. The 11 affected co-twins are those 10 previously regarded as concordant plus a further twin (MZ 23B), a devotee of the occult with grandiose and other odd ideas, who was thought by most judges to be developing an insidious schizophrenia. All 33 DZ probands were diagnostically acceptable as at least probably schizophrenic, and so were three of their partners; thus concordance here remained at 9%.

Going further and omitting cases of *doubtful* schizophrenia, we were left with 20 MZ pairs. The two further probands now omitted are: First, a girl (MZ 13A) whose alternative diagnoses are behavior disorder, inadequate personality, manic-depressive or atypical psychosis. Her twin was acceptable to all judges as schizophrenic or (in one instance) probably so; and in fact the illnesses of these two twins were of a similar schizoaffective kind. The second of these probands was a case of alcoholic hallucinosis (MZ 9*A*), whose neurotic twin sister was never considered as schizophrenic. From the affected co-twins the grandiose occultist previously added, must now be omitted again, as must another co-twin who was only doubtfully schizophrenic. This second co-twin is a woman (MZ 6*B*) whose alternative diagnosis is a somewhat atypical psychotic depression in a schizoid personality leading to partial invalidism—her illness was similar to, if somewhat milder than, that of her twin. MZ concordance for *definite* schizophrenia in both twins is thus *reduced* to 8 out of 20 pairs or 40%. But two pairs with similar, essentially schizophrenic-like illnesses are thereby counted as technically discordant. In the DZ pairs, on the other hand, the number of acceptable pairs was reduced from 33 to 31 (DZ 27 and 32), but the number of concordant pairs remains at 3, so that the concordance rate here was *raised* slightly to 10%. The inclusion of ?S gave better discrimination between MZ and DZ pairs than restriction to cases of definite schizophrenia; the MZ : DZ concordance ratios were 5.5 vs. 4.1.

Individual Judge Concordance Rates

Table 5.3 presents the results obtained by asking what percentage of pairs are concordant when a proband is diagnostically acceptable to a particular judge. Concordance has been calculated using two different criteria for schizophrenia—(a) that restricted to S or those cases classified as definite or first-choice schizophrenia, and (b) schizophrenia inclusive of qualified cases, i.e., ?S. The same criterion was applied to both twins.

There is no consistent relationship between conservatism of diagnosis, judged by the frequency of using schizophrenia, and the concordance rates in either MZ or DZ pairs. The judges at either extreme, JB with 31% and 12% and PM with 58% and 24%, secure less discrimination as indicated by the MZ : DZ ratio than the other four judges. Birley's attempt to apply the objective criteria of the Present Psychiatric State interview to the summaries we provided did not succeed in isolating a nucleus of schizophrenics with an overwhelming genetic contribution (i.e., with a very high MZ concordance rate). Perhaps this was partly due to the unsuitability of our summaries compared to a face-to-face interview of a hospitalized patient. We tried to provide the clearest examples of delusions, hallucinations, and thought disorder during the patient's lifetime. The lack of distinction between MZ and DZ rates is painfully glaring if only the (a) criterion for Birley in Table 5.3 is accepted; he rejected 9 MZ pairs as not containing an unqualified schizophrenic, and, of the remaining 15, only three or 20% were concordant on the same criterion. There were five others where the second twin had been hospitalized for a psychosis which he preferred to call manic-depressive and only possibly schizophrenic from the evidence in the summaries. Eleven DZ

TABLE 5.3

Direct Pairwise Judge and Consensus Concordance Rates

Judge	Criteria[a]	MZ		DZ		p[b]	MZ/DZ ratio
ES	a	7/17	41%	3/30	10%	*	4.1
	b	9/18	50%	3/31	10%	***	5.2
PM	a	13/24	54%	7/33	21%	*	2.6
	b	14/24	58%	8/33	24%	**	2.4
KA	a	8/16	50%	2/27	7%	***	6.8
	b	9/16	56%	2/27	7%	***	7.6
JB	a	3/15	20%	3/22	14%	N.S.	1.5
	b	5/16	31%	3/25	12%	N.S.	2.6
LM	a	8/17	47%	2/23	8%	**	5.4
	b	10/20	50%	2/29	7%	***	7.2
JP	a	8/22	36%	3/31	10%	*	3.8
	b	11/23	48%	3/33	9%	***	5.3
Consensus of 6	a	8/20	40%	3/31	10%	*	4.1
	b	11/22	50%	3/33	9%	***	5.5
Previously diagnosed		10/24	42%	3/33	9%	**	4.6

[a]Definite schizophrenia in both twins (a) and schizophrenia inclusive of ?S in both twins (b), in pairs where a *proband* has met the diagnostic requirement.

[b]χ^2 (Yates corrected, one-tailed) test of association between zygosity type and level of concordance; *, $\leqslant$ .05; **, $\leqslant$ .01; and ***, $\leqslant$ .001.

probands were also rejected by this judge, all from discordant pairs, but he accepted both twins in the three DZ pairs considered concordant by the other judges. Birley's DZ concordance rate is 14% compared to his 20% MZ rate; however, none of the discordant DZ co-twins were judged by him to have a psychosis, unlike the corresponding MZ co-twins. The lack of contrast clearly does not do justice to the facts and suggests the operation of a methodological artifact. Thanks to the care used by Birley in recording his diagnoses, we can see what would happen if we include as affected anyone whom he considered just possibly schizophrenic (??S). When this was done, his rates were 10/21 or 48% MZ and 3/30 or 10% DZ, quite in accord with the consensus diagnoses.

Going now to the most liberal usage of the diagnosis of schizophrenia by our judge Meehl, we find that this does not help either in successfully detecting an overwhelming genetic factor. Meehl's (1972) expectation would have been, as would Heston's (1970), that by combining the information of the summaries, verbatim abstracts of tape-recorded mental status interviews, and the MMPI when available, he would have identified virtually all of the MZ co-twins of schizophrenics as being borderline schizophrenics, schizotypes, or markedly schizoid. In fact, his MZ concordance rate was the highest at 58%, but it was at the price of lowering the MZ : DZ ratio because his DZ rate rose to 24%. In addition to the three consensus-concordant DZ pairs, his standards led him to diagnose four DZ co-twins as borderline and one as a schizotype. Meehl's rates of 58% vs. 24% do not take into account the fact that he could discriminate chronic and acute schizophrenia from his other categories; if we choose to restrict the calculation of concordance rates to Meehl's chronic and acute cases we obtain 11/22 or 50% MZ and 3/33 or 9% DZ, again in accord with the consensus.

The best discrimination between MZ and DZ pairs seems to originate from applying diagnostic standards which might be regarded as rather broad and loose by some European lights, but as rather narrow and conservative by the great majority of United States psychopathologists.

The View of a Scandinavian Psychiatrist—Erik Essen-Möller

We consider ourselves extremely fortunate to have enlisted the cooperation of Erik Essen-Möller in the blind diagnosing of our 114 twins. His diagnostic formulations could not be fitted into the same framework as the other judges'. We chose to use Essen-Möller as a criterion for a Scandinavian point of view; he is one of the elder statesmen of Swedish and world psychiatry. References were made in Chapter 2 to his own classical twin study of schizophrenia, but less well known perhaps is his scholarly interest in personality theory (especially Sjöbring, 1963) and his work on the concept of schizoidia (1946). Our objective was to examine the effect on concordance rates when a Scandinavian psychiatrist makes a distinction between schizophrenia and schizophreniform psychosis, and further, to discover how successful he would be, compared to other clinicians, in identifying conditions that might be related genetically to schizophrenia. Our summaries could not of course provide the data Essen-Möller preferred for assessing characterological defects such as facial tonicity and emotional accessibility.

The twins were diagnosed according to his personal preference (e.g., "affective fluctuations with confusion and ? early brain lesion") and also allocated to one of six categories: 1, true schizophrenia; 2, schizophreniform psychosis; 3, other diagnosis but possible (??) schizophrenia; 4, schizoid or possibly schizophrenia-related personality; 5, other diagnosis; and, N, normal. In our 11 consensus concordant MZ pairs, all 22 received one of the first four category diagnoses. The 11 consensus discordant MZ pairs seemed to fall into two clusters: five pairs (MZ 3, 8, 11, 14, 16) in which both twins were coded for one of the first four categories, and six pairs (MZ 4, 5, 9, 18, 20, 24) in which only one or neither were thought likely to be schizophrenia-related. Category diagnoses for the MZ and DZ twins made by Essen-Möller are given in Appendix D. It is interesting that none of the probands in the latter six discordant pairs was coded as a true schizophrenic.

MZ Pairs

Twelve probands were coded "1" and 7 of their co-twins or 58% were concordant at that level. By the addition of two co-twins with psychoses with schizophrenic-like features, coded "3," the probandwise concordance becomes 9/12 (75%). Pairwise the rates were 4/9 (44%) and 6/9 (67%). Essen-Möller found every one of the remaining co-twins to have personality traits that might have a genetic relationship to schizophrenia. Although the concordance rate for typical schizophrenia was higher than he had found in his own twin study, the conclusion he had reached about the co-twins was reaffirmed—MZ co-twins of typical schizophrenics when not themselves affected, revealed characterological abnormalities of a schizoid kind. Up to now, Essen-Möller's clinical assessments represent the most successful attempt ever made to identify the schizotype or spectrum disorders.

DZ Pairs

That this success did not result from a high usage of categories 1-4 will now be shown. Concordance for true schizophrenia in the DZ pairs was 2/19 (11%) probandwise and 1/18 (5.6%) pairwise. Only one of the nineteen co-twins of probands called "1" had a schizophrenia-related personality, besides the two overtly schizophrenic ones. The concordance rates for being affected, in the spectrum sense, then become 3/19 or 16% probandwise and 2/18 or 11% pairwise. Neither of these latter figures gives strong support to Heston's (1970) idea that schizoid disease is due to a dominant gene with virtually complete penetrance. However, in Essen-Möller's own sample of 24 DZ pairs (1941b), 12 to 15 were significantly abnormal when assessed in person.

If we transform Essen-Möller's diagnoses by loosening his standards to include his categories 1-4 as schizophrenia, we obtain an MZ pairwise concordance rate of 13/20 (65%). No MZ rate in Table 5.3 is higher. The corresponding rate in DZ pairs becomes only 3/32 (9%) by this loosening.

For the record, it is of interest to report what was found in pairs with a schizophreniform diagnosis in a proband. Three MZ probands were so diagnosed. The co-twin of one of them, MZ 6B, was called an affective psychosis in a schizoid personality, while MZ 4B was normal and MZ 24B was called a depressive episode in a normal personality. Of six schizophreniform DZ probands, all had normal co-twins except one, DZ 10A, whose co-twin was schizophrenic (cf. Appendix D). Essen-Möller's use of the diagnosis "schizophreniform psychosis" seemed to be applied mostly to acute paranoid schizophrenia.

Contaminated Diagnosis, Too Much or Too Little?

It is generally supposed that the most likely bias to occur if diagnosis of the twins is not made independently, or if the summaries on which blind diagnoses are made are themselves contaminated, is a spuriously high concordance with respect to the same diagnosis. In a study of schizophrenia this could arise through calling a twin schizophrenic or ? schizophrenic who might otherwise have been regarded as either "within normal limits" or given some different diagnosis. The former possibility, that of calling a normal person schizophrenic, can be largely excluded in our series since it was clear from the summaries that the co-twins previously diagnosed schizophrenic had all required hospitalization, many of them for lengthy periods and on more than one occasion. Only in one pair is concordance due to a consensus diagnosis of ?S in a twin (MZ 23*B*) who had never seen a psychiatrist. The few cases of doubt are all a matter of differential diagnosis between schizophrenia on the one hand and a personality disorder or affective illness with schizophrenic-like features on the other.

However, there is also the possibility, perhaps equally likely, that the summaries in some cases were not sufficiently detailed for a judge to make a diagnosis of schizophrenia which he would have made had certain known facts been presented to him. Such details could have been omitted either because they had never been recorded by the treating psychiatrist or, if recorded, had been omitted by us inadvertently or in the interests of conciseness when we prepared the summaries. In some pairs our summaries may even have been too lengthy and the relevant observations submerged with irrelevancies or misleading details in our efforts to be thorough and objective, so that they might sometimes have been overlooked by a judge. Birley, who believes that schizophrenia is often misdiagnosed, and who made a deliberate attempt to base his own diagnoses on recorded symptoms and not on the diagnostic opinion of others, thinks that influences such as these may partly account for the fact that he diagnosed schizophrenia less frequently than the other judges. Meehl might have diagnosed more cases as borderline schizophrenia or as schizotypes had detailed interview protocols and psychological test results been available in more cases. Slater, when reviewing MZ pairs in which, after diagnosing blind, he was found to have disagreed with the consensus opinion that the pair was concordant for schizophrenia, considered, when he had all the information before him and not just the summary, that in three cases (MZ 6*B*, 13*A*, and 15*A*) he might well have diagnosed schizophrenia rather than something else; this was on account of reading the complete abstracts of hospital notes and the suggested evidence of thought disorder provided by the full transcript of the taped interviews.

More often than a possibly unjustified diagnosis of schizophrenia in the co-twin of a known schizophrenic, there is what we may call "negative contamination." We see in the treating psychiatrists for this series a great reluctance to make the diagnosis and a strong desire to give the other twin the benefit of any doubt. In one MZ pair (MZ 17*A*) the Maudsley consultant preferred to make a diagnosis of "hysterical pseudodementia," in another (MZ 19*B*) "sensitive personality and *folie à deux*," and in a third (MZ 21*B*) "possible encephalitic Parkinsonism," rather than schizophrenia; and in DZ 10*A* the Maudsley diagnosis was "paranoid state," although the co-twin was known to be a deteriorated chronic schizophrenic. Particularly in the MZ pairs, the follow-up and the consensus diagnosis based on it left no serious doubt that both twins in these pairs were indeed schizophrenic.

Diagnosis based on summaries is probably as likely to result in *missing* schizophrenia through lack of relevant information included in the summaries as in its overdiagnosis. While we believe it is important to mention the possibility of a bias in the direction of underestimating concordance through the absence of relevant details in the summary, we do not in fact believe that such a bias occurred to any significant extent with respect to the consensus diagnosis, certainly not so far as chronic or acute schizophrenia is concerned.

SUMMARY AND DISCUSSION

Our introductory remarks on the subject of diagnosis stressed our opinion that schizophrenia is much more than a semantic convenience.

The aims of the experiment described in this chapter were (1) to achieve an independent diagnosis of each twin, uninfluenced so far as possible by a knowledge of the zygosity of the pair or the clinical state of the other twin, and (2) to compare the effect on MZ and DZ concordance rates of different criteria for the diagnosis of schizophrenia.

Summaries of the history of each of the 114 twins were given to six judges representing a wide range of psychiatric opinion from three different countries. They were invited to diagnose the histories according to their preferred practices. Their diagnoses were classified as Schizophrenia, ?Schizophrenia, Other diagnosis, or Normal. An analysis of variance by Slater suggested that all judges were attempting to measure the same things but had different cutoff points for doing so. Cases diagnosed as chronic or acute schizophrenia but not borderline by the judge with the broadest concept of schizophrenia corresponded very closely with those diagnosed as schizophrenia including "??Schizophrenia" by the judge with the narrowest concept.

The judges at each extreme achieved the poorest discrimination between MZ and DZ concordance rates. In this experiment the best discrimination was obtained by standards that would be regarded as rather broad in Europe but as rather narrow in the United States.

A consensus diagnosis was reached without difficulty. It resulted in the exclusion of two MZ pairs because they did not include a schizophrenic proband. When the six cases of ?Schizophrenia were included, pairwise concordance for schizophrenia according to the consensus diagnosis was 11/22 (50% ± 11) in MZ pairs and 3/33 (9% ± 5) in DZ pairs.

As a distinguished representative of Scandinavian psychiatry, Essen-Möller also diagnosed the summaries. Only 34 twins were regarded as strict schizophrenics compared with 43 by our next most conservative consensus judge. Given an MZ twin who was a strict schizophrenic, Essen-Möller was blindly able to assess every co-twin as either schizophrenic, ?schizophrenic, or as having a personality possibly related genetically to schizophrenia; he did this without a high "false positive" rate among the twins of DZ schizophrenics. Even when his standards were broadened for probands excellent discrimination between MZ and DZ twins was still obtained (65% vs. 9% concordance). The statement above about measuring the same thing but with a different cutoff point seems also to apply to Essen-Möller.

Reasons were given for believing that the summaries were not likely to have been contaminated in such a way as to lead to misleadingly inflated consensus concordance rates. Diagnoses based on the summaries of histories are as likely to result in missing cases of schizophrenia. Some psychiatrists are extremely reluctant to diagnose schizophrenia in the co-twin of a known schizophrenic.

Unless otherwise stated, we shall base our further analysis on the consensus diagnosis. An index to the twin pairs by consensus diagnoses is given as Appendix E.

6

CLINICAL GENETIC ANALYSIS

SEVERITY AND CONCORDANCE

Viewed in retrospect, many schizophrenic twin studies reveal a tendency for severely affected probands from MZ pairs to have affected co-twins more often than probands whose illnesses were relatively mild. Severity is a vague and ambiguous term, and it has been used by the investigators themselves or by recent reviewers, including ourselves, to refer to different, if related, characteristics. In the Luxenburger (1928) study, hebephrenic and catatonic cases could be regarded as severe; paranoid cases as mild. In the Kallmann (1946) study, severity was defined by degree of deterioration. In reanalyzing the Slater (1953) case histories, Rosenthal (1959) applied the Phillips Scale (1953) of premorbid adjustment, a scale that correlates with outcome. Inouye's (1963) "chronic progressive" and "relapsing" cases could be regarded as severe and his "chronic mild or transient" cases as mild. Kringlen (1967) used a global rating of severity of psychopathology, similar to those used by Alanen (1966), Mosher *et al.* (1971), and Wender *et al.* (1968). It would seem appropriate to use more than one criterion for severity in any one study so long as the construct remains incompletely defined.

In our provisional report (Gottesman & Shields, 1966a) we used three simple and objective measures for our sample. Two of them related to length of hospitalization and the other to outcome. At the Puerto Rico conference (Gottesman, 1968b) a more refined analysis of the data was presented, taking into account length of observation and percentage of follow-up time spent in hospital. Findings were reported, both for our Grade I concordance (diagnosed schizophrenic and hospitalized) and for "Judge A's" (Slater's) blind diagnosis of schizophrenia. The previously reported association between severity of the

TABLE 6.1

Relationship between Severity and Concordance in MZ Pairs

Analytic criteria of "severe" and "schizophrenia"	Concordance for schizophrenia in co-twins of mild probands	Concordance for schizophrenia in co-twins of severe probands	Significance of difference (one-tail) (p)
1. Not working, or in hosp. within past 6 months. (Grade I diag.)	17% (2/12)	75% (12/16)	$< .005^a$
2. As (1) above (consensus diag.)	27% (3/11)	80% (12/15)	$< .025^a$
3. In hosp. $\geqslant 1$ year (consensus diag.)	33% (3/9)	70% (12/17)	$< .10^a$
4. In hosp. $\geqslant 2$ years (consensus diag.)	38% (5/13)	77% (10/13)	$< .10^a$
5. Classified as hebephrenic or catatonic by Judge A (consensus diag.)	33% (5/15)	91% (10/11)	$= .004^b$
6. Classified as strict Sc by Scand. psychiatrist (consensus diag.)	20% (2/10)	81% (13/16)	$= .003^b$
7. First hosp. at 25 or later (consensus diag.)	47% (7/15)	73% (8/11)	N.S.^{a}
8. Mean global psychopathology rating $\geqslant 5.5$ (consensus diag.)	42% (5/12)	71% (10/14)	N.S.^{a}

^{a}Chi square, Yates corrected.
^{b}Fisher exact test.

proband and diagnosis of schizophrenia in the co-twin held up in this refined approach.

We now present a similar analysis, employing several different criteria of severity. Data are in terms of probands and their co-twins, since the two twins in pairs where both were probands sometimes differed in severity. In the main, diagnosis of schizophrenia is determined by consensus of our six judges; the overall MZ probandwise concordance rate is 58% (15/26). However, in order to illustrate the effect of using consensus diagnosis, the first row in Table 6.1 is based on Grade I concordance.

Outcome

The criterion of severity in rows 1 and 2 is one which showed the biggest difference in concordance between the co-twins of mild and severe probands in our previous analysis. It uses outcome to assess severity. Any proband who, at the time of follow-up, had not been in hospital during the past 6 months and was working or running a home was regarded as mild, and all others as severe. Row 2 applies the same criterion of severity to the consensus diagnosis of schizophrenia. The omission of MZ 10 and MZ 12 and the transfer of MZ 23—a mild proband according to all criteria—from the discordant to the concordant group has the effect, both here and generally, of slightly reducing the contrast between the rates.

Length of Hospitalization

Rows 3 and 4 relate to the total time hospitalized for psychiatric disorder at time of follow-up, probands hospitalized for less than 1 or 2 years, respectively, being called mild. The possibility arises that cases assessed as severe, using length of hospitalization as the criterion, might have been observed longer than mild cases. This could have led to the classification of pairs as mild and discordant when they would eventually have become severe and concordant. To avoid any possible artifact of this kind, concordance was therefore calculated (Gottesman, 1968b) using cutting points for severity of at least 15%, 20%, 25%, 30%, 50%, and 75% of time spent in hospital since first admission. Results were remarkably consistent, but the contrast was diminished. There still remains the possibility that the difference might be further exaggerated through the inclusion of pairs followed up for a relatively short time. For instance, a concordant pair followed for, say, only the first 5 months after first hospitalization of a proband who had spent only 1 month in hospital would yield a severe case according to 20% time in hospital as criterion; if the proband's remission proved permanent, longer follow-up would have yielded a pair that was mild and concordant. When pairs followed up for less than 5 years were omitted, the severity-concordance relationship was not thereby reduced. The relationship still held, if anything more strongly, when the diagnosis of one of our more conservative judges was used.

Workers who use "hard" data, such as length of hospitalization, to measure severity are criticized on the grounds that such measures do not take clinical features into consideration and that reasons for hospitalization may be extrinsic to the illness per se. On the other hand, clinical assessments are criticized as unreliable and biased. We therefore report our findings both ways.

Clinical Classification

Row 5 uses Kraepelinian subtype classification in an attempt to divide the consensus-diagnosed probands into nuclear or process (severe) and peripheral or reactive (mild) groups. Probands diagnosed blind by Slater as hebephrenic or catatonic nearly all had twins who were schizophrenic according to the consensus diagnosis. The severity-concordance association is significant here beyond the .005 level. A Scandinavian psychiatrist's diagnosis of a proband as a strict schizophrenic also discriminates very well indeed between the concordant and discordant consensus-diagnosed pairs (row 6).

Phillips Scale

It is customary in clinical and experimental work on the process-reactive dimension (Garmezy, 1968) to use objective premorbid life-history data rather than clinical symptoms as a prognostic index. The Phillips Scale, originally intended to predict response to ECT, has been found useful in this regard. On this scale marital status carries more weight than any other item (Klein & Klein, 1969). In our study, where outcome is already known, there may appear to be little purpose in using a cutting point on the Phillips Scale to dichotomize cases into mild and severe. Nevertheless, since the scale is so widely used, we asked a graduate student to score the twins from the (not very extensive) premorbid information given in the diagnostic summaries. Counting, as is usual, those scoring 15 points or over on Sections A-F as "premorbid poor" and hence as likely to be process-schizophrenics, there were too few "premorbid goods" among the schizophrenics for the score to be useful in the present context. Only 4 MZ probands were "premorbid goods." Two MZ and seven DZ nonschizophrenic co-twins were "premorbid poors." Except for an excess of DZ pairs where the *A*-twin was a "poor" and the *B*-twin a "good," the test did not effectively distinguish MZ pairs from DZ pairs. We do not wish to stress these negative findings, which were not unexpected in the circumstances. Later in this Chapter (p. 251) we show that an MZ pair is more likely to be concordant if the premorbid personality of the proband was schizoid than if it was not.

Age at Hospitalization

Taking age at first hospitalization as an indicator of severity, row 7 shows the usual association between severity and concordance. The two probands not hospitalized till after age 40 had schizophrenic-like illnesses associated with chronic alcoholism (MZ 9) or thyroid disorder (MZ 20), and their co-twins were not schizophrenic.

Severity Ratings

We turn now to global ratings of severity. Judge LM was invited to rate each case blind on a scale of severity with which he had previous experience (Mosher *et al.*, 1971). The scale is shown in Table 6.2. Judge PEM was asked to rate the twins on the same scale. Good agreement was obtained ($r = .88$). In the case histories (Chapter 4) we have reported the mean of the global psychopathology ratings of both judges. These have been used to produce row 8. The cutting score of 5.5 divides the probands into two nearly equal groups and, once again, shows higher concordance when the proband is severe. In general, findings in DZ pairs show the same tendency. For instance, using criterion 8, when the proband was mild, 5% (1/18) of DZ co-twins were concordant; when severe, 19% (3/16) –difference nonsignificant (Fisher). We could also replicate Kringlen's (1967) pairwise method of analysis. Where the more severely affected member of an MZ pair had a global rating of less than 6.0, 33% (5/15) of pairs were concordant for schizophrenia. In pairs that included a case as severe as or more severe than 6.0, concordance was 86% (6/7)–the difference is significant ($p = 0.0256$, Fisher). The relationship between severity and concordance shows up in yet another way. The mean severity rating of schizophrenics from concordant pairs, whether probands or not, was higher than that of schizophrenics from discordant pairs, and similar in both MZ and DZ pairs (Table 6.3).

TABLE 6.2

Rating Scale for Severity of Global Psychopathology

In comparison with your overall clinical experience, how "sick" or "mentally ill" is this person? Guidelines given should be used only as guides, not absolutes. This should be an overall estimate based on history, mental status, and interpersonal, occupational and social functioning.

0. Supranormal, more than one S.D. above "norm of overall functioning"

1. Normal, not ill at all

2. Questionably mentally ill ("Classifiable, but not diagnosable." A few psychologic or psychosomatic complaints)

3. Mildly ill ("Diagnosable, no treatment indicated." Distinct psychiatric symptoms, or sees himself as sick)

4. Moderately ill ("Diagnosable, treatment indicated." Clearly neurotic or character disordered with functional impairment)

5. Markedly ill (Probably psychotic at some time, reactive schizophrenia)

6. Severely ill (Prolonged psychosis, process schizophrenia)

7. Among the most severely ill (Unremitting severe psychosis, deterioration)

TABLE 6.3

Mean Global Psychopathology Ratings of
Schizophrenics

	MZ	N	DZ	N
From concordant pairs	5.6	22	5.7	6
From discordant pairs	5.1	11	5.2	30

This may be the appropriate point at which to report on the severity ratings more generally. Table 6.4 shows the distribution of the schizophrenics and is similar in MZ and DZ probands, thus supporting the representativeness and comparability of our two subsamples. Rosenthal (1961) pointed out that "the twins that get into a sample define and limit the population to which generalizations can be made"; and he demonstrated (1959) that in Slater's series it was the more severely ill twin who became the proband in concordant pairs with only one proband. His criticism was that Slater missed mild cases. The same tendency does not hold in our series. If anything, the schizophrenics who were not probands, i.e., the secondary cases, tended to be more severe than the probands. This is understandable. DZ 10B, for example, could not have entered our sample of Maudsley probands, for he had been chronically hospitalized elsewhere since before 1948.

Conclusion

Without belaboring the point further, we may conclude this section by saying that in our sample consistent differences in concordance were found between the MZ twins of severely and mildly affected probands, whether severity was assessed by outcome, length of hospitalization, age at first hospitalization, Kraepelinian subtype, or on a mean rating of global psychopathology. The possible meaning of the higher concordance rates found when schizophrenia in a proband was severe will be discussed in Chapter 10.

MORBID RISK

In trying to assess the risk of the twin of a schizophrenic becoming affected in the course of time, we shall use the consensus diagnosis as the best available starting point. This includes as concordant one pair (MZ 23) in which the co-twin has never had a psychiatric consultation. If his inclusion be thought to overestimate the concordance, it must also be remembered that there are three probands (MZ 9, 20; DZ 27), all with nonschizophrenic partners, where the

TABLE 6.4

Distribution of Global Psychopathology Ratings

Consensus schizophrenics	<5	5–	6–	7	Total	Mean
MZ probands	8	11	5	2	26	5.4
DZ probands	11	16	7	0	34	5.3
Secondary cases	1	4	4	0	9	5.5
Total	20	31	16	2	69	
Rejected probands (MZ)	2	–	–	–	2	4.5

schizophrenia was probably purely symptomatic. We shall attempt to correct the casewise proband rates for age. We do not expect in this way to obtain single figures for MZ and DZ co-twins of schizophrenics that will be correct in any absolute sense. To do so would be unrealistic. We do, however, hope thereby to gather an impression of the order of risk for such individuals, as compared with individuals from the general population.

In order to assess the risk for a twin, one has to bear in mind what is known of the distribution of the ages of onset of schizophrenia in the general population, and what is known of the interval between onsets in concordant pairs, both in other studies and in our own. For the United Kingdom it has been shown by Slater and Roth (1969) that by age 35 men have passed 62% of the risk and women 47% of the risk of their being admitted to mental hospital with a diagnosis of schizophrenia in the course of their lives. By age 45 the corresponding percentages are 82% and 68%, and by age 55, 92% and 83%. It is clearly incorrect to suppose, particularly in the case of women, that the risk of being hospitalized for schizophrenia nowadays is virtually over by the age of 40. Although ages at first admission differ in men and women, the total lifetime expectancy for schizophrenia, as calculated from United Kingdom national statistics for the years 1952-1960, is approximately equal in the two sexes —1.064% for men and 0.988% for women, with age 55 as end point.

The risk for the MZ twin of a schizophrenic is, however, related to the age at which the psychosis first became manifest in his or her twin, as well as to the chronological point reached. It would be misleading to suppose that the same risk was still to be incurred by a normal co-twin of 45 when his brother's schizophrenia had started at 15 rather than 40. According to Luxenburger (1936) 24% of concordant MZ pairs fell ill within 1 year of each other, 42% within 2 years, 84% within 6 years, and 95% within 11 years. Kallmann (1946) found no difference in age at first admission in 26% of concordant pairs; 55% were admitted within 4 years, and 89% within 8 years. In Slater's (1953) series, half the pairs fell ill within 3 years, but differences in age at onset of 20, 22, and

31 years were observed in 3 of the 28 pairs. Kringlen (1967) considered that 24% of his 17 concordant pairs had onsets within a year, though in only one pair were both twins hospitalized in the same year. Within 5 years, 50% of 16 pairs in which both twins were hospitalized had been admitted, and within 10 years 88%. Intervals of 11 and 23 years separated the admissions of Essen-Möller's (1970) two strictly concordant MZ pairs.

Of our ten pairs in which both twins were hospitalized and schizophrenic, admission occurred within a year in four, within 2 years in 5, and within 3 years in 8 pairs. In the remaining two, the intervals were 5 and 12 years. Abe (1969) has reanalyzed Slater's data, combining it with our former Grade I data. He worked on the assumption that each pair had been observed from the time of onset in the first twin. Ages of onset were estimated clinically by Abe from a study of the original case notes. He found that the probability of onset in the second twin fell constantly with the lapse of time in a logarithmic way. An exception to this tendency was an excess of pairs in which the twins fell ill within 2 years of one another. Abe attributed this displacement of the curve to the influence of one twin on the other.

Although findings in the above studies vary, it seems clear that, even after five years' follow-up, there still remains a considerable proportion of the risk of falling ill for MZ co-twins. However, this generalization is much less true for older co-twins. Kringlen's case histories show that all 14 of his concordant pairs were concordant by age 43 at latest. Among pairs where neither twin was hospitalized till age 35 or later, 3 were concordant, the second twin in these pairs being hospitalized in the same year and 2 and 5 years later. There were 10 discordant pairs where first hospitalizations of the probands occurred between the ages of 38 and 49, and in these the co-twin had been observed for a further 8

TABLE 6.5

Age and Length of Follow-up of Discordant MZ Co-twins
from First Hospitalization of Proband—Consensus Diagnosis

MZ co-twin	Sex	Age	Follow-up
14	m	20	3 years 7 months
4	f	39	2 years 11 months
11	m	32	9 years 9 months
18	f	37	6 years 7 months
24	f	40	7 years 9 months
5	f	42	13 years 4 months
16	m	44	27 years 2 months
8	m	46	16 years 11 months
3	m	50	16 years 1 month
20	f	46	3 years 0 months
9	f	47	3 years 5 months

to 21 years without falling ill. It would therefore appear that MZ co-twins aged 40 or over and followed up for 5 years or more from the first hospitalization of the proband incur little risk of becoming schizophrenic.

Table 6.5 shows the essential information for each of our 11 nonschizophrenic MZ co-twins—their age, and the length of time that has elapsed since first hospitalization of the proband. The intervals tend to be longer if estimated age at onset is used, and shorter if age when a final diagnosis of schizophrenia was made is used, instead of first hospitalization, but the differences are not of sufficient consequence to merit separate analysis. A compensatory advantage of a small sample size is that a table such as Table 6.5 may permit a general picture to be obtained at a glance, which may be as valid as any attempt to apply conventional statistical procedures. In view of the relationship between severity and concordance already mentioned, we also need to give some weight to the nature of the schizophrenia in the proband.

MZ 14B is most at risk—he was only 20 years old when reliable information about him was last obtained, and at that time his MMPI had a profile psychotic in shape. Though the proband in this pair first fell ill at 15 and was hospitalized at 16, he was 20 before he received a formal diagnosis of schizoaffective psychosis. The remaining discordant co-twins have survived much more of the risk period. MZ 4B is still under 40 and it is barely 3 years since her sister fell ill; however, her sister's illness was mild and atypical, which may reduce the likelihood of the co-twin's breaking down. MZ 11B and 18B are still relatively young and (so far as one knows) may be at risk for a paranoid form of illness later on. With MZ 24B and 5B one becomes increasingly more confident that they will remain nonschizophrenic, and even more so with MZ 16B and 3B. MZ 9B and 20B, though both over 45, have been followed up for less than 4 years from hospitalization of the two probands; but there are factors peculiar to the illnesses of the two A-twins that make it unlikely that the schizophrenia potential of the B-twins will be elicited.

TABLE 6.6

Objective Alternative to Age-Correcting:
Effect on Concordance Rates if Additional MZ Co-twins Become Schizophrenic

Additional co-twins becoming schizophrenic	Concordance for schizophrenia			
	Pairwise (uncorrected)		Casewise (proband method)	
None	11/22	50%	15/26	58%
One more	12/22	55%	16/26	62%
Two more	13/22	59%	17/26	65%
Three more	14/22	64%	18/26	69%

Table 6.6 shows the rates that would obtain if up to three more MZ co-twins should become schizophrenic in the course of time. A reasonable guess—and it cannot be more than a guess—is that not more than two of the above co-twins will become diagnosable schizophrenics. In that event 65% would be the morbid risk for MZ co-twins, to be compared with 1% for the general population. If one wishes to consider MZ 23B as not yet firmly diagnosable as schizophrenic, and to omit MZ 9A and 20A as phenocopies, very similar rates to the above are obtained. To give a broad range, one may say that, based on this impressionistic but informed assessment, the morbid risk will lie somewhere between 60% and 70%, probably nearer the former figure than the latter.

The conventional shorter Weinberg method, with age 40 taken as the end of the risk period, gives a risk of 15/19.5 = 77%. The method reduces the denominator of 26 co-twins by one-half of those over age 15 who have not yet reached the age of 40, on the assumption that they will, on the average, have survived half the risk. An even higher figure than 77% would have been obtained if age 45 had been taken as the end of the risk period, as is sometimes done. Kallmann and Slater also found that the shorter Weinberg method, when applied to MZ twins and in conjunction with the proband method, gave figures that were too high. Slater (1953) developed an alternative method which took into consideration the differences in age at onset within pairs. Using the weights which he derived from his own material, which may, of course, not be entirely applicable to ours, we calculate that our 11 discordant MZ co-twins have survived 9.68 risk-lives. Morbid risk is then 15/24.68 = 60.8%, which seems a more reasonable estimate. It is close to what would obtain if only one more co-twin fell ill. Abe's (1969) method is equivalent to the longer Weinberg method, which is similar to that used in actuarial life insurance tables. (An example of the longer Weinberg method appears in Slater and Cowie, 1971, Appendix D.) Instead of using chronological age, Abe used the time elapsed from the onset in the first twin. Abe calculated for us that 70% of our 24 former Grade I MZ pairs might be expected to be concordant if they were followed up for 25 years, as compared with the 42% presently known to be concordant. A considerably higher figure would be obtained if the casewise proband method were used, along with our consensus diagnosis of schizophrenia. Such a figure would certainly be an overestimate and would seem to imply that some co-twins would fall ill after the age of 60. Our material is too small for the reliable application of life table procedures. Until more satisfactory methods for age correction in MZ twins are developed, we prefer our more conservative, if more commonsense, estimates of 62% or 65% for the lifetime risk of an MZ co-twin also becoming a schizophrenic.

In concordant DZ pairs, ages at onset are probably not so closely correlated as in MZ pairs. In DZ 10, for instance, the twins were hospitalized at ages 25 and 49, a difference of 24 years. Chronological age is therefore likely to be a better yardstick for estimating the proportion of risk survived than it is in the case of MZ pairs and the shorter Weinberg method to be correspondingly more reliable. There are, in any event, too few concordant DZ pairs for using the type of procedure we have applied to MZ pairs. Taking age 40 as the end of the risk period for the purposes of the shorter Weinberg method, the uncorrected probandwise rate of 4/34 (12%) is increased to 4/23 (17%). The latter rate would be reached by the uncorrected rate if only two further co-twins were to fall ill (6/34 = 17%). Since 9 unaffected co-twins are under age 30, and 14 have been followed up for less than 10 years from first hospitalization of the proband, it is not unreasonable to envisage the latter possibility. We would not, however, like to hazard a guess as to which DZ twins are most in danger. If only one more were to fall ill the rate would be 15%, very close to the Kallmann and Slater figures. As a check on the shorter Weinberg method, we may also apply Strömgren's method. Strömgren (1935) presented tables showing the proportion of risk passed for male and female sibs separately according to chronological age and age of onset in the proband. His tables were based on Bavarian population data of the time and the correlation in age at onset in sibs (.19) found in Schulz's large material of 660 schizophrenic probands. Correcting only for the age differences in the 30 discordant pairs of DZ twins gives a corrected concordance rate of 11% (3/26.98) pairwise and one of 14% probandwise.

In conclusion, taking the more conservative of our estimates, casewise rates of concordance, corrected both for mode of ascertainment (proband method) and for age, would be of the order of 16/26 = 62% ± 10 in MZ pairs and 5/34 = 15% ± 6 in DZ pairs. In Chapter 9 we shall compare these estimates with the corresponding rates found by Slater, Fischer *et al.*, and Kringlen.

NONSCHIZOPHRENIC ABNORMALITIES IN CO-TWINS

Compared with 14 *B*-twins (11 MZ and 3 DZ) who had a consensus diagnosis of schizophrenia or probable schizophrenia, 16 were diagnosed as having other (O-group) psychiatric abnormalities and the remaining 27 were normal. Omitting MZ 10*B*—a psychopath who was matched with a proband considered to be psychopathic rather than schizophrenic—6 of the O-group co-twins were from MZ pairs and 9 from DZ pairs. Five were male, 10 female. Seven received psychiatric treatment, 5 of them as inpatients. The judges differed amongst themselves as to how they formulated the abnormalities in co-twins. According to our own review of their opinions, they comprise: 5 personality disorders, 3 neurotic depressions, 1 recurrent depression, 1 postpartum depression, 1 adolescent depression, 3 anxiety neuroses, and 1 juvenile delinquent. None of the depressions were of a bipolar, manic-depressive type (Perris, 1966). Numbers are small, and we have no controls to enable us to decide whether our findings are specific for schizophrenia. Nevertheless, it will be worth looking rather more carefully at what we have found, first clinically and later, in Chapter 7, psychometrically.

First, how far are these abnormalities related to schizophrenia genetically? It seems that four co-twins at most would qualify as schizoid/eccentric psychopaths in Kallmann's (1938) sense, or as possible borderline schizophrenics or inadequate personalities of the kind that Kety *et al.* (1968) would place on a spectrum of schizophrenic disorder with high confidence. MZ 14*B* could be described as a spectrum-type inadequate personality, and his MMPI has a psychotic profile with a score on the schizophrenia scale ($T = 94$) that equals his brother's. MZ 16*B* is a paranoid personality. DZ 14*B* is a colorless, relatively inadequate, depressive personality who might be considered as having a spectrum disorder. DZ 25*B* had personality traits of a schizoid kind, though it was by virtue of his juvenile delinquency that he was rated as abnormal; the latter could be regarded as a normal reaction to his environment. The other 11 abnormal co-twins do not fit in readily with the usual concept of the schizoid—they are not socially withdrawn, emotionally cool, paranoid, or eccentric. Neurotic affective disorders predominate.

Table 6.7 shows the diagnoses of both twins arranged according to the kind of schizophrenia in the *A*-twin. From this array two further possibilities emerge. Some of the abnormalities may represent genetic elements other than specifically schizophrenic ones which the co-twin shares with the proband or, in the case of DZ co-twins, with other family members. Other abnormalities may require a more environmental interpretation. Schizophrenias with affective and paranoid features predominate among the *A*-twins in Table 6.7. There are also three schizophrenias associated with epilepsy or drugs and occurring in premorbidly existing abnormal personalities. There are few nuclear schizophrenics. Those that are, have twins who are schizoid (DZ 25; MZ 16), or who suffered from a mild transient reactive disorder, occurring in a normal personality (DZ 11, 20).

The abnormal co-twins of the paranoid cases tended to have depressions. MZ 24*A* and DZ 17*A* were themselves depressed as well as paranoid, and it may be noted that other markedly paranoid probands with late onset (DZ 2*A*, 10*A*) had previous episodes of depression. Thus we are led to suggest that elements common to the development of both affective and schizophrenic disorders coexist in some paranoid schizophrenics and that some of their co-twins show the affective elements only.

In the case of the three schizophrenias associated with epilepsy or drugs, their co-twins suffered from disorders related to personality characteristics which they shared with the proband. In MZ 5 these were anxiety tendencies, which in the case of the *B*-twin were exacerbated by marital stress. In MZ 9, both twins may have been somewhat character disordered. In DZ 27 both twins

TABLE 6.7

Pairs Where the Co-twin Had a Nonschizophrenic Diagnosis

Pair	Sex	Diagnosis of A-twin	Diagnosis of B-twin
MZ 10	m	Personality disorder (psychopath)	Personality disorder (psychopath)[a]
MZ 14	m	Schizoaffective	Personality disorder (inadequate, hypochondriacal)
MZ 18	f	Schizoaffective	Personality disorder (hysterical), anxiety[a]
DZ 24	f	Schizoaffective	Anxiety neurosis (with depression and ? alcoholism)[b]
MZ 24	f	Para schizophrenia, affective features	Post-partum depression[a]
DZ 5	f	Para schizophrenia	Anxiety neurosis (with depression)[a]
DZ 14	m	Para schizophrenia	Personality disorder (neurotic, depressive)
DZ 17	f	Para schizophrenia, "invol. depression"	Neurotic depression[a]
DZ 21	f	Para schizophrenia	Recurrent depression[a]
MZ 5	f	Schizophrenia, barbiturate addiction	Neurotic depression, anxiety[b]
MZ 9	f	? Schizophrenia, alcoholism	Neurotic depression, anxiety
DZ 27	f	? Schizophrenia, epilepsy	Personality disorder (psychopath)
DZ 25	m	Cat schizophrenia, ? psychopath	Juvenile delinquency, schizoid personality
DZ 11	f	Cat schizophrenia, psychopath	Anxiety neurosis (transient)
DZ 20	m	Schizophrenia, acute undifferentiated	Adolescent depression (mild)[b]
MZ 16	m	Pseudoneurotic schizophrenia	Personality disorder (paranoid)

[a]Psychiatric IP.
[b]Psychiatric OP.

had hysterical and sociopathic disorders which were in accord with the disturbed environment.

PSYCHIATRIC ABNORMALITIES IN PARENTS, SIBS, AND CHILDREN

Twelve parents and 16 sibs had what may be called serious psychiatric abnormalities, including five cases of schizophrenia. One child was schizophrenic. These 29 abnormalities occur in the following 22 pairs: MZ 2, 4, 7, 13, 18, 20; DZ 2, 3, 5, 6, 7, 8, 10, 13, 15, 16, 20, 22, 23, 27, 28, 32. As criteria for serious abnormality we have used suicide and psychiatric treatment or its equivalent, such as 100% pension for "shell-shock" after World War I. Eighteen of the above relatives are known to have been hospitalized for psychiatric disorder and a further two committed suicide. We have not included as abnormal those relatives with minor personality deviations such as the overprotective mother of MZ 1 who later became a spiritualist. The total abnormality count may therefore be an underestimate. Pairs MZ 10 and 12, where the proband was not schizophrenic according to the consensus diagnosis, are omitted from the present calculations.

Parents

On account of illegitimacy or adoption there is no information about four fathers and two mothers. Of the remaining biological parents, eight out of 51 fathers (16%) and four out of 53 mothers (8%) were abnormal in the above sense, making 12% in all. In no pair were both parents abnormal. Five out of 42 parents of MZ twins were abnormal as were 7 out of 62 parents of DZ twins.

Only one parent was definitely schizophrenic, the mother of DZ 7, who was a chronic paranoid schizophrenic by the age of 34. In this family both twins and a brother were also schizophrenic. Applying the shorter Weinberg method of age correction and counting parents under age 40 as having, on average, survived only half a life-risk, the morbid risk for schizophrenia in the parents is only 1/100.5, or 2/109 (1.8%) if, as is customary and theoretically correct, the proband method is used. In either case the risk is lower than that generally found—4.4% in pooled data from the genetic literature according to Slater and Cowie (1971). Hallgren and Sjögren (1959), however, found only one case of definite schizophrenia in the parents of 247 schizophrenics, but the risk of psychosis of any kind was 8.6%.

Sibs

Not counting twins, 108 full sibs were over age 15 or, in one instance (DZ 32), had been treated psychiatrically before that age. Ten out of 56 brothers (18%) and 6 out of 52 sisters (12%) were abnormal, making 15% in all. Abnormality was more frequent in the sibs of DZ twins (12/69 or 17%) than in those of MZ twins (4/39 or 10%). The difference, not found in other studies, is probably due to chance. The MZ twins in our sample had relatively few sibs. Of 47 sibs of the same sex as the twins, 6 were abnormal (13%).

Three brothers and one sister were schizophrenic. In two pairs, DZ 13 (male) and DZ 23 (female), the affected sibs were of the same sex as the proband and in two (DZ 5, 7) of opposite sex. The age-corrected risk for schizophrenia in the sibs is 4/84.5 (4.7%) or 5/89.5 (5.6%) probandwise. The latter is close to the corresponding rate of 5.4% in Slater's (1953) twin study but less than that in the pooled data from the literature (8.5%). In our sample the risk of schizophrenia for sibs is less than that for DZ twins.

A maternal half-sister of DZ 15 may have been schizophrenic.

Children

To complete the picture, a son of the female proband DZ 24 was schizophrenic. Two other children (MZ 18; DZ 33) came to psychiatric attention for what were probably minor troubles and are not considered here. Since only 8 of the 25 children of probands were over 15 and none was older than 28, there is no point in reporting the percentages affected. Combining DZ co-twins, sibs, and children, morbid risk for schizophrenia (probandwise) is 10 out of a corrected total of 116.5 or 8.6%.

Sex of Affected Relative

We have already seen that there is no more abnormality in female than in male relatives. This is unlike the finding in some earlier studies, which led to doubts being expressed as to the extent of the role of genetics in schizophrenia (Jackson, 1960; Rosenthal, 1962b). Nor is there a relative excess of pairs of like sex, unlike what is predicted by the sex-identification theory. Pairing the sex of the 29 abnormal first-degree relatives with that of the twins, there are 11 pairs of the same sex (6 of them female) and 18 pairs of opposite sex.

Nonschizophrenic Illnesses in Relatives

Only 6 of the 29 abnormal parents, sibs, and children had schizophrenia. As with the co-twins, the nonschizophrenic abnormalities span a wide range. The alleged paranoid illnesses of one or possibly two brothers of MZ 18 may indeed be schizophrenic. Some, such as the epilepsy of the brother of DZ 6, are doubtless independent of the schizophrenia of the proband. In others, such as the chronic neurotic disorders of the fathers of DZ 22 and 32, it would be unwise to assert dogmatically how they have contributed genetically and environmentally to the illnesses of their daughters. Some of our tentative interpretations at the case history level have been indicated in Chapter 4. Manic-depressive psychosis and related recurrent depressions occur only in families, such as MZ 20 and DZ 28, where there is some doubt as to whether the twin had a "true" schizophrenia.

SIMILARITIES AND DIFFERENCES WITHIN CONCORDANT PAIRS

To those who are unfamiliar with the literature on human genetics, the term *concordant*, as applied to schizophrenic twins, may at first conjure up an image of a pair of twins who, if not both schizophrenic at birth, probably fall ill together, are both hospitalized, and remain overtly schizophrenic for the rest of their days with the same symptoms. A moment's reflection causes one to guess that the term might comprise a much wider range of possibilities, as indeed it does. One twin might have a severe, deteriorating psychosis with early onset, and the other, a mild schizophrenic episode of late onset. Or both twins might have episodic illnesses; and these might or might not coincide, so that at a given point in time both might be well, both might be schizophrenic, or one might be actively schizophrenic and the other recovered. Jacobs and Mesnikoff (1961) described what they termed "alternating psychoses" and expressed the view that in some pairs at least there might be a psychodynamically determined tendency for twin *A* to be well when twin *B* is sick and *vice versa*.

The genetic investigators have, of course, paid considerable attention to resemblances and differences within pairs that were technically concordant. Luxenburger (1928), who was the first to gather a series systematically, was impressed by the comparative rarity of the photographic similarity in concordant MZ pairs, which the reports of single cases led him to expect. Rosanoff *et al.* (1934) were concerned with what they called qualitative and quantitative discordance. Findings have varied somewhat from study to study, at least in emphasis. While Slater (1953) and, even more so, Essen-Möller (1941b) were impressed by differences in outcome for pairs in which both twins had schizophrenic symptomatology, Kallmann (1946) concluded that both members of all concordant MZ pairs had deteriorated. Slater (1953) compared pairs of MZ twins and other relatives where both were schizophrenic and found a significant resemblance for particular features. Kringlen (1967) reported almost perfect resemblance for subtype in concordant pairs (13/14). The Genain quadruplets (Rosenthal *et al.*, 1963) suffered from schizophrenia ranging from severe to mild. They were similar, however, with respect to their constricted activity and speech and in a general pattern of unfolding of subtype syndromes. They and their father all had a similar EEG abnormality.

We have already indicated that in all but one of the pairs we are calling concordant both twins were hospitalized. We shall now look at age at onset, symptoms and subtypes, course and outcome.

Age at Onset

A precise date for the onset of schizophrenia is difficult if not impossible to pinpoint. Age at first psychiatric hospitalization is a useful objective measure, but it may not coincide with the onset of illness. In some cases the illness may have lasted for years, diagnosable but undiagnosed before hospitalization which may indeed never occur. It should be obvious that hospitalization is determined by a host of factors such as rural-urban environment, family tolerance, demands on the labor supply, and number of psychiatric beds available. In others the true schizophrenic nature of the illness may not be apparent on first hospitalization, though in retrospect it would be misleading to say they were then suffering from a different disease. For each schizophrenic we have recorded three measures.

1. Estimated age when the first serious symptoms related to the eventual schizophrenia were reported by the patient or noted by relatives—This is the measure recorded as "onset" in the case histories of Chapter 4. In several instances the event that was picked upon was inevitably somewhat arbitrary. Nevertheless, the attempt was made, if only to indicate that first hospitalization is not necessarily to be equated with age at onset.

2. Age at first psychiatric hospitalization, whether or not a diagnosis of schizophrenia was made during the course of that hospitalization—Many experienced psychiatrists could not bring themselves to make such a diagnosis until more unambiguous signs and symptoms were present than those elicited at first admission.

3. Age at first diagnosis of schizophrenia.

Examining the records of the 68 hospitalized cases accepted as schizophrenia, we find that the interval between measures 1 and 2 was, in terms of age at last birthday, 0 in 24 cases, 1 year in 17, and 2 years or more in 26 (maximum 15 years); in one case 1 was later than 2. Intervals between 1 and 3 were 0 in 18 cases, 1 year in 17 and 2 years or more in 33 (maximum 15 years). Intervals between 2 and 3 were 0 in 42 cases, 1 year in 9, and 2 years or more in 13 (maximum 13 years); four cases were diagnosed schizophrenic 1, 2, 8, and 15 years before hospitalization. Turning to the 11 consensus-concordant MZ pairs, estimated age at onset occurred at the same year of age or 1 year apart in 7 pairs and in four after intervals of 4, 5, 10, or 16 years. Of the 10 pairs where both twins were previously diagnosed schizophrenic, the diagnoses were made at the same age in 6 pairs and in the others at intervals of 2, 3, 5, and 12 years of age.[1]

Whereas neither 1 nor 3 can always be precisely dated, this is not so in the case of 2. In Table 6.8, we report in detail the within-pair differences for first

---

[1] As pointed out previously (Gottesman, 1968b), there was no tendency for pairs in which both twins were probands to have been diagnosed schizophrenic within shorter periods of time than the others.

hospitalization, including DZ pairs and pairs where twins were hospitalized but are not schizophrenic. In four MZ pairs concordant for schizophrenia the twins were hospitalized within a year of one another. Indeed in one pair hospitalization occurred on the same day and in two other pairs the twins were admitted to different hospitals within a week. The proportion of pairs hospitalized so close together is higher than in other series and the tendency should not be generalized. Four further pairs became concordant for hospitalization within 2½ years, and the longest interval in our MZ sample was 12½ years. In one DZ pair the interval was practically 24 years. Column (4) of Table 6.8 gives a good impression of the correlation in age at first hospitalization, which was .83 for all pairs in Table 6.8.

We also show whether the twins were living together or apart at the time of the first admission of the first twin to be hospitalized. According to these data there is no evidence that those hospitalized within short periods of one another were more often living together. If we go on to ask whether the discordant MZ pairs were more often living apart than those where both twins have been hospitalized, there is no sign of this either. At first hospitalization of A the twins were living together in 5 pairs (MZ 3, 11, 14, 16, 12) and apart in 5 (MZ 4, 5, 8, 9, 20). This does not deny the possibility of one twin influencing the time of hospitalization of the other. We believe this to have occurred in MZ 17 and 19 and perhaps others such as MZ 7 and DZ 4 and 7.

Symptoms and Subtypes

We shall not attempt a formal analysis aimed at showing which symptoms show the greatest resemblance, such as was attempted by Slater (1953), since our material is not large enough. Furthermore, the interview based assessment of the presence or absence of a particular symptom in a given case, such as thought disorder, is known to be considerably less reliable than the diagnosis of schizophrenia itself (Kreitman, 1961; cf. Shepherd et al., 1968). If this is so when several psychiatrists interview a patient at the same time (or review a videotape) for the purposes of research, assessment as to whether a particular symptom was ever present must be even more difficult to make reliably, since it depends on whether it was elicited and adequately recorded at the time.

We are well aware of the unreliability of Kraepelinian subclassification and other subtypes of schizophrenia and of the mixed and changing subtype patterns shown by many patients throughout their history. Two of our judges attempted a Kraepelinian subclassification of the summaries and we were not surprised that they failed to show close agreement. However, the classification by Judge A (Slater), when made blind, confirmed the previous reports of at least a moderate

TABLE 6.8

Intrapair Differences in Time of First Psychiatric Hospitalization

(1) Pair No.	(2) Sex	(3) Interval between dates of first hospitalization	(4) Ages (last birthday) at first hospitalization	(5) Twin first hospitalized	(6) When first hospitalized twins living apart or together
			Both twins schizophrenic		
MZ 19*	f	Same day	31–31	–	Together
MZ 17	m	3 days	22–22	B	Apart
MZ 7*	f	6 days	18–19	A	Together
MZ 22*	f	7 weeks	19–19	A	Apart
MZ 13	f	13 months[a]	14–15	A	Together
MZ 2	m	24 months	31–33	A	Apart
MZ 1	m	27 months	27–30	B	Together
MZ 15	m	28 months	34–36	B	Apart
MZ 6	f	5 years, 7 months	27–32	B	Apart
MZ 21*	m	12 years, 7 months	17–30	A	Together
MZ 23	m	—[b]	22–	(A)	Apart
DZ 7*	f	5 months	22–22	A	Together
DZ 4	f	6 years, 10 months	27–34	A	Apart?
DZ 10	m	23 years, 11 months	25–49	B	Together?
			Not both schizophrenic		
MZ 10[c]	m	3 months[d]	17–17	B	Apart
MZ 24	f	24 months	31–33	B	Apart
MZ 18	f	9 years, 6 months	20–30	B	Apart
DZ 17	f	7 years, 3 months	52–59	A	Together
DZ 21	f	9 years, 8 months	23–32	B	Together
DZ 5	f	10 years, 3 months[e]	31–41	A	Apart

*Both twins are probands.

[a]Admission to Remand Home here regarded as equivalent to hospitalization. Twin A was not admitted to mental hospital till age 15, 13 months after twin B.

[b]Twin B, age 26, never hospitalized.

[c]Neither twin schizophrenic (consensus diagnosis).

[d]Sentence to Detention Center by Court here regarded as equivalent to hospitalization. Twin A was first admitted to mental hospital at 23, 5 years, 9 months after twin B was first admitted to a psychiatric clinic.

[e]Twin A may have been hospitalized at 16, 25 years before twin B.

243

subtype resemblance in concordant MZ pairs (see Gottesman, 1968b). We present the findings here, as we believe they are a reasonably independent demonstration of within-pair phenotypic similarities that reflect a mixture of schizophrenic-specific and background gene effects. We incorporate additionally Slater's opinions reached when he reviewed his diagnoses for those consensus-concordant pairs in which he had disagreed with the consensus diagnosis of schizophrenia. Table 6.9 presents his subtype diagnoses in the 11 MZ and 3 DZ pairs that we are treating as concordant for schizophrenia. Eight out of 14 pairs were allotted the same subtype when classification was made blind. On review a further two were regarded as the same subtype. We note that all four remaining pairs include a paranoid schizophrenic matched with an atypical or ? latent schizophrenic or, in the case of a DZ pair, with a hebephrenic co-twin. A lesser degree of familial resemblance for the paranoid as compared with the other Kraepelinian subtypes is what Kallmann found, as noted by Rosenthal (1963).

TABLE 6.9

Subtype Similarity[a,b]

	No. pairs
Same	
Both twins hebephrenic: MZ 1, 2, 7*, 17, 22*†; DZ 7*	6
Both twins catatonic MZ 6 on review (originally Cat − Other Dx); DZ 4	2
Both twins paranoid MZ 19*	1
Both twins atypical (schizoaffective) MZ 13 on review (originally Other Dx − Atyp)	1
Different	
Atypical-Paranoid MZ 15 on review (originally Other Dx − Para); MZ 21*	2
Paranoid − ?latent Sc MZ 23 (unchanged on review)	1
Paranoid-Hebephrenic DZ 10	1

[a]Consensus-concordant Sc pairs. Blind diagnosis of predominant subtype by Judge A (Slater), including his revised diagnosis in pairs where Sc was not his first choice primary diagnosis of both twins.
[b]*, Both twins are probands. †, Both recurrent heb.

Course, Duration, and Outcome

Slater (1953) found pairs of schizophrenic relatives including MZ twins to be significantly alike in whether they had more than one attack. Nowadays the tendency is for a schizophrenic to have more than one spell of treatment in the course of his illness. Our concordant MZ pairs were alike in whether or not they had fewer than 4 spells as defined in Chapter 4. In 5 pairs both twins had fewer than 4, in 4 pairs both had 4 or more, and in 2 pairs one twin had fewer than 4 and the other 4 or more spells. The similarity is not entirely due to the younger pairs having had fewer spells. Pairs MZ 13 and 22, at ages 19 and 28, respectively, are liable to recurrent episodes, for instance. It is relevant to note that Inouye's (1963) highest MZ concordance rate by subtype was for "relapsing schizophrenia," 86% (6/7). Perris (1966), studying the genetics of recurrent affective disorders, found that "unipolar depression" (defined as three or more attacks of depression) appeared to be a specific genetic disorder that bred true and was distinct from recurrent illnesses with mania and depression. The course of a psychotic illness may turn out to be independent of the psychosis, genetically conditioned, and an indicator of "psychic disease resistance"; or, it may be part of the psychosis itself and then furnish pathognomic clues to genetic heterogeneity.

From our finding, reported earlier in this chapter, that probands from concordant pairs tended to suffer from relatively severe illnesses as measured by duration of hospitalization, it comes as no surprise to learn that in 6 MZ pairs out of 10 where both twins were hospitalized for schizophrenia both accumulated more than 2 years as inpatients. In 2 of these (MZ 7 and 17) both twins have been virtually chronic hospital invalids from the start. However, it is of greater interest to note the differences. In MZ 2 one twin has been in hospital 622 weeks longer than the other, in MZ 15 469 weeks longer. In MZ 19 there is a difference of 227 weeks, in MZ 1, 150 weeks, and in MZ 21, 139 weeks, to mention only those with differences of over 2 years.

The variability of the schizophrenias and other hospitalized psychiatric illnesses is seen in Fig. 6.1. Restricted to spells of hospitalization, insofar as these can be reproduced on such a scale, this oversimplification of the twins' psychiatric histories does, we believe, succeed in highlighting salient factors concerning the age at first admission and patterns of remission and relapse, both within and between pairs. Concordant pairs are shown first. The shaded areas indicate spells in hospital reckoned from first admission of the first twin to be hospitalized (e.g., at age 27 in MZ 1*B*) to the end of observation (death of MZ 1*A*, 35 years later).

We have no instance of consistently alternating psychoses, in the sense of Jacobs and Mesnikoff (1961); perhaps MZ 13 is the nearest. Numerous other

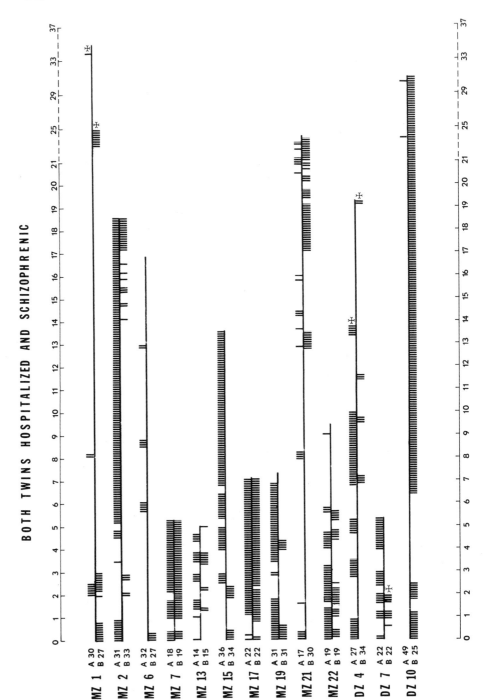

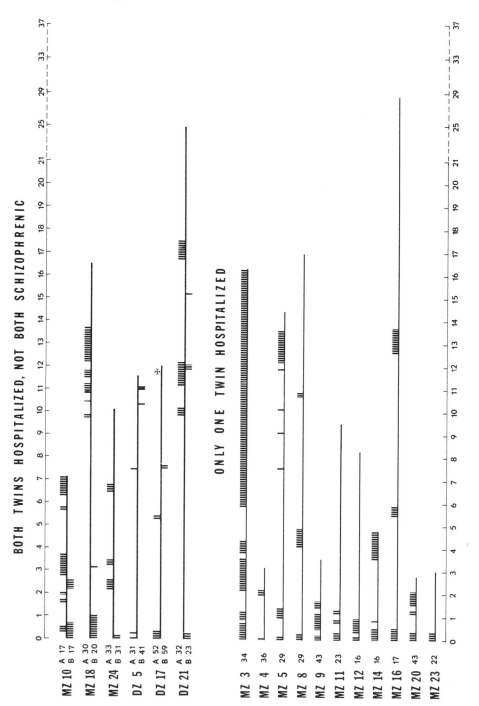

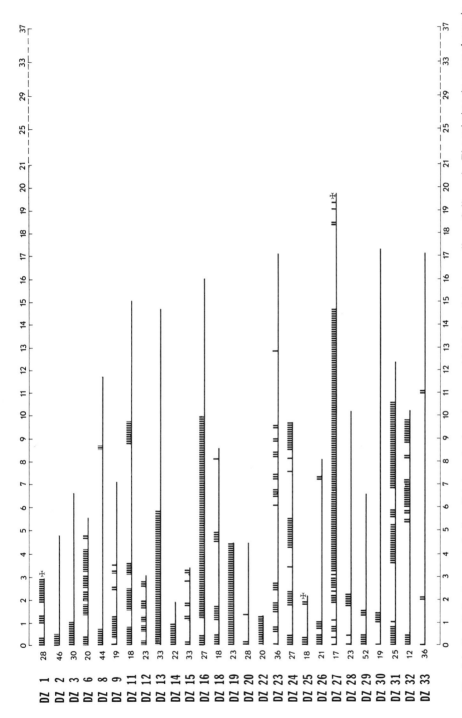

Fig. 6.1. Hospitalization periods for all twins who have been psychiatric inpatients, regardless of diagnosis. Elapsed time in years shown in horizontal scale above and below each section; hospitalized intervals of 5 weeks and less shown as one vertical stroke; fact of death indicated by a crown.

248

patterns are seen. Besides chronicity in both twins (MZ 7 and 17 mentioned above), there are pairs in which both twins remit (e.g., MZ 22) or in which one has a long chronic illness and the other acute attacks of late onset (DZ 10). In a few pairs, the second twin was admitted only after the last known discharge of the first twin (MZ 6 and 15), though more often periods of illness overlapped. The first to fall ill was not consistently the one who was hospitalized longer (cf. MZ 6, 15, and 21). The longest differences in hospitalization have already been mentioned.

If we compare the illnesses in the concordant MZ pairs with those of the *A*-twins in the discordant MZ pairs, the latter will readily be seen to include a higher proportion of milder cases. The discordant DZ pairs, however, show a more representative distribution of mild and severe cases.

Most of the twins from MZ concordant pairs were either in hospital at the time of follow-up or had died. The chart cannot tell the whole story. The following comments refer to some of the exceptions. In MZ 6, *A* was a home-invalid; *B* was one of the few in this group who had been holding a job for several years, but like MZ 15*B*, her adjustment may have depended on family support. MZ 19*A* was only recently discharged from hospital; her sister, though still paranoid, had apparently been getting by for the past three years. MZ 22*A* had had a minor relapse, while her co-twin was probably in the best shape of all those from concordant pairs; but we have since heard that MZ 22*B* has had two acute relapses. There is some hope that the young schizoaffective sisters of MZ 13 may improve. But on the whole, long-term outcome unfortunately seems rather similar in this mostly hard-core group of schizophrenics. None have been improved by their illness; all seem resigned to a lesser station in life than might have been the lot of their childhood social peers. Nevertheless, the fact that the paths of schizophrenic twins can differ considerably continues to offer a challenge and hopes of being able to halt or control the process effectively.

PREVIOUS PERSONALITY

This section deals with the premorbid personality of schizophrenics, schizotypy in nonschizophrenic co-twins, and possible genetic and environmental heterogeneity in the schizophrenic disorders.

It would be unrealistic to demand detailed, sensitive, reliable and unbiased prospective accounts of the premorbid personalities of all probands and co-twins. It is, however, important to try to assess as best we can the kinds of personality in which the potentiality for schizophrenia developed or failed to develop in our twins. It is not sufficient to ask how many of the nonschizophrenic MZ co-twins were in some sense schizoid personalities or schizotypes, or to use the observed characteristics of these co-twins to define the schizoid. We should also ask whether the schizophrenic probands themselves would have been recognizably schizoid or schizotypic before they were manifestly psychotic. If some schizophrenias develop in a normal personality or in a personality that is abnormal but not notably schizoid, there would seem to be no compelling reason for supposing that all discordant MZ co-twins should now be recognizably schizoid. On the other hand, if we hypothesize from nontwin data that the "true" schizophrenic psychosis is an environmentally determined outcome of a highly genetically determined manifest characterological defect which we may call schizotypy, then all MZ co-twins of "true" schizophrenics should be schizotypes insofar as we can validly detect the condition in its compensated form. If they are not (ceteris paribus), we may suspect that in such cases we may be dealing with a schizophrenia in the proband of a basically different kind.

Using all available clinical information, we have attempted to classify the MZ schizophrenics into those who were premorbidly schizoid, those who were otherwise abnormal, and those whose personalities were apparently within normal limits. We also grouped the personalities of the nonschizophrenic co-twins into the same categories. We fully recognize the possible errors in our judgments, arising from the retrospective nature of the enquiries and the ill-defined criteria of the schizoid and the abnormal. Schizoidness is a quantitative trait and overlaps to an unknown extent with "normal" introversion. We have not called a subject schizoid if somewhat schizoid characteristics were over-shadowed by abnormal traits of other kinds. Table 6.10 shows the often

TABLE 6.10

Preschizophrenic Personalities of MZ Twins[a]

Case No.	Twin A	Twin B	Case No.	Twin A	Twin B	Case No.	Twin A	Twin B
Concordant for schizophrenia			21*	S	S	11	N	N
1	N	?N	22*	A	A	14	?N	S
2	S	N	23	S	S	16	S	S
6	?N	?N	Discordant			18	A	A
7*	S	S	3	S	?S	20	N	N
13	A	A	4	N	N	24	N	N
15	A	A	5	A	A	Discarded		
17	S	S	8	N	N	10	A	A
19*	S	S	9	A	?A	12	?S	N

[a]Symbols: *, Both twins are probands; S, schizoid personality; A, otherwise abnormal personality; N, within normal limits.

tentative or insecure rating we give to each MZ twin. Decisions were based on a synthesis of judges' comments and our clinical hindsight.

From Table 6.10 we see that the previous personality of a schizophrenic is not always obviously schizoid; this indeed was common knowledge (cf. Bleuler, 1911; Slater & Roth, 1969). Only 13 of 33 schizophrenics were judged schizoid with a further 9 probably otherwise abnormal. If we accept the genetic factors in schizophrenia and assume that they are similar in most cases, the observation suggests that it is not merely a tendency to be schizoid that is inherited and in some way causally related to schizophrenia, but something more besides. In addition it suggests that not all genes get "switched on" until certain psychosomatic states are reached. The schizoid personality is clearly an important contributory factor; and, however central to or independent of the main diathesis we may regard it, it is one which is itself probably under considerable genetic control. Table 6.11 shows that concordance is highest in pairs where the previous personality of the *A*-twin was schizoid and lowest when it was normal. Since premorbid personality is well-established as a prognostic index, the finding reflects the relationship between severity and concordance discussed earlier.

The above analysis tends to support a multifactorial view of schizophrenia with different though related combinations of genetic and environmental factors contributing to the etiology in each case. Perhaps the strongest support for a contrary view comes from the assessments which Essen-Möller made on our material (see pp. 219-220 and Appendix D). These could be interpreted as support for the view of a single major etiological factor for "true" schizophrenia, a factor which is usually manifest at a characterological level, but which is absent in the balance of basically unrelated schizophrenic-like psychoses of various origins. In the 11 consensus-discordant MZ pairs there were 5 in which both twins were suspected of a genetic connection with schizophrenia, and these included the only three pairs in this group where the proband was thought to have had a "true" schizophrenia. The other six consensus-discordant pairs (MZ 4, 5, 9, 18, 20, and 24) were the only ones to include a twin not suspected by Essen-Möller

TABLE 6.11

Premorbid Personality and Concordance in MZ Pairs

Concordance for schizophrenia	*A*-twin (or, as the case might be, proband) with premorbid personality assessed as:			
	Schizoid	Otherwise abnormal	Normal	Total
Pairwise	6/8	3/6	2/8	11/22
Probandwise	9/11	4/7	2/8	15/26

of being at least schizoid and they included no twin considered to have had "true" schizophrenia.

It is interesting that the probands in the latter 6 MZ pairs mentioned above are those in which in our comments on the case histories we were most tempted to suspect etiological heterogeneity among the probands as a major reason for the discordance.

As we shall suggest later, it may be possible to account for most of the heterogeneity with a model which supposes some of the genes in a polygenic system and some of the environmental contributing factors to be relatively rarer than the others but at the same time to have a greater effect on the psychosis or on some aspects of the psychosis than the others. Where one of these relatively major factors is present, a subject will require fewer of the other individually common enhancing factors, such as schizoidia or anxiety readiness, before succumbing to a schizophrenic-like psychosis. Clinicians at different centers will tend to have different personal thresholds for calling a patient "true" schizophrenia rather than schizophreniform as a function of the weight given these major factors. With Slater, we personally would exclude the alcoholic hallucinosis of MZ 9A and the symptomatic psychosis of MZ 20A as readily as we would exclude the nonconsensus schizophrenias of MZ 10A and MZ 12A. With the other Essen-Möller discordant pairs (MZ 4, 5, 18, 24) where neither twin was at least schizoid, we have varying doubts as to whether they should best be excluded from the schizophrenias or included as relatively peripheral cases in which the genetic diathesis was less marked than usual and the stresses undergone by the proband correspondingly extreme or unique ones.

SUMMARY

In our series of twins the simple, directly observed, pairwise concordance rate for schizophrenia or probable schizophrenia (consensus diagnosis) was 50% for 22 MZ pairs and 9% for 33 DZ pairs. The present chapter attempted to expand this simplified statement.

In MZ pairs the co-twin was more likely to be schizophrenic if the illness of the proband was severe, as judged by a number of different clinical and objective criteria. Consistent with this finding was an analysis showing that concordance was high, given a proband whose previous personality was assessed by us as schizoid.

The association between severity and concordance, found also in other studies, is one of the limitations of any general estimate of the risk of the twin of a schizophrenic being likewise affected. A further limitation arises from the difficulty of allowing for the fact that some as yet unaffected co-twins may still become schizophrenic. Of many age-corrected estimates which may be made from our material, morbid risks of 62% ± 10 for MZ co-twins of schizophrenics and 15% ± 6 for DZ co-twins of the same sex seem to be reasonable ones; they correspond closely to the casewise rates that would be obtained by the proband method if one more twin of each type were to fall ill. The above risk for MZ twins is lower than those obtained in the older studies of Kallmann (1946) and Slater (1953), but higher than those obtained in recent Scandinavian population studies. There is less variation in the reported DZ morbid risks.

While 11 MZ and 3 DZ *B*-twins were schizophrenic, 6 MZ and 9 DZ *B*-twins of "consensus" schizophrenics had other psychiatric abnormalities, mostly personality disorders and various kinds of depression other than bipolar manic-depressive psychosis. Probably only 4 of the 15 *B*-twins had disorders that could be regarded as part of a specifically schizophrenic spectrum (Kety *et al.*, 1968). We mentioned some ways in which these disorders might have occurred without their being closely related to schizophrenia genetically.

Twelve parents, 16 sibs and one child had what were regarded as serious psychiatric abnormalities. Only one parent, four sibs and one child were definitely schizophrenic. The morbid risk for schizophrenia in sibs was 5.6%, lower than in most other studies. There was no tendency for proband and seriously abnormal relative to be of the same sex, contrary to the recently popular hypothesis that identification with a sick relative of the same sex is an important etiological factor in schizophrenia.

How alike are "concordant" pairs? In our material there was a close correlation within concordant pairs in estimated age at onset and in age at first psychiatric hospitalization. There was no tendency for twins to be in hospital either at the same time or at different times. The in-and-out-of-hospital history of each twin was presented pictorially. While the data were not considered suitable for an analysis of resemblance in respect of specific symptoms, we were able to demonstrate by means of Slater's independent subtype classification of the individual histories, and in other ways, that there was often a very considerable resemblance in the form the psychosis took in a pair of schizophrenic twins. Illnesses of a paranoid kind were an exception to this tendency.

Of the 10 concordant schizophrenic MZ pairs where both twins were hospitalized there were five in which one twin had spent more than 2 years longer in hospital than the other. Active physical methods of treatment and easy discharge facilities led to a better prognosis than obtained in an earlier era. However, outcome was far from favorable in schizophrenic twins from concordant pairs.

A personality of a kind which may be described as schizoid is almost certainly largely genetically determined. It is nothing new to show that such a personality is a poor prognostic index in schizophrenia, and it is not surprising that we have shown that the MZ co-twins of such schizophrenics are frequently schizophrenic themselves. However, not all schizophrenics develop from a schizoid personality, even in concordant pairs of twins, and not all nonschizo-phrenic MZ co-twins are diagnosable as schizoid. We have argued that it is not merely a tendency to be schizoid which is inherited and in some way causally related to "true" schizophrenia. Some genes may be implicated which only get "switched on" once certain psychosomatic states have been reached. Different though often related genetic and environmental factors may contribute to the etiology of each case. This introduces heterogeneity of a kind which may make it largely a personal matter whether a particular "schizophrenic-like" case is regarded as part of the group of schizophrenias or as distinct from it.

7
PSYCHOMETRIC CONTRIBUTIONS TO GENETIC ANALYSIS

To the extent that psychometric tests of personality and thinking styles have validity, they have the potential for making an important contribution to the search for phenotypic indicators of some kind of genetic predisposition to schizophrenia. A new dimension of objectivity is added to studies of schizophrenic twins and their relatives by removing the subjectivity of the experimenter in assessing the kind and amount of psychopathology, by introducing a normative frame of reference whereby the experimental subjects can be compared with previously studied, large criterion groups of normals or patients, and by systematically collecting responses to the same stimuli. The distribution of test scores permits working hypotheses about whether some of the data on some traits or clusters are in fact dichotomous and thus amenable to simple Mendelian explanations; if some score or configuration of scores, for example, should only occur in 50% of DZ co-twins, 50% of proband parents, and in all of the MZ co-twins of schizophrenics, we would have confirmation of a simple dominant gene theory and a way of identifying the carriers of the gene before they decompensated. As might have been expected, we are nowhere near such an ideal state of affairs. Personality tests are much less reliable and valid than, for example, intelligence tests (Cronbach, 1970), and they vary among themselves along these dimensions and others (e.g., projective versus objective and clinical versus statistical combining of raw data to reach a decision, cf. Meehl, 1954; Sines, 1970).

It was in the hope that psychological test scores would provide a heuristically useful set of phenotypes for genetic analyses that we used the Minnesota Multiphasic Personality Inventory Test (MMPI) and the Goldstein-Scheerer Test [hereafter called Object Sorting Test (OST)] of Concept Formation (as described by Lovibond, 1954).[1] We were buoyed by the example in research on twins of diabetics (Then-Bergh, 1938) which reported concordance rates of 43% and 14% for overt diabetes in MZ and DZ pairs; when abnormal glucose-tolerance test results were accepted as indicators of the diabetic genotype, the rates rose to 100% and 40%, respectively. Pyke *et al.* (1970) reported less striking results in their MZ co-twins, but Cerasi and Luft (1967) found all their MZ co-twins to have a diabetic-like plasma insulin response to a glucose infusion test.

The concurrent validity of a test, that is, its ability to detect overt schizophrenics, is not so much at issue here as is its ability to show a difference between the clinically normal relatives of probands and relatives of normal controls or nonschizophrenic but otherwise psychotic controls. Of course, if the number of false positives is too high, the number of controls identified as having a schizophrenic potentiality will be too high to be credible for a disadvantaging genetic characteristic. This might mean, for example, that fewer than say 10% of controls unrelated to a schizophrenic should be detected as "abnormal"; Deming (1968) has shown that in the general population about 2.1% of living adults over age 15 have or have had a diagnosis of schizophrenia[2] in hospitals; our figure of 10% allows for a ratio of four potential schizophrenics for every one schizophrenic detected clinically.

In Lovibond's original development of an OST scoring procedure (1954) for quantifying Rapaport's (1946) clinical observations, 4/45 controls showed "impairment," but only three (7%) were as bad or worse than 77% of hospitalized schizophrenics. Using similar criteria, Rosman *et al.* (1964) later found 57% of parental pairs with normal offspring to have at least one high-scoring member on the OST (7 or more points), while McConaghy and Clancy (1968) found 75% of parents of normals selected for their high scores also to have high scores. Thought disorder, or cognitive slippage as it is called (Meehl, 1962) in its milder forms, as well as such traits as schizoidia, may stem from many different sources, only some of which are genetically related to schizophrenia. Brain damage, for example, has long been demonstrated to impair sorting kinds of tasks (Vigotsky, 1934; Kasanin, 1946; Tutko & Spence, 1962).

Individuals with MMPI profiles suggestive of clinical schizophrenia have not yet been followed long enough to see whether they develop into overt cases. However, Hathaway *et al.* (1970) have partially described the 4.75% of adoles-

---

[1] Our intentions (Gottesman & Shields, 1966b) to apply Meehl's Schizotype Checklist, to have the OST scored blindly by Margaret Singer, and to utilize the information on the Briggs History Form proved unworkable for various reasons.

[2] Only first admission diagnoses were used so that Deming's figure for New York state is an underestimate that does not take into account individuals whose diagnoses were changed to schizophrenia on a second or later admission (cf. Clark & Mallett, 1963; Astrup & Noreik, 1966).

cent boys and girls (age 14) from a cohort of some 11,000 with scores higher than two *SD*'s above the normal mean on the Schizophrenia or Paranoia scales and followed to age 25. Although only one had a psychiatric hospitalization, the selected "children and their careers are suggestively similar to [retrospectively] reported data on schizophrenia patients as children and adults." For example, they were disproportionately of lower class origin and showed no or downward social mobility. Peterson (1954) studied 33 false negative Veteran's Administration outpatients who were *later* hospitalized and diagnosed as schizophrenics; their outpatient MMPI's with primary elevations (see below, p. 260) on $Sc(85)$, $Pt(82)$, and $D(81)$ were like those of true positives $(82')$ and unlike those of true negatives $(123')$. The MMPI was constructed so that no more than 2.5% of normals obtain standard scores (T) greater than 69 on any one scale.

A number of other investigators have been concerned with the psychometric characteristics of the relatives of schizophrenics (e.g., Mednick & Schulsinger, 1968; Rosenthal *et al.*, 1968). Their focus has often been on the tests as an aid to describing the quality of interaction within the families (e.g., Alanen, 1966), rather than as an indicator of a schizophrenic genotype. The most ambitious and successful programmatic research has been conducted by Singer and Wynne (1965, 1966; Wynne, 1968) who have studied 280 families including 92 where the index case was an adolescent or adult schizophrenic. Their test battery has included Rorschach, TAT, MMPI, OST, and other instruments, but their scoring procedures go beyond detecting "schizophrenics" to the assessment of "schizophrenogenicness" from verbatim protocols. The latter permits rating features of communication that have a "transactional impact" on the other family members, and that would be expected to "induce difficulties in focusing attention and handling meaning." Wynne (1968) reported that 85-95% of the families were correctly differentiated with regard to offspring diagnosis using this framework. Lidz's Yale group (Wild *et al.*, 1965) improved the discrimination between parents of schizophrenics and controls after shifting to the same approach. In this fruitful focus on communication deviance, the investigators are no longer using the test (e.g., Rorschach) psychometrically, but as a means of eliciting less guarded verbal behavior. Such findings tell us much more about phenomenology than about etiology (cf. Mishler & Waxler, 1968a,b); the deviance may be the result of having a schizophrenic in the family, rather than the cause. Hirsch and Leff (1971), using the 41-category Rorschach scoring procedure developed by Singer and Wynne, found considerable overlap between the performances of neurotics' and schizophrenics' parents with 40% of schizophrenic parental couples showing total group median or lower deviance scores; both the neurotic and schizophrenic offspring had been hospitalized (at the Maudsley-Bethlem and two other hospitals) for acute illnesses. Mothers of the two groups did not differ from each other, whereas the difference between fathers was traced to the greater word usage for schizophrenics' fathers; Hirsch implicated a form of experimenter bias (cf. R. Rosenthal, 1966) during the Americans' inquiry period to account for his findings contradicting those of Singer and Wynne. Earlier, Schopler and Loftin (1969) had shown that the differences in OST performance between parents of psychotic children and controls was diminished by the context of testing; apparently less anxiety-arousing circumstances led to less thought disorder so that parents of psychotics did not differ significantly (30% versus 12%, N.S. χ^2) from parents of normal controls. Mishler and Waxler (1968b) found that disruptive communication (speech fragmentation and repetitions) were *more* frequent in normal than schizophrenic families.

The psychometric assessment of schizophrenics' relatives for the express purpose of clarifying genetic factors does not have a long history. Meehl (1962) believed that ordinary family studies, based as they were on formal psychiatric diagnoses, threw away valuable information about relatives who were neither normal nor overtly psychotic. He called for priority to be given to the development of high-validity indicators of "compensated schizotypy," both psychometric and neurophysiological. McConaghy's (1959) work was an important step in the right direction. Using scores on the OST he showed that at least one parent of a schizophrenic with both OST and clinical thought disorder in each of 10 pairs showed appreciable thought disorder (7 or more points) although clinically "normal." Since only 9% of 65 normals had scores so high, he was inclined to believe that the OST could identify the heterozygote carrier of the recessive (sic) gene for schizophrenia. A possible link with neurophysiology had been provided earlier by his colleague Lovibond (1954) who, following Pavlov, posited a failure of cortical inhibition in schizophrenics which led to the "loosening of association," and thus poor concept formation.

None of the subsequent attempts to replicate achieved such unambiguous results, whether using the OST or other tests (Lidz *et al.*, 1962; Rosman *et al.*, 1964; Phillips *et al.*, 1965; Stabenau *et al.*, 1965; Romney, 1969). Various reasons have been adduced including differences in age, sex, social class, intelligence, anxiety, absent or unknown high score in index cases, and, by McConaghy himself (McConaghy & Clancy, 1968), that loosening of associations or "allusive" thinking is normal and ubiquitous. Each reason has some merit.

In a dissertation by Toms (1955), focusing on the personality and attitudes of the mothers of long-stay paranoid schizophrenics in a Veterans' Hospital, the mean MMPI profiles of 25 experimental mothers and 30 controls were given; the controls were not well selected and the test information about five additional mothers who had invalidated their tests by Lie scores over 70 was omitted. The mean profiles of both groups were unremarkable and within normal limits. A configural approach to profile analysis (Sullivan & Welsh, 1952) which we shall use later on was more revealing. Three "signs" characterized the profiles of schizophrenics' mothers—Hypochondriasis, Psychopathic deviate, and Psychasthenia were each comparatively more elevated than Hypomania within the profiles. Such results can parsimoniously be explained by assuming that the patients' mothers were relatively more depressed than mothers of normals. Bradford (1965), while assessing the perceptions of mothers by male schizophrenics and by their normal brothers, was able to obtain the MMPI's from 13/20 patients' brothers and 15/15 normal controls' brothers. Again the mean profiles from both groups of siblings were unremarkable and within normal expectations.

Heston (1966) found that the mean scores on the Menninger Mental Health Sickness Rating Scale (MHSRS) (Luborsky, 1962) differentiated the offspring of schizophrenic mothers from matched controls with normal mothers; mean scores were 65.2 and 80.1, respectively, on the 100-point scale. The excellent discrimination was a result of the very low scores of 26 of the 47 experimentals rather than to a general lowering for each one. The mean scores on the MMPI did not distinguish between the two groups, but 28 of the 47 experimental subjects were not tested including the most disabled and schizophrenic.

MMPI IN THE MAUDSLEY TWIN STUDY

We used the MMPI because of our previous experience with it in twins (Gottesman, 1963b, 1965), its network of relations with other psychological variables, its validity in normal and psychiatric populations (Dahlstrom, Welsh, & Dahlstrom, 1972; Marks & Seeman, 1963; Butcher, 1969), and its ease of administration. Our concern that it might be inappropriate or rejected by British subjects proved unwarranted; very few items were too "American" and the profiles generated agreed with our experience in the United States.

Description of the MMPI Scales

Although Hathaway and McKinley (1951) derived their inventory within a Kraepelinian framework and labeled their scales according to the diagnoses of their criterion groups, advances in interpretation have led to the substitution of numbers for scale names and a dependence on the configuration of all scales in the profile rather than single-scale elevations for diagnostic inferences. A tremendous wealth of clinical and research knowledge about the MMPI has accumulated since its beginnings before World War II. The inventory consists of 550 self-descriptive statements printed in a booklet which the subject is asked to assess as *true* or *false* by marking an answer sheet; a *cannot say* category is provided by allowing the subject to omit an item, but he is encouraged to do this only rarely. The content areas sampled by the items include social, sexual, and religious attitudes, mood, morale, general health, phobias, and masculine-feminine interests. Although many of the items elicit admissions of obvious symptoms, many are subtle and have a "projective" component to them (Meehl,

TABLE 7.1

MMPI Scale Numbers, Abbreviations, and Names

Validity scales	Clinical scales
? Cannot say	1 *Hs* Hypochondriasis
L Lie	2 *D* Depression
F Rare endorsement	3 *Hy* Hysteria
K Correction	4 *Pd* Psychopathic deviate
	5 *Mf* Masculinity-femininity
	6 *Pa* Paranoia
	7 *Pt* Psychasthenia
	8 *Sc* Schizophrenia
	9 *Ma* Hypomania
	0 *Si* Social introversion-extroversion

1945). Unlike earlier personality tests, the answers to items are not necessarily taken at face value as surrogates for behavior. Items were selected empirically from a large pool because they were endorsed significantly more frequently in a given direction by the abnormal criterion group, say hysterics or schizophrenics, than by the normal control group. In this fashion, scales of varying length were built, for example, the Schizophrenia scale has 78 items while the Paranoia scale has 40. After a simple scoring procedure, four validity scales and ten clinical scales are plotted on a standard profile sheet and the points connected. The scales with their numbers and abbreviations are set out in Table 7.1 for mnemonic convenience.

Regardless of the number of items per scale, each has been standardized to a mean score of 50 and a standard deviation of 10 in the normative population. Standard or *T* scores above 69 are only obtained by 2.5% of normals and provide an arbitrary level above which abnormality may be suspected. We will not give *T* scores for individual twins, but they can be reconstructed with good accuracy from the profile codes in Chapter 4 which used the Welsh (1948) system; for example, 8′ (see p. 71) would be equated to a *T* score of 75 on the Schizophrenia scale knowing that the actual score could not be lower than 70 or higher than 79. Sex differences have already been provided for before coding takes place.

The Validity Scales

The raw score on *?* is simply derived from the number of items omitted; up to 30 items can be omitted if done randomly with no important consequences. It was our practice in our use of the MMPI to score all omitted items in the "sick" direction for each scale on which they appeared so as to maximize the amount of psychopathology revealed. In the few cases in which this was necessary, it resulted in striking confirmation of the wisdom of the procedure; in DZ 13*A*, for example, who left out 107 ($T = 72$) items it was obvious that items were not omitted randomly since he was delusional, agitated, and clinically paranoid. His profile went from a 8-420169/375: FKL- to a 8*4″29′063-571/ F*LK-. Whatever error in the twins' MMPI's may have been introduced was independent of zygosity. The 15-item L scale allows the subject to claim various social virtues, too many indicating a tendency to put oneself in a favorable light; four items are modal and give $T = 50$. High L scores also permit inferences about personality traits such as rigidity, naivete, and no insight. The F scale consists of items answered in the same direction by at least 90% of normals. It is loaded with bizarre statements such as "My soul sometimes leaves my body" and only three items yield a $T = 50$. High scores suggest psychosis, malingering, poor reading ability, or even random answering. K is both a correction scale and a sign of certain attributes. It is much more subtle than L in detecting the effort to look good psychologically. Five scales, *Hs, Pd, Pt, Sc,* and *Ma,* are corrected by the addition of portions of K. Like $F+\%$ on the Rorschach, too much or too little K suggests psychopathology.

The Clinical Scales

Scale 1 (*Hs*) was derived from the responses of a criterion group of 50 hypochondriacs and measures preoccupations with bodily health and functions. Scale 2 (*D*) indicates depression, having been derived from responses by 50 cases of reactive depression and unipolar depression. The criterion group for Scale 3 (*Hy*) consisted of conversion reactions; the scale measures psychological immaturity, repression, and denial. Scale 4 (*Pd*) was derived from 100 young adults with histories of delinquency and indicates impulsiveness, inability to anticipate the consequences of behavior, and emotional shallowness. The criterion group for Scale 5 (*Mf*) consisted of a small group of male sexual inverts; scale elevations indicate interests similar to those of the *opposite* sex. Although male homosexuals often obtain high scores, the number of false positives is too high in Western cultures for such scores to be sufficient evidence of homosexual practices. Scale 6 (*Pa*) was derived from a mixed group of patients who were suspicious, oversensitive, and plagued by delusions of persecution; diagnoses were usually paranoia, paranoid state, or paranoid schizophrenia. High 6 scores readily permit the inference of paranoid tendencies, but low scores do not permit ruling them out. The criterion group for Scale 7 (*Pt*) was composed of 20 patients with severe phobias or compulsive behavior who could be subsumed under Janet's term "psychasthenia" (a contemporary Scandinavian term would be *anankastic*); high scorers are characterized by obsessional ideation, perfectionistic tendencies, feelings of guilt, and anxiety. An additional criterion for item selection (cf. McKinley & Hathaway, 1942) led to additional saturation with "general maladjustment" variance; scores above 70 led Gilberstadt and Duker (1965) to diagnose "anxiety state." Scale 8 (*Sc*) was derived from 50 criterion patients diagnosed as schizophrenic. The scale is central to our concerns but it is a weak one in the identification of schizophrenics all by itself with a hit-rate of 60% on cross-validation, i.e., 40% of diagnosed schizophrenics obtain T scores less than 70. However, the false positive rate is quite low with only 2% of the normative sample obtaining T scores higher than 70. Hypomania (*Ma*) or Scale 9 was derived on a small criterion group characterized by the pattern of overactivity, emotional excitement, and flight of ideas. The scale items are quite heterogeneous and its reliability over time is .56, the lowest of any of the scales; the nature of the construct being measured conspires against reliability with the most hypomanic patients refusing to be captured and tested. The last of the standard clinical scales appearing on the MMPI profile sheet is called Social introversion (*Si*) or Scale 0. It differs from the others described thus far because it was derived in a nonpsychiatric setting. The criterion group consisted of 50 college students who were the highest third on an inventory (Minnesota T-S-E) of social introversion (Evans & McConnell, 1941) constructed rationally to measure three components of introversion—social, thinking, and emotional. The items selected from the MMPI to form the new scale differentiated a contrast group of 50 students who were in the lowest third of scorers on the social introversion component of the T-S-E. Subsequent validity studies showed that Scale 0 high scorers were not participants in social activities and were shy, brooding, and self-depreciatory; high scorers in psychiatric settings were often called schizoid.

Experimental Scales

Literally scores of additional experimental scales have been derived from the MMPI, but few meet the tests of time and utility. One group of scales that has particular interest for us is that derived by Rosen (1962). Rather than trying to discriminate psychiatric criterion groups from normals, he set himself the much more difficult task of discriminating five clinically homogeneous groups from a random sample of psychiatric patients in the same hospital. The five groups and scales were the following—Conversion reaction (*Cr*), Somatization reaction (*Sm*), Depressive reaction (*Dr*), Anxiety reaction (*Ar*), and Paranoid schizophrenia (*Pz*). We shall report our findings on the clinical and these experimental MMPI scales below.

We do not attempt, in this work, to defend the MMPI against the kinds of criticisms that have been leveled against it; the interested reader will find the ideas of Block (1965), Dahlstrom (1969), and Wiggins (1969) informative in these respects. There is no denying that personality measurement is an infant science, but all is not bleak. There are dim lights on the horizon and we may be able to "bootstraps" our way toward them (cf. Meehl, 1972).

MMPI PROFILES OF PROBANDS, CO-TWINS,
AND ALL FIRST-DEGREE RELATIVES

Table 7.2 shows the sources of the MMPI data we analyzed. We were able to obtain the MMPI from 77% (34/44) of the consensus MZ schizophrenics and their co-twins, 67% (44/66) of the corresponding DZ twins, 11 mothers (plus two adoptive mothers, MZ 22 and DZ 11), 8 fathers, 5 nonschizophrenic siblings, and one adult offspring (a total of fifty first-degree relatives counting the DZ *B*-twins).

In a number of respects our personality test data are unsatisfactory, but their uniqueness and heuristic value outweigh their shortcomings. We had to endure a number of obstacles to unambiguous interpretation of the various results. We could not be sure that untested or dead or illiterate twins were like those finally tested. Since the MMPI profile varies with clinical condition, the assessment should ideally have been done at a uniform stage of the schizophrenic process, such as time of first admission for each twin; ideally, both members of a pair would have retests when both were in remission as well as premorbidly, etc. However, no control was possible over the clinical condition or over concurrent environmental events. One DZ co-twin was tested 26 years after his first admission, having spent virtually all that time in hospital. Other twins were tested at various times in the course of their illnesses, sometimes in hospital, most often not. Probands in varying degrees of remission, either spontaneous or with an indeterminate amount of support from phenothiazine medication and a meliorative environment, will decrease the pathology measurable by the MMPI and make contrasts with normal or compensated co-twins less impressive. The conditions prevailing at the time of testing may be surmised from the case histories in Chapter 4. Original probands who did not meet the criteria for a consensus diagnosis of schizophrenia were removed from the personality test analyses along with their co-twins and other relatives (cf. Gottesman & Shields, 1968). MMPI's were obtained from 33 pairs of twins and 12 unpaired twins; the unpaired twins may be either probands or co-twins. Small sample sizes for many

TABLE 7.2

Twins and Relatives Tested with the MMPI

	From 22 MZ pairs	From 33 DZ pairs
A-twins	18	19
B-twins (of which schizophrenic)	16 (8)	25 (1)
Both twins of a pair	15	18
Other first-degree relatives		25	

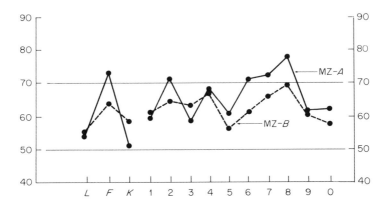

Fig. 7.1 Mean MMPI profiles of 18 MZ *A*-twins and 16 MZ *B*-twins.

of our analyses necessitate an emphasis on trends suggestive of interesting hypotheses and a consequent casual treatment of statistical significance, i.e., a context of discovery rather than one of justification (Reichenbach, 1951).

MZ Proband and Co-twin Contrasts

Of the 22 consensus MZ pairs, we tested 18 *A*-twins, all of whom were probands by definition, and 16 *B*-twins, of whom three were probands in their own right (one of the *B*-twins in our four double-proband MZ pairs was not tested, all *A*-twins in these pairs were tested). Figure 7.1 shows the mean MMPI profiles for the 18 *A*-twins and the 16 *B*-twins; 8 of the latter were consensus schizophrenics.[3] The figure portrays a simple pairwise view of our MMPI data and constitutes objective test support for the clinical analyses reported in Chapter 6 based on clinical diagnoses and our interpretation of events in the twins' case histories. In the "eyes of the MMPI" the mean profile of MZ *A*-twins is clearly psychotic in both shape and elevation; furthermore, the first-choice diagnosis (independent of MMPI) obtaining for this kind of profile in other psychiatric settings is expressly a schizophrenic one (Marks & Seeman, 1963; Gilberstadt & Duker, 1965). The co-twins' profile has one score outside of normal limits, *Sc* with *T* = 70, and with a configuration quite similar to that of the *A*-twins. The similarity of shape is conveyed by the fact that the four highest scales in the co-twin profile (8, 4, 7, 2) are found among the five highest scales in the *A*-twin profile; further, the pattern of the validity scales, *L, F,* and *K,* is similar and typical for a psychiatric population.

---

[3] Standard deviations for all mean scores used in profiles and contrasts are given in Appendix F.

The demonstration of personality test similarity above does not speak to the issues of the origin of the similarities (genetic or environmental) or to the usefulness of the MMPI in detecting schizotypes; for the former, we will also require our data on DZ co-twins, while the latter requires the exclusion of MZ co-twins who are already diagnosed as schizophrenic, or a look at the data from schizophrenics in remission. All of these issues will be dealt with in due course.

The effects of removing all co-twins with a consensus diagnosis of schizophrenia from the *B*-twin profile in Fig. 7.1 are to reduce the sample size to eight and to eliminate the strong similarity between profiles. We can contrast an MZ mean profile consisting of *all* with a consensus diagnosis of schizophrenia, $N = 26$, with the profile of the eight normal or "other" diagnosis (see Table 6.7) twins.

	L	*F*	*K*	1	2	3	4	5	6	7	8	9	0
All MZ consensus schizophrenics	54	72	52	59	70	59	68	60	70	71	78	62	62
Normal and "other" MZ co-twins	55	57	61	64	62	65	63	52	55	62	63	60	53

If the latter profile had been generated by a much larger number of nonschizophrenic identical co-twins of schizophrenics, it should describe the MMPI phenotype of those individuals with the genetic predisposition for developing schizophrenia, whether they be twins or not (cf. Rosenthal & Van Dyke, 1970), but who do not develop schizophrenia. Samples of discordant MZ pairs would be biased in favor of the less genetically loaded cases and those which were phenocopies. The co-twin profile above does not readily lend itself to a specific clinical description and information is lost by the averaging process, but the profile is clearly too uniformly elevated to be called "normal."

The mean profile of the schizophrenics in the above tabulation is very similar to that of the *A*-twins in Fig. 7.1, although based on 8 more cases. The fact points to the stability of the schizophrenia-associated profile across *A* and *B* schizophrenics.

In Chapter 6, we found that the probands in discordant pairs tended to be milder schizophrenics than those in concordant pairs. According to the MMPI, the mean profile of the probands in the discordant pairs tends to confirm the clinical evaluation in that the discordant probands scored lower on *F* and three scales of the psychotic tetrad, *Pt, Sc,* and *Ma*, and they had a profile shape that suggested an acute illness while the concordant probands' profile suggested a chronic process. The small sample sizes involved, together with intragroup variability associated with clinical state, require a cautious acceptance of these data, but the trends observed do highlight an important strategy for future studies which use clinically defined homogeneity to search for test-defined homogeneity which, in turn, can be used to *bootstraps* toward a search for, say, biochemical homogeneity. Halevy, Moos, and Solomon (1965) examined the whole blood serotonin levels of psychiatric patients; the MMPI's of the lows were very high with leading scores on scales *D, Pt,* and *Sc* (a schizophrenic-like profile) while the highs had low profiles. The authors favored an explanation in terms of a relationship between severity of psychopathology and levels of serotonin; extensions of this kind of research with attention to diagnostic specificity would be valuable. The contrasts between the 9 *A*-twins from concordant pairs and the 9 from discordant pairs are given in the following tabulation; there are no marked changes in the contrasts when the 3 tested *B*-twins who were probands in their own right are added in.

	L	F	K	1	2	3	4	5	6	7	8	9	0
MZ *A*-twins from concordant pairs	56	76	52	63	67	57	67	64	70	74	81	65	61
MZ *A*-twins from discordant pairs	52	70	50	56	75	59	69	58	72	70	75	58	64

Sullivan and Welsh (1952) have provided a method for objectifying some of the aspects of the configural analyses performed by clinicians judging patterned responses and arriving at decisions about group membership. The technique is more advanced than the one pioneered by Meehl (1950), which uses patterning of test *items*, in that its units are the scales of the MMPI; both approaches sought to derive empirical indexes or "signs" which differentiated identified criterion groups from suitable control groups. The focus here is on the *intratest* comparisons of scales rather than the traditional one of intertest comparisons with t tests for significances between group means. In brief, for each subject in the group under consideration, say, for instance, our 26 consensus MZ schizophrenics, a matrix is formed of the relative rank of each of the scale-comparison pairs within each twin's MMPI profile using the coded profiles in Chapter 4; there are 45 different scale-comparison pairs for each twin, 1 with 2, 1 with 3, etc., each yielding a plus(+) when the first is greater than the second, a minus(−) when less, and an equals sign (=) when tied, resulting in a 26 × 45 matrix of pluses, minuses, and equals for our criterion group. A similarly derived matrix, 8 × 45, is generated for the contrast group of 8 MZ co-twins who do not have a diagnosis of schizophrenia. Tests of significance (χ^2, 1 df) between the two groups were then applied to the totals of pluses and minuses for each of the scale comparisons to obtain a list of tentative signs (equals were combined with pluses in our analyses). Without cross-validation, the signs cannot be other than suggestive.

Given 45 scale comparisons, we would expect 2 to be significant at the .05 level by chance alone in each contrast between two groups. The values of χ^2 required for significance at the .05, .01, and .001 levels were 3.84 (*), 6.63 (**), and 10.83 (***). The signs which differentiated consensus schizophrenics from the nonschizophrenic identical co-twins [i.e., the tabulation on page 265] were the following.

	χ^2
4 > 3	6.91**
6 > 1	5.19*
6 > 3	10.62**
6 > 4	4.09*
8 > 3	6.27*

Since the technique of Sullivan and Welsh makes no use of absolute elevation, its strength lies in its power to discriminate *relative* elevations within a profile; the two kinds of information supplement each other when maximizing the meaning in profiles. The above findings may point to factors which contribute to the release of schizophrenia. They can be looked at in two ways—the schizophrenic had too much of the traits reflected in the scales on the left—aggressive, paranoid, and schizoid; and/or he had too little of those traits reflected on the right—largely defense mechanisms of repression and denial. The converse would then hold for the co-twins who have the same diathesis—their predisposition to schizophrenia was suppressed by the relative absence of elements on the left and/or the presence of those on the right. Support for these speculations will be sought in other contrasts subsequently. It may be well to keep in mind that scales 1, 2, and 3 have been termed the "neurotic triad" and scales 6, 7, 8, and 9 the "psychotic tetrad" (Dahlstrom *et al.*, 1972). However, these ways of looking at the profile differences tacitly assume that the MMPI scales are a valid reflection of premorbid personality. This may not be so. It could be that the schizophrenic illness had increased or decreased the personality traits in question rather than that the traits had released or suppressed the schizophrenia; only a longitudinal study of schizotypal subjects and their MMPI's would unravel the problem.

DZ Proband and Co-twin Contrasts

Of the 33 consensus DZ pairs, we were able to test 19 *A*-twins and 25 *B*-twins; none of the *B*-twins were probands, but one, DZ 10*B*, was a consensus schizophrenic. Figure 7.2 shows the mean MMPI profiles for these DZ twins. The high profile is almost a perfect match with that of the corresponding MZ twins in Fig. 7.1, objectively demonstrating that the MZ and DZ probands were drawn from the same population of schizophrenics, a demonstration depending largely

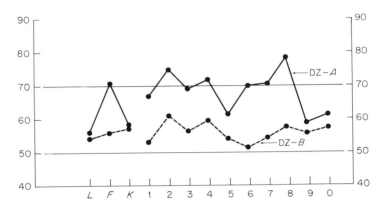

Fig. 7.2. Mean MMPI profiles of 19 DZ *A*-twins and 25 DZ *B*-twins.

on clinical judgments in other twin studies. Only the two signs expected by chance came out of the Sullivan-Welsh pattern analyses of the MZ versus DZ proband profiles. The DZ co-twin profile is remarkably within normal limits; with the omission of 10*B*, no *T* score in the profile is greater than 60. The configuration of the validity scales does not have the characteristic inverted V for psychiatric patients, but rather is like that of normals. The Sullivan-Welsh technique elicited nine signs that significantly differentiated the DZ probands from their co-twins.

	χ^2		χ^2
1 > 9	4.63*	7 > 9	5.17*
2 > 0	5.59*	7 > 0	7.75**
6 > 3	5.25*	8 > 9	9.12**
6 > 9	4.17*	8 > 0	4.28*
6 > 0	5.68*		

As might have been expected from looking at the differences between the two profiles in each of Fig. 7.1 and 7.2, the DZ co-twins were much more frequently distinguished from their schizophrenic twins with whom they shared only half their genes (on average), than were the MZ co-twins from their schizophrenic twins with whom they shared all their genes. The simplest explanation for the DZ proband/co-twin contrast from the Sullivan-Welsh configural analysis is that the probands were psychotic and the co-twins were not.

We can now return to one of the questions raised earlier about the possible origins of the personality test similarities between the MZ probands and their co-twins. If the similarity were mainly due to shared environments from the time of ovum fertilization until they went their own ways in late adolescence, we would expect the similarities between DZ probands and their co-twins to closely approximate that observed between the MZ pairs. The contrast between the two mean co-twin profiles involved shows how far wrong such an expectation would be.

	L	*F*	*K*	1	2	3	4	5	6	7	8	9	0
MZ co-twins of probands (*N*=20)	56	66	57	62	66	63	67	57	64	67	73	62	59
DZ co-twins of probands (*N*=26)	54	58	57	54	62	58	60	54	53	55	59	57	58

A standard deviation or more in the "healthy" direction separates the DZ co-twins from their MZ counterparts on Paranoia, Psychasthenia, and Schizophrenia.

However, the MMPI was not able to distinguish between the mean profiles of co-twins in the above contrast once the consensus schizophrenics had been removed, either by mean differences or by Sullivan-Welsh searching; the techniques should be tried with larger samples of phenotypically nonschizophrenic co-twins than 8 MZ's and 24 DZ's.

Sex Differences in the MMPI

It would be largely repetitive to report the results of the analyses of MMPI mean profiles by sex; the results are virtually the same as those we reported earlier (Gottesman & Shields, 1968). The findings presented above for observed differences and similarities in MZ and DZ mean profiles were paralleled when same-sex male twins were looked at separately from same-sex female twins. Across the male samples of MZ and DZ consensus schizophrenics, the Depression scale was much more elevated than across females, giving the males the appearance of more chronicity than females. Across the female samples of MZ and DZ schizophrenics, the Paranoia and Psychopathic deviate scales were relatively more elevated in their profiles than for the males, giving them the appearance of a paranoid-agitated clinical form of disorder.

The apparent effect of schizophrenia on sexual identification, as inferred from the *Mf* scale, is worth noting. In all four subsamples of females (MZ and DZ probands and their co-twins) the mean *Mf* score was about 50, i.e., they were "averagely" female in their identification regardless of the presence of schizophrenia and matched the general population of normals in this respect. The male probands, however, regardless of zygosity, had a mean score of 70 on *Mf*, two standard deviations above the mean of the normative male population. The means for the MZ and DZ male co-twins were 64 and 57, respectively, but the former would have been only 60 without consensus schizophrenics. It may be that same-sex sexual identification is disrupted for males only, when schizophrenia appears. Similar observations about the *Mf* scores of schizophrenics and their twins were made by Mosher *et al.*, (1971), but with different interpretations.

Intrapair Profile Similarity and Intrapair MMPI Scale Similarity

The degree to which the entire profiles of each pair of twins share a similarity in shape or configuration can be inferred from the rank-difference correlation coefficient (ρ) for each of the paired MZ profiles, $N = 15$, and each of the paired DZ profiles, $N = 18$ (cf. Gottesman, 1963b). The values for ρ are given in Table 7.3 and can be interpreted as if they were Pearsonian correlations; we have eight degrees of freedom for the ten clinical scales *Hs* through *Si*, which requires a correlation of 0.6 for significance at the .05 level. The ordinarily high correlations within MZ pairs, anticipated from the literature on intelligence and personality tests, are only found for those pairs where both have a consensus diagnosis of schizophrenia; even then we have two exceptions, MZ 7 and MZ 15, perhaps explainable by the fact that they were discordant for their socioclinical condition at the time of testing. The genetic isomorphism between SO and SN identical twin pairs is apparently masked at the phenotypical level by the presence of schizophrenia now or earlier in the proband. The phenomenon is even more striking in the DZ pairs who share half their genes in common, on average, and typically correlate about .5 for measures of such continuously

TABLE 7.3

Rank-difference Correlations for Paired MMPI Profiles

MZ pair	Consensus	ρ	DZ pair	Consensus	ρ
2	SS	.69	10	SS	−.32
6	S?S	.53	11	SO	−.11
7	SS	.10	14	SO	.03
13	?SS	.83	17	SO	.53
15	SS	.19	20	SO	.09
21	SS	.68	2	SN	−.37
22	SS	.59	6	SN	.04
23	S?S	.71	8	SN	.19
5	SO	-.02	9	SN	.05
9	?SO	.12	12	SN	−.23
14	SO	.39	13	SN	.29
24	SO	.41	18	SN	.44
4	SN	.10	19	SN	−.04
8	SN	.13	22	SN	.16
20	SN	.05	28	SN	.59
			30	SN	.09
			32	?SN	−.27
			33	SN	.50

distributed traits as height and IQ. None of the 18 correlations differ from zero; in fact, half the correlations are zero or negative. The one consensus concordant pair, DZ 10, was clinically discordant at the time of testing. All rhos were calculated from the Welsh codes in Chapter 4. Earlier work with the MMPI profiles of 34 pairs each of normal adolescent MZ and DZ twins resulted in more typical similarity with median rho values of .4 and .3, respectively.

When we look at the individual MMPI scales by means of the intraclass correlation coefficient, we again gather evidence suggestive of phenotypical dissimilarity for personality for the DZ pairs and a diminution of personality similarity for the MZ pairs, perhaps as both a cause and a result of schizophrenia in the probands. Table 7.4 shows the intraclass r's for the MMPI scales calculated from the same paired profiles used for Table 7.3 together with the pooled values (Vandenberg, 1967) obtained from 120 normal MZ pairs and 132 normal DZ pairs tested in three other independent studies of the MMPI profiles of twins (Gottesman, 1963b, 1965; Reznikoff & Honeyman, 1967). The small number of schizophrenic proband pairs requires a correlation of .5 for statistical significance (.05 level); all normal pair correlations of .17 and higher are significant. Recalling that 7/15 of the tested MZ schizophrenic proband pairs were discordant for schizophrenia, their intraclass correlations may be the net result of opposing forces and hence undeserving of interpretation. Yet it is disturbing to note the low r's obtained for *Pd, Pa,* and *Sc,* given the importance of those scales to a schizophrenic's profile. The DZ schizophrenic proband pairs lend themselves to explanation since all but one were discordant for schizophrenia. The r's are consistent with each other in that none are different from zero, but the wave of

TABLE 7.4

MMPI Scale Intraclass Correlations for Schizophrenic and Normal Proband Pairs

Scale	MZ		DZ	
	Sc	Normal	Sc	Normal
L	.19	.46	.36	.17
F	.14	.40	−.21	.38
K	.22	.35	.32	.20
Hs	.46	.41	−.27	.28
D	.55	.44	−.18	.14
Hy	.61	.37	−.39	.23
Pd	−.11	.48	−.10	.27
Mf	.66	.41	.04	.35
Pa	−.02	.27	−.30	.08
Pt	.50	.41	−.37	.11
Sc	.14	.44	−.35	.24
Ma	.54	.32	.15	.18
Si	.30	.45	−.13	.12

negative signs was unexpected; the 9/13 negative r's could be a chance effect, or a statistical artifact (i.e., unmet assumptions underlying correlation for discordant pairs), perhaps the effect of the schizophrenic process itself in destroying the personality trait similarity observed in normal DZ pairs. The data on profile similarity in Table 7.3 appear to be consistent with those on scale similarity in Table 7.4; the analyses do need repeating with larger samples homogeneous for their consensus status.

Clinical Comments on Individual Twin Pair and Family MMPI's

A large amount of rich clinical fare is obscured by averaging MMPI profiles and the true amount of variability existing around each mean profile is given short shrift. It was a price we were temporarily willing to pay in order to maximize the nomothetic utility of our data; now we can turn to the idiographic with no apologies. A look at selected pairs and families of our collection of profiles highlights many of the difficulties of combining personality assessment, a relatively crude art, with genetic theorizing, a relatively elegant science. A brief digression is necessary in order to bolster the thesis that test-defined taxonomic classes, i.e., MMPI code types, have validity for identifying the schizophrenic phenotype, and that such classes are a step in the process of identifying the schizophrenic genotype. Marks and Seeman (1963), Gilberstadt and Duker (1965), Sines (1966), and Meehl (1973) have all made important contributions to this problem area.

It is possible to form 45 different two-point code types and 120 different three-point code types with the ten standard MMPI scales (e.g., 27 = 72 and 278 = 872). However, nowhere near such numbers of combinations are actually generated by patients admitted to psychiatric facilities. Marks and Seeman found that only 16 code types (two and three points), strictly specified by different sets of rules, were able to classify 78% of their patient population. By relaxing the specification rules, Briggs *et al.* (1966) were able to classify 84% of the patients in a different hospital setting with the same 16 code types. Less glowing results are to be expected in further cross-validation but these examples point out that there are in fact meaningful clusters, a fraction of the number possible, which supports that idea of taxonomic specificity, more or less. Table 7.5 was constructed from data in Appendix A in Marks and Seeman (1963). It shows those nine of their sixteen code types for which a diagnosis of schizophrenia was *modal*. In addition, the table shows the percentage of each code type receiving that diagnosis, the percentage diagnosed as in a schizophrenic "spectrum," defined here as schizophrenia plus paranoid state, and schizoid or paranoid personality, and finally the total percentage receiving any psychotic diagnosis. The data are to be taken as a demonstration of relative specificity rather than as something more definitive at this stage. The relevant base rates against which to evaluate the specificity demonstrated are the following (based on 350 females

TABLE 7.5

MMPI Code Types with Diagnostic Specificity for Schizophrenia

Code type	% Schizophrenic	% Schizophrenic "spectrum"[a]	% Any psychosis[b]
278	37.5	41.7	58.4
28	50	55	70
46	45	60	55
482	62.6	66.8	70.9
83	33.3	42.8	47.6
86	54.5	68.1	68.1
89	65.2	65.2	69.5
96	60	75	85
K+	33.4	33.4	47.7

[a]Spectrum defined here as including persons diagnosed paranoid state, schizoid and paranoid personality, plus schizophrenics, of course.

[b]Other psychoses were affective; K+, 28, and 86 profiles were diagnosed organic brain syndromes at rates of 23.8%, 15%, and 13.6%, with none of the latter rates included in this column.

only): schizophrenia, 30.8%, "spectrum," 34.8%, and psychotic, 45.7%. All the code types refer to T scores of 70 or higher. All but one (K+) of the codes features either the Schizophrenia scale or the Paranoia scale, our minimal expectation for face validity. Only 10% of diagnosed schizophrenics obtained code types not found in Table 7.5.

In Chapters 4, 5, and 6, we were concerned about qualitative versus quantitative dissimilarity between pair members and about the "noise" added to research on schizophrenia by symptomatic cases; we again treat these problems here and in Chapter 8 when searching for reasons to explain the discordance for consensus diagnoses in pairs of MZ twins. The reader should think of the information in this section on MMPI profiles as supplementing that in the other chapters. Figure 7.3 shows the profiles of MZ 21 *A* and *B*, one of our consensus SS pairs; the remarkable degree of similarity can be traced mostly to the fact that they were tested at a time when both happened to be matched for their clinical state. They were tested within 1 week of each other: *A*, 4 weeks after discharge and 2 weeks before readmission, *B*, 3 months before discharge. Still, there are important differences; only *A* is an overt homosexual and this is reflected in the difference in *Mf* elevations; *B*'s score on *Pa* is enough higher than *A*'s to warrant only him being labeled paranoid schizophrenia. The psychometric discordance for Kraepelinian subtype further argues against such subtypes having different etiologies (cf. Table 6.9). The profile similarity in this 40-year-old pair cannot, in our opinion, simply be the result of both being schizophrenic; Table 7.5 shows that the odds of two random schizophrenics having the same profile are 1/81.

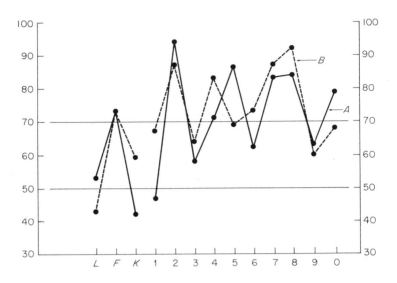

Fig. 7.3. MMPI profiles of MZ 21A and B, consensus concordant schizophrenic 40-year-old males tested in similar clinical states.

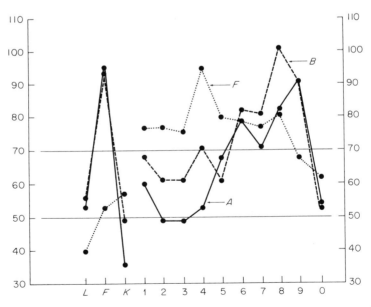

Fig. 7.4. MMPI profiles of MZ 13A and B at age 19 and their father (F) at age 40, all three similar for a schizoaffective disorder.

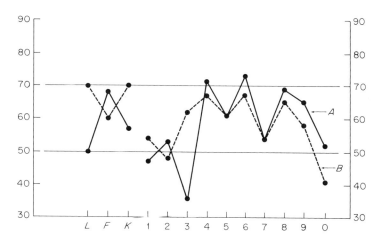

Fig. 7.5. MMPI profiles of MZ 23*A* and *B*, 26-year-old males when *A* was in remission and *B* was consensus ? schizophrenic, although never hospitalized.

Equally striking is the degree of MMPI similarity for MZ 13 *A* and *B*. We also show the profile of father tested at age 40 in Fig. 7.4 to support our speculation in Chapter 4 that a gene with a large effect for schizoaffective psychosis may be segregating in this family. Twin *A* has the less psychotic profile at the time of testing which was 6 months after psychiatric discharge, with markedly less thought disorder (i.e., *Sc*) and less anger (i.e., *Pd*). Twin *B* was also tested at age 19, but had been admitted 9 days previously to a locked ward with a relapse of her schizophrenia; her profile is an excellent match to the clinical description at the time (see p. 105). Although the father was diagnosed as a hysterical psychopath on his one inpatient admission, both his profile and his life style place him on a schizoaffective continuum near his daughters. The genetic and the environmental contributions to the twins' personalities from the father are reflected in the "triplet" profiles, but the fact that the similarities prevail despite differences in sex and age implicate genetic factors more heavily.

When the 26-year-old male twins in MZ 23 were tested, they were qualitatively dissimilar for a diagnosis of schizophrenia despite the qualitative and quantitative similarity shown in their profiles in Fig. 7.5. The proband was in remission while the co-twin, unbeknownst to us at the time, would end up as consensus ?S (hence concordant), despite never having been treated in a psychiatric facility. It leads us to suggest that variations of a 468 profile type, even when not elevated over 70, should be scrutinized as an indicator of a schizophrenic genotype.

A clear example of qualitative dissimilarity for diagnosis but with quantitative similarity for personality at one point in time and dissimilarity at another is

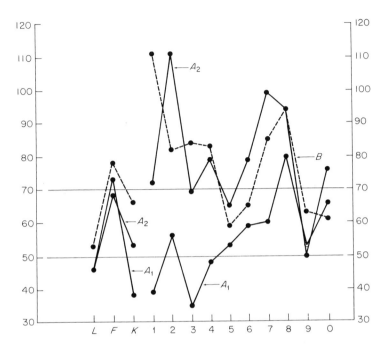

Fig. 7.6. MMPI profiles of MZ 14*A* and *B*, 20-year-old males showing qualitative dissimilarity for diagnosis, quantitative dissimilarity for personality after ECT (A₁), and quantitative similarity for personality without treatment (A₂).

provided by the profiles of MZ 14 *A* and *B* in Fig. 7.6. On consensus, the twins are SO, with the *B* twin who has never been seen in psychiatric treatment having a personality disorder (inadequate, hypochondriacal) at age 20; one consensus judge (PEM) called *B* a pseudoneurotic schizophrenic while Essen-Möller diagnosed him a schizoid personality. Twin *A*'s profile when first tested is the lowest in the figure with *D* = 56 and *Sc* = 80, while B's at the same time is the dashed line coded 18* etc. The degree of psychometric discordance seen here is due to the fact that *A* was tested the afternoon of the morning on which he received his seventh ECT in a course. It is obvious that the main effect of electroshock on *A*'s symptomatology was to markedly reduce the affective (depressive, hysteroid, and aggressive) and obsessional components but leave the schizophrenia itself intact. Twin *A* was next tested 2 months later after transfer to the Maudsley at a time when he had been taken off all psychotropic medication for a 4-day period so as to get a clearer picture of his disorder. The profile at that time is clearly a schizophrenic one coded 278* etc. and surprisingly similar to B's; *A* was again tested 6 months later, on the same hospitalization, but then on chlorpromazine plus chlordiazepoxide hydrachloride, 6 months prior to discharge to an aftercare hostel. Twin *A*'s third MMPI is closest of all to *B*'s:

				L	*F*	*K*	1	2	3	4	5	6	7	8	9	0
No. 3	MZ	14	*A*	63	68	62	82	89	89	71	63	67	83	96	55	44
No. 1	MZ	14	*B*	53	78	66	111	82	84	83	59	65	85	94	63	61

Our experience with this pair leads us to support the view (Kety, *et al.*, 1968) that at least some so-called inadequate personalities do indeed belong in the schizophrenic spectrum. The phenomenology of the pair also call to mind some of H. S. Sullivan's insights about schizophrenia. He noted (1962, p. 335) that some patients alternated between a chronic paranoid state and a chronic hypochondriacal state; in MZ 14, we may have an instance of the simultaneous occurrence of equivalent disorders in the same genotype. The *B* twin in this pair is our personal candidate for the MZ co-twin most likely to become an overt schizophrenic of those who were 0 or N on consensus.

The profiles of MZ 22 *A* and *B* provide a clear example of the problems facing users of psychological tests of personality to detect "concordance." When *A* was first tested, she was in a good clinical remission, it having been almost 3 years since discharge as an inpatient. Her MMPI profile 0'2, etc., was suspicious but would not have led us to forecast her acute decompensation and readmission just seven weeks later. The *B* twin was also seen and tested while in a good remission, it having been a little past 3 years since her last discharge as an inpatient. The two remission profiles are remarkably similar and are among our candidates for the kinds that may be a clue to the carriers of the schizophrenic genotype. The high profile in Fig. 7.7 belongs to the *A* twin 10 days after her admission to a day hospital and is clearly a schizophrenic profile. Had we been unfortunate enough to test one twin only in remission and the other only during an acute phase of her illness, we would have missed another instance of psychometric equivalence and clues to the forms taken by schizotypes. It is important to note that the mean *discharge* profiles of two of the code types in Table 7.4 as reported by Marks and Seeman (1963) were 0'2 and 0'68 belonging with the 278 and 86 types, respectively. Again, we have support for the proposition that there is some method in madness.

In the Comments section of Chapter 4 (pp. 96, 126), we expressed our reservations about the consensus schizophrenic diagnoses of two MZ probands, 9 *A* and 20 *A*, suggesting instead that both had symptomatic schizophrenias; the identical twins of such cases would not be expected to be schizophrenic without identical exogenous stressors. The MMPI codes of the *A* and *B* twins involved are set out here to show some psychometric support for our *ex cathedra* opinions.

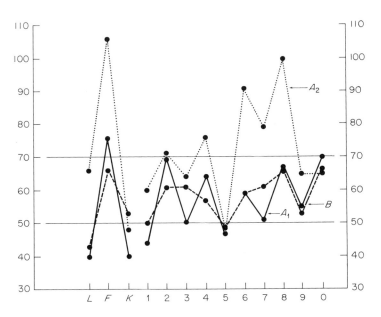

Fig. 7.7. MMPI profiles of MZ 22*A* and *B,* 28-year-old females concordant for schizo-phrenia in remission (A₁ and B) and after *A* decompensated (A₂).

Both *A* twins' profiles are more psychopathic than anything else and, although MZ 9 *A*'s is admittedly quite an exaggerated condition, it is compatible with a view of her as a symptomatic schizophrenia in a chronic alcoholic (she was probably not sober at the time of testing). Neither of the *B* twin profiles is suggestive of psychosis, schizoid personality, or even subtly reminiscent of the kinds of profiles singled out so far for schizotypy.

	A		B	
MZ 9	4*89″<u>67</u> 2′0-31/:5# (F″′-L/K:)		239 <u>67</u> 0/48*5* 1 (KL/F:)	
MZ 20	4′-<u>31</u> <u>8679</u>/25:0# (KL-/F:)		93/<u>581</u> <u>702</u> 64 (KL-/F:)	

Although Fig. 7.2 demonstrated that the DZ co-twins as a group were unusually normal on their MMPI's, the averaging process hid the differences that might exist between consensus 0 and consensus N twins. In Table 6.7, we showed the diagnoses of the nine DZ co-twins who were 0; anxiety and/or depression characterized seven of them while only one was called schizoid. The 6 MMPI's we had on this subset did not reflect such clinical homogeneity, partly because the diagnoses often referred to a transient condition or one not present at the time of testing; according to the MMPI, four, at most, of the six tested co-twins could be seen to have a noteworthy degree of schizoidia (17 *B*, 21 *B*, 14 *B*, and 25 *B*). If we add to these four the only two N co-twins whose MMPI's suggested a schizoid component, 23 *B* with a 201', etc., and 32 *B* with a 0'8, etc., we would have among the 33 DZ co-twins, three overt schizophrenics, one diagnosed schizoid personality (25 *B*), one suspected schizoid personality not detected by the MMPI or consensus (31 *B*, see Comments, p. 202), and five co-twins where a schizoid element is suggested by the MMPI. The *maximum* pairwise concordance rate for "schizoid disease" (Heston, 1970) using all this information would be 10/33 or 30%, vs an expectation of 50% on a dominant gene with complete manifestation hypothesis.

The small amount of MMPI data we did obtain on parents and nontwin sibs of the probands was tantalizing and we wish we had more. We believe the strategy of testing whole families with personality tests is to be encouraged. In one completely tested family, DZ 22, both parents, the proband, and an older same-sex sib had abnormal MMPI's with only the DZ co-twin appearing to be normal; the father was 9'; mother, 27'; proband 8<u>26</u>*, and the sib, 2'0. Except for the proband and sib code types, the others are not particularly schizophrenically oriented. In another family, DZ 18, we tested the entire family (see p. 173) which included an older and a younger same-sex sib; both parents and all the boys except the proband had MMPI's within normal limits; the co-twin's highest score, however, was Social introversion (68) and the youngest (age 24) brother's code led with a 278 (*Sc* = 63). Following up this particular family would be rewarding. Three additional fathers and one mother produced MMPI's that could be related to schizophrenia although none of these parents have been in psychiatric care: fathers, DZ 6 (7″08'), DZ 9 (2'0), and DZ 30 (6'45); mother, DZ 12 (6'<u>813</u>). In the latter cases, the testing of personality provides unique information. Since it was our practice not to do mental status examinations with parents whom we used as informants and to score the MMPI's after contact with a family was completed, we could not follow the leads to probable abnormality provided by the tests.

When we pooled the MMPI profiles of the nontwin first-degree relatives, we found a mean profile very much like that of the DZ co-twins in Fig. 7.2; the largest difference between the means of any scale was three points. We therefore combined the MMPI results of all first-degree relatives, that is, DZ co-twins

	L	*F*	*K*	1	2	3	4	5	6	7	8	9	0
Mean	53.7	55.0	57.6	54.0	60.4	56.2	58.6	54.0	53.7	54.5	56.9	54.6	55.9
SD	9.7	7.3	8.8	10.0	11.2	8.1	11.1	9.3	9.4	10.1	9.3	10.7	10.2

($N = 25$) and others ($N = 25$) to yield a mean reference profile for the first-degree relatives of schizophrenics. We shall return to the significance of these data on first-degree relatives for the mode of transmission of schizophrenia in Chapter 10.

MMPI'S OF SCHIZOPHRENICS IN REMISSION

Seldom do investigators in the field of schizophrenia research pay close attention to schizophrenics who are not actively psychotic. Remitted schizophrenics should be able to provide many important clues to the variable psychopathology of schizophrenia which can be linked to etiological issues. Are remitted schizophrenics qualitatively or "merely" quantitatively different from actively psychotic patients? Does the remitted schizophrenic give us a picture of what "schizophrenics-to-be" look like or is there seldom complete restitution? The nature of our sampling was such that we could divide our consensus schizophrenics into working, recent hospital discharge, and non-working (cf. Chapter Four) and then use the MMPI as a dependent variable. The mean profiles of these three groups, disregarding proband and zygosity statuses, is given in Fig. 7.8. The similarity of the recent hospital discharge profile to the nonworking profile led us to combine them into one "actively schizophrenic" profile ($N = 24$) to contrast with the "schizophrenia in remission" profile ($N = 22$). Both the between-profile differences and the similarities are impressive. The lowest profile reinforces the earlier theme that differences in clinical condition are important to MMPI interpretation; the remission profile has only one scale as high as $T = 70$ and that is Sc. Nevertheless, the overall

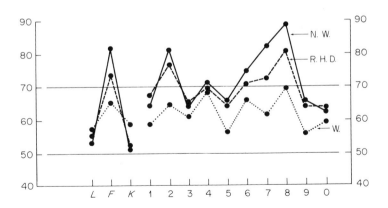

Fig. 7.8. Mean MMPI profiles of three subgroups of consensus schizophrenics: nonworking (NW), recent hospital discharge (RHD), and working (W).

"sawtooth" shape of the profile is far from normal and is malignantly coded as 8'46273-. There appears to be only a slight quantitative difference between the recent hospital discharge profile and the non-working one; we are tempted to posit that a slow healing process is depicted in Fig. 7.8 with the recently discharged patients "moving" toward the remission profile.

The Sullivan-Welsh technique was applied to determine the signs that distinguished between the active and the remission profiles. Seven signs characterized the contrast between the intraprofile configurations of the active versus the remission profiles:

	χ^2
2 > 4	5.54*
7 > 1	6.96**
7 > 3	8.51**
7 > 4	7.06**
8 > 3	3.93*
8 > 4	4.32*
8 > 6	4.68*

Recalling the specificity of the 278 code type for a diagnosis of schizophrenia, the signs which distinguish the active from the remitted schizophrenic profiles can be seen as reflecting a difference on a dimension of schizophrenicity. It is not just belaboring the obvious to say that the active schizophrenics are more schizophrenic than the remitted ones; the implication of the signs is that the difference is primarily a quantitative one. Equally attractive as an interpretation of the sign differences between active and remitted schizophrenics is that the former have been potentiated by the effects of anxiety and the latter have had anxiety removed, thus permitting remission (cf. Mednick, 1958; Meehl, 1972). We made our inferences about the role of anxiety in the MMPI from the specificity of *Pt* as an indicator of anxiety and of the *D Pt* code type as an indicator of an anxiety reaction (e.g., Gilberstadt & Duker, 1965). The role of anxiety as a potentiator of the schizophrenic genotype in some predisposed persons, if confirmed, would go a long way toward explaining the beneficial effects of some anti-psychotic medication and some anxiety reducing effects of psychotherapy for some schizophrenics.

Although we did not have enough data to obtain stable results, we would have liked to have seen an analysis of the MMPI's of the co-twins of active versus remitted schizophrenics to see if there were meaningful differences, especially in MZ pairs. Such an analysis would cast light on the question of whether this difference was an instance of etiological heterogeneity for "process" versus "reactive" cases. Based both on our data in Chapter 6 and that in this section, we are inclined to believe that such distinctions are best seen as ends of a continuum of degree of schizophrenicity.

EQUIVALENCES BETWEEN ESSEN-MÖLLER DIAGNOSES
AND MMPI PROFILES

There were many other ways in which we might have analyzed our MMPI results with the personality test data being used sometimes as an independent variable and sometimes as a dependent variable. Considerations of space allocation force us to defer all but one more unique use of these data in this book. By using Essen-Möller's diagnoses from Chapter 5 as independent variables and the mean MMPI profiles of his classes as dependent variables, we can gain some insight to the possible equivalences of the two kinds of data, we can initiate the process of construct validity for such clinically useful constructs as schizophreniform psychosis and schizoid personality, and, finally, get a feeling for the validity of our combining his categories 1–4 (see p. 219) so as to obtain a pooled concordance rate for "schizophrenia."

The MMPI code types for the mean profiles associated with each of Essen-Möller's diagnostic categories are given in Table 7.6 together with the total number of twins assigned each diagnosis, regardless of zygosity. Again, a certain amount of information is sacrificed by using mean profiles, but the reader can perform his own analyses by using the information in Appendix D in conjunction with the MMPI data in Chapter 4. The code types (and their plotted profiles) clearly show a great deal of communality for the first four categories in Table 7.6. Closest of all to each other are the true schizophrenics and the schizoid personalities; in fact, except for lower elevations on F and Sc, the schizoid personalities as diagnosed from a Scandinavian point of view appear to overlap greatly with individuals called overt schizophrenics in other countries. If this idea is supported by actual cross-national diagnoses of the same patients, it could help explain some of the apparent discrepancies between concordance rates in Scandinavian MZ twins compared to other areas (see p. 310). Although the schizophreniform psychosis profile is only based on ten cases, it does not appear to be other than what its label says; the MMPI phenotype for it is a form

TABLE 7.6

MMPI Code Types for Essen-Möller's Diagnostic Categories

	Diagnosis	N	Code type of mean profile	
1	True schizophrenia	21	8"274'650 139-	F'-LK/
2	Schizophreniform psychosis	10	8'213 67 40-95/	F-KL/
3	Possible schizophrenia	6	8'67 9 45-20 13/	F'-L/K:
4	Schizoid personality	10	2847' 316 50-9/	FK-L/
5	Other diagnosis	15	482 3 76-901 5/	K-F L/
6	Normal	16	24 930 85 71/6:	KFL/

of the schizophrenic profile type. The profile for possible schizophrenics looks more like a schizophrenic one than anything else, while the elevation and shape of the remaining profiles are clearly not suggestive of schizophrenia with the normals in fact looking like normals to the MMPI, and the "other diagnosis" resembling a mild character disorder profile.

RESULTS OF MMPI EXPERIMENTAL SCALES

Rosen's (1962) method of scale construction contrasted his criterion diagnostic groups with a mixed group of *psychiatric* inpatients instead of with normals. This gave us high hopes for his *Pz* (Paranoid schizophrenia) scale being more powerful than the *Sc* scale for detecting schizophrenics. He reported that the mean *Pz+K* score (standard or *z*) for his cross-validation group of schizophrenics was 60; it was 50 of course for the General Abnormal sample, and 46 for his criterion diagnostic groups with anxiety reaction and depressive reaction. Using a cutting score of 60 to examine the differentiating power of the *Pz+K* scale, Rosen reported that 53% of schizophrenics equalled or exceeded that score whereas only 39% of them exceeded the same score on *Sc+K*. Although Rosen only worked with male United States veterans, we used his norms for our samples of males and females; we found no differences between the means for males and females on any of his five scales. For our 46 MMPI-tested consensus schizophrenic twins, we too found 54% to equal or exceed a score of 60 on *Pz+K*; however, 85% of them exceeded that score on the *Sc+K* scale. Our results with the *Sc* scale may have been the result of more stringent criteria for a diagnosis of schizophrenia. In Table 7.7, we show the mean scores on Rosen's five experimental scales for several of our subgroups and for two of his in order to add to the construct validity data for the scales. Rosen did not report any data from giving the scales to normals; we show that both the group of twins

TABLE 7.7

Mean Rosen Scale Scores

Group	N	Pz+K	Ar	Dr	Sm	Cr
Rosen's schizophrenics	51	60	48	50	48	46
Our MZ consensus schizophrenics	26	64	43	50	52	47
Working schizophrenics	22	61	44	54	57	53
Nonworking schizophrenics	14	71	45	48	48	42
First-degree relatives	50	49	45	56	56	57
Essen-Möller normal twins	16	49	44	56	56	57
Rosen's anxiety reactions	83	46	57	52	52	51

diagnosed as normal by Essen-Möller and the group of first-degree relatives of schizophrenics have scores close to that of the General Abnormal sample. Such data suggest that the Dr (depressive reaction), Sm (somatization reaction), and Cr (conversion reaction) scales are quite weak, but the data support the idea that schizophrenia has a great deal of specificity among the groups of psychiatric conditions in the table.

The $Pz+K$ mean score for those eight MZ co-twins shown on p. 265 was only 53 contrasted with a mean of 64 for MZ consensus schizophrenics; the T scores on $Sc+K$ for these two groups were 63 and 78, respectively. The 14 nonworking schizophrenics in Fig. 7.8 did, appropriately, obtain the highest score on Pz, 71, but they also obtained the highest score on Sc of 89. For our purposes, we will have to search further for a simple, unitary indicator of the schizotype.

OST IN THE MAUDSLEY TWIN STUDY

Our experiences in administering, scoring, and interpreting the Goldstein-Scheerer Test of Concept Formation as a mode of formally and objectively assessing some aspects of schizophrenia-related thought disorder left us with more questions than answers. The OST was an object lesson in a number of senses, but most importantly it forced us to confront the realities of spreading ourselves too thin for a project requiring a major effort rather than an adjunct one and specific expertise beyond our resources. We gained an appreciation of the multifaceted amorphousness that could be subsumed under the term *thought disorder* and of how many facets were *not* tapped by the OST. In addition to the kinds of invalidators to interpreting OST scores mentioned at the beginning of this chapter, a number of others involving the effects of drugs and the effects of process versus reactive are now known; definitive reviews are available in the papers by Payne (1966, 1970).

We had assumed that thought disorder or the loosening of association (Bleuler, 1911) was central to the diagnosis of schizophrenia and that the detection of more subtle forms such as overinclusiveness (N. Cameron, 1938) would aid in the precision of labeling relatives of probands as "affected." No tests are needed for the more gross aspects of thought disorder such as delusions, hallucinations, neologisms, fragmentation, or clang associations—they usually are part of the clinical record or easily elicited on interview. More subtle disorders called variously cognitive slippage, underinclusiveness, concrete, constricted, amorphous, wooly, impoverished vagueness, and interpenetration, overlap little, if any, with overinclusiveness and are not usually scored (e.g., Lovibond, 1954) from the OST protocol. Also, Watson (1967) reported that overinclusiveness was not a unitary phenomenon and could not be explained in terms of stimulus generalization.

The groundwork has been laid for reporting our largely negative findings from using the Lovibond scores for overinclusiveness on the OST protocols of 83 twins and 16 biological parents; we were also fortunate to have Loren Mosher rescore all of the protocols of families where at least one parent was tested using the Singer-Wild orientation. The concurrent validity (and hence construct validity) of the OST was first undermined by our observation that only 53% of our consensus schizophrenic A-twins were abnormally overinclusive, using a criterion of 7 or more Lovibond points; if overinclusiveness were the archetypal characteristic of schizophrenics regardless of clinical condition (i.e., remission or acute or chronic), we should have found all A-twins to be abnormal. Inspection of the protocols revealed that it was too easy to avoid accumulating points for this aspect of thought disorder by simply saying, "I don't know" or by giving a very concrete or constricted answer to a test item. Other consensus schizophrenics obtained normal scores apparently because they were in a good remission, often with the help of phenothiazine drugs. We found no important difference in rates of abnormality on the OST in MZ B-twins compared to DZ B-twins—23% versus 17%. However, in line with the results of many other investigators, we found that 44% of the 16 tested parents had abnormal scores.

We looked for a relationship between high OST scores and scores over 70 on MMPI scales F, Pt, Sc, and Pz and found none of the χ^2 to be significant; when we grouped according to the profile types modal for schizophrenia (Table 7.5) we did find a significant relationship with high OST scores. The demand characteristics of the two tests are quite different from each other and permit the MMPI to be more powerful as a detector of abnormality given a modicum of cooperation from the subject; the MMPI asks for self-description in the format "Are you now or were you ever X?" whereas the OST demands that right now you demonstrate your ability to perceive and conceptualize optimally, without penalizing you for silence or withdrawal to a concrete answer.

At the clinical level, we were impressed with a number of anomalies. Contrary to the literature, many of our most chronic or process schizophrenics obtained very abnormal scores instead of being concrete and thus did not resemble brain injured patients (DZ 7 A, 34; MZ 16 A, 26; DZ 2 A, 27; MZ 17 A, 19; and MZ 17 B, 23). MZ 9 A was tested while intoxicated and obtained a score of 24. Within the range of intellect of our subjects, we did not note any relationship with ability to perform the sortings accurately; in fact, we were surprised to find that two "normal" fathers with high professional status obtained high scores of 13 (DZ 9) and 7 (DZ 20); the proband son of the former only obtained a score of 3 while being markedly deviant on the MMPI (827*)—he appears to be a dramatic example of the power of 30 mg a day of trifluoperazine to rectify thought disorder. A few of the high OST scores obtained could be accounted for in terms of concurrent anxiety as determined from both the interview and the MMPI; a few others seemed best accounted for by the hypothesized ubiquity of allusive thinking (McConaghy & Clancy, 1968).

In sum, the assessment of overinclusive thinking with the OST did not lead to the detection of subtle signs of schizotypy in the relatives of our probands. In fact, our results suggested that, in general, the OST was a poor predictor of a consensus diagnosis of schizophrenia. In future efforts to measure the presence of thought disorder in probands' relatives, we would extend the search to more facets (cf. Fish, 1967; Broen, 1968) than overinclusiveness.

SUMMARY

We described the hopes of those who advocate a psychometric approach in family studies of mental abnormality, with special reference to the MMPI and the Object Sorting Test of thought disorder in schizophrenia.

The early promise of the OST as a means of identifying gene carriers or rating "schizophrenogenicness" was not confirmed. Unfortunately the OST proved of little value in our study since at the time of testing barely half of our consensus schizophrenics were abnormal, using the original criterion of overinclusive thinking. OST score was not associated with scores on any of the most likely MMPI scales. Like other investigators, we found nearly half the parents of schizophrenics to be high scorers, but these were not always the parents of schizophrenics with overinclusive thought disorder. Only some 20% of co-twins were high scorers, and there was no suggestion of a significant MZ : DZ difference. In our hands the test did not predict a consensus diagnosis of schizophrenia, let alone lead to the detection of subtle signs of schizotypy.

With the MMPI we explored new territory. The mean profile of consensus diagnosed schizophrenics, whether probands from MZ or DZ pairs, or secondary cases, was clearly schizophrenic in shape and elevation, thus justifying the use of the test in this sample. The schizophrenics were, however, tested at varying stages of illness or remission, making intrapair comparisons difficult to interpret.

The mean profile of MZ co-twins was similar in shape to that of the probands and was highest on the *Sc* scale. It was very much more elevated than that of DZ co-twins, which was not notably schizophreniclike. The MZ : DZ difference was accounted for almost entirely by the consensus diagnosed schizophrenics. The 8 nonschizophrenic MZ co-twins did not as a group have a schizophrenic or schizoid looking profile, nor one which could be distinguished from that of the DZ co-twins. Thus, in this small sample the averaged MMPI profile failed to identify a group with a genetic predisposition to schizophrenia.

The profiles of the schizophrenic and nonschizophrenic MZ twins were compared, using the Sullivan and Welsh (1952) method of configural analysis. Scales associated with schizoid, paranoid, and aggressive traits were higher than those reflecting neurotic mechanisms of repression and denial significantly more often in the schizophrenics than in the nonschizophrenics. The finding may give a lead as to which traits favor or hinder the development of schizophrenia in predisposed persons. However, as in all tests of schizophrenics, it is impossible to know how far findings reflect premorbid characteristics or the effects of the illness. A comparison of schizophrenic and nonschizophrenic DZ twins elicited a larger number of significant, distinguishing configural signs, which may reflect no more than that the probands were psychotic and the co-twins not.

The general findings held when the sexes were analysed separately. The elevation of the *Mf* scale suggested that schizophrenia may have an unhealthy effect on sexual identification in males only.

When the rank order of scores on the ten clinical MMPI scales was correlated between twins in pairs where both were tested, significant correlations were found in concordant MZ pairs only. Schizophrenia in one twin appears to have masked many resemblances in personality found in nonpsychotic twin pairs. In the majority of concordant MZ pairs the correlations may be considered as reflecting the personality and clinical resemblances noted in Chapter Six.

Intrapair correlations calculated for the separate scales tended to be non-significantly positive in MZ pairs. For *Sc* r was only .14. The mixture of concordant and discordant pairs may have distorted the findings, leading to lower correlations than those reported in normal twins. In DZ pairs, all but one discordant for schizophrenia, the correlations were frequently negative. Among possible explanations, the destructive effects of the schizophrenia on personality resemblance is in our view the most likely.

The MMPI clinical scales were developed by comparing different diagnostic groups with normals. The experimental scales of Rosen compared diagnostic groups with a mixed group of psychiatric inpatients. In our sample the Rosen *Pz* (paranoid schizophrenia) scale was less powerful than the *Sc* scale as an indicator of schizophrenia or its predisposition.

So far, the findings might suggest that, beyond confirming some of the clinical diagnostic findings, the MMPI has been of no value in providing a quantitative dimension for genetic analysis, or in identifying high risk genotypes. However, our averaging of MMPI profiles has obscured information of interest.

The profiles of 5 MZ pairs were presented to illustrate or supplement points made in this and other chapters. These include

1. Basic similarities and differences, revealed when schizophrenic twins are tested in a similar stage of their illnesses
2. The effect of temporary variations in clinical state upon profile elevation, including two pairs in which one twin was tested more than once
3. A previously hospitalized father whose profile resembled that of his concordant twin daughters
4. Nonelevated profiles as possible indicators of a schizophrenic genotype
5. A pair showing qualitative dissimilarity in psychiatric diagnosis (inadequate hypochondriacal personality vs. schizo affective psychosis) combined with close similarity in personality measures

Psychometric evidence is presented in support of the opinion that the schizophrenias of two consensus probands developed on a nongenetic organic basis (symptomatic schizophrenia).

Among nonschizophrenic DZ co-twins there were at most 6 whose MMPI's suggested a schizoid personality component. Our information from parents and sibs indicated that the psychological testing of whole families is to be encouraged. The pooled MMPI profiles of these other first-degree relatives resembled that of the DZ co-twins. The means and S.D.'s of the combined data (N = 50) were presented, and they were within normal limits.

Advantage was taken of the presence in our sample of 22 schizophrenics who had been out of hospital for at least six months and were working. Their mean MMPI profile was compared with that of the 24 other more "active" schizophrenics tested. Differences were marked but of a quantitative nature. From differences on the Psychasthenia scale, inferences were drawn about the role of anxiety in determining remission and relapse.

In diagnosing the histories for Chapter 5 Essen-Möller distinguished between true schizophrenia, schizophreniform psychosis, probable nonschizophrenic psychosis which could possibly be schizophrenia, and schizoid personality (whether suffering from a psychosis other than true schizophrenia or not). While the MMPI's of all four groups had much in common differentiating them from Other diagnosis and Normal, the closest to each other in code type were the true schizophrenics and the schizoid personalities. These are the two which Essen-Möller himself believes to be similar genotypically.

CONCLUSIONS

Selective factors in our MMPI-tested subsample were not great enough to invalidate the finding that the MZ co-twins of schizophrenics are as a group more schizophrenic in their objective test scores than the DZ co-twins. This predicted support for the psychiatric diagnostic findings of Chapter 5 argues against the hypothesis that the latter were seriously contaminated by what was included in the summaries on which the consensus diagnosis was based.

We were less successful in our hopes that the MMPI would identify virtually all MZ co-twins and a high proportion of DZ co-twins as persons with a high risk of developing schizophrenia. Our use of the test here has been more that of hypothesis building than hypothesis testing. Ideas about the specificity for schizophrenia or the schizoid of code types such as 4-6-8, or 2-7-8, or 0-2 require further testing. If such code types were found with equal frequency in the close relatives of patients with affective disorders, or if prospective studies of these code types in the normal population revealed no increased risk of schizophrenic-like disorder, our tentative hypothesis would be refuted. Our ideas about the importance of anxiety for remission and relapse in persons predisposed to schizophrenia will likewise require confirmation, but are consistent with some observations in the physiological and pharmacological fields.

Our observations point to two important limitations in the psychometric search for stable underlying constitutional variables in a disorder like schizophrenia. On the one hand, correlations between proband and relative (e.g., on the *Sc* scale) may be obscured by intra-individual fluctuations; not unexpectedly, findings highlighted the effects of schizophrenia itself and the clinical state of the subject at time of testing, and similar considerations may apply to other kinds of tests and to other disorders (cf. Coppen, Cowie, & Slater, 1965; Shields, 1972). On the other hand, when abnormalities are found in relatives they may be less pertinent than expected, since features in which schizophrenics tend to be abnormal may have little or no relevance to schizophrenia when they occur in other groups (e.g., "allusive" thinking in schizophrenia and university students, McConaghy and Clancy, 1968).

Perhaps we should not be surprised at our comparative lack of success, when we recall that in Huntington's chorea extensive search has not yet succeeded in identifying gene carriers *in choreic families.* And in epilepsy, studies of the EEG of relatives do not tell the whole story of the genetic predisposition; although specific abnormalities are found, it is not through persons with such abnormalities in particular that epilepsy is transmitted, and the EEG is of only limited value in genetic counseling (Harvald, 1954). Whereas one may be sufficiently encouraged to continue examining relatives of schizophrenics by means of whatever tests seem most promising, failure to identify the predisposition would not be surprising. There may be no such monolithic stereotype—at least in terms of behavioral traits. But identification of factors that contribute to the predisposition to schizophrenia would be progress indeed, as much in understanding its pathogenesis as in formal genetic analysis.

8
ENVIRONMENTAL FACTORS

If the genes are necessary but not sufficient for the development of schizo-phrenia, it must also be freely acknowledged that the environment is necessary but not sufficient for the development of schizophrenia. Such obvious state-ments remain trite and trivial clichés unless they are coupled with attempts at specification and quantification. Although the design of our twin study focused primarily on genetic aspects of schizophrenia, this chapter examines the nature of environmental factors that contributed to the concordance or discordance for schizophrenia in our twin pairs. The fact that the concordance rate for schizo-phrenia is very far from 100% in many studies is probably the best single piece of evidence there is pointing to the general importance of environmental fac-tors—as yet unknown.

Many suggestions have been made in the literature concerning the variables responsible for the predisposition to, or precipitation of, the disorder. We were unable to make any broad generalizations; no one or two types of environmental input such as a chaotic or rigid family, or marital schism or marital skew, appeared to be predominant let alone universal. A few families such as MZ 7, DZ 7, and DZ 27 typify those reported in the psychodynamically oriented litera-ture. In another family, MZ 4, a deviant mother treated only the future schizophrenic in an irrational manner, but such instances were exceptional. In other pairs such as MZ 14, nongenetic but biological determinants—a large birth weight difference and forceps delivery—distinguished the proband from his schizotypic but so-far compensated co-twin. But it is easy to point to contrary observations. MZ 2B was 4 pounds lighter at birth and not expected to survive; he was ridiculed by his father for a speech impediment; yet both twins became schizophrenic with the more advantaged A-twin spending 12 years *longer* in hospital than his brother. The influence of the psychosis in one twin in precipitating schizophrenia in the other could be noted in a few pairs (MZ 17, 19, DZ 4). In the case histories of Chapter 4 we commented on additional experiential contributions to pathogenesis.

COMMON BACKGROUND VARIABLES
IN CONCORDANT PAIRS

The more important stressors are for the development of psychopathology, the more apparent should be the differences between the family and other shared features for concordant pairs contrasted with the discordant. Kringlen (1967) reported a trend for concordant pairs to come from a lower social class and to have experienced economic hardships in childhood. They also had less social contact with other children and a closer twin bond; the latter facts probably reflect a more schizoid premorbid personality as much as the lack of opportunity for socialization. Furthermore, the trend was not significant statistically, despite the availability of 21 concordant and 34 discordant MZ pairs. We were unable to confirm his findings regarding social class, but we had only 6 MZ pairs from social classes 4 and 5 combined. We did note a tendency for greater concordance in premorbidly schizoid twins (p. 251).

In an attempt to examine possible stressors more thoroughly, we analyzed the frequency of minority group membership, illegitimacy and adoption, broken homes (both institutional rearing and parental loss), and the presence of psychiatrically abnormal relatives in the home of rearing. None of these variables, singly or in combination, were more frequent in concordant pairs, including concordance at the level of "any abnormality in the co-twin." In the combined sample of 57 pairs there was only a slight trend for concordant schizophrenic pairs to have been exposed to more stressors than those in which the second twin had another diagnosis; the same was true for the latter pairs compared with the completely discordant ones. We would not like to defend conclusions based on this analysis of our small sample. We also looked at birth order, maternal age at twin birth, and season of birth—we were not surprised at the absence of positive findings. Fischer's (1972) findings and conclusions on her Danish sample were essentially the same as ours with respect to "environment."

DIFFERENTIAL PREMORBID HISTORIES
IN DISCORDANT PAIRS

Investigators have usually set out a detailed table showing which twin or in how many pairs the affected or sicker was the first born, more dominant, more fiery, etc. All have to some extent come up against the difficulty of too few cases per cell for stable results, minuscule absolute differences, uncertain retrospective information, and missing data. The low yield from such efforts has convinced us to take a more clinical approach. However, the heuristic value of the approach leads us to comment on some of the variables most commonly analyzed by the other twin researchers: birth order, birth weight, global assessment of who was physically stronger in early years, premorbid intelligence, and intrapair dominance. Such variables and others can be obtained from the case histories in Chapter 4 by those wishing to perform additional analyses on our twins.

Particular attention was paid to the discordant MZ pairs (consensus diagnosis), omitting MZ 9 and MZ 20 where exogenous factors in adulthood were crucial to their breakdowns, and to those pairs differing by one full point in mean global severity of psychopathology rating (p. 229). MZ 21 defies easy categorization and has been omitted; although the *B*-twin was the more severe (6 vs 5), *A* was hospitalized 13 years before *B*.

Early Physical Differences

As in the other studies birth order was irrelevant to which twin became schizophrenic. Much attention has been focused on birth weight and perinatal hazards. If we look at the published data from the twin studies with representative sampling by Slater (1953), Tienari (1963), Inouye (1963), and Kringlen (1967), none of them found lower birth weight to distinguish the future schizophrenic or the more severe of a concordant pair; totaled, the ratio was 36 : 37, lighter : heavier. Our own findings for nine discordant pairs plus five differing by one point in severity bring the totals to 41 : 41. DZ pairs also fail to show a trend.

The data above include pairs where both twins were of normal birth weight and several in which the difference was small. If birth weight were of major importance for developing schizophrenia, we would expect concordant pairs to contain two individuals under 5 pounds and for within-pair differences to be larger in discordant than in concordant pairs. Neither of these expectations was confirmed in our small subsets.

Although there was a tendency for the twin with greater birth complications to be more at risk for schizophrenia in other studies, our evidence, when available, was inconclusive.

When we looked at who was stronger early in development or who had more childhood illness, findings were similar to those for birth weight with which these variables appear to be correlated.

Intelligence and Achievement

Inconsistency marks the findings in this area too. Luxenburger was inclined to think that the brighter of a pair was more at risk because he encountered more challenges and therefore more exposure to failure. Other investigators have suggested the opposite, that the less competent is more at risk. These opposing views might be resolved if the relationship between ability and risk of decompensation were curvilinear with too much *or* too little intellectual competence increasing the risk but for different reasons. Slater found the schizophrenic MZ twin to be more intelligent, but Tienari and Kringlen found the reverse. Only 6 of our 14 pairs under consideration differed in ability; in only three was the future schizophrenic the brighter. For our three pairs with the biggest discrepancy in achievement, MZ 2, 11, and 14, the brighter had the worse outcome in the first and last. In the MZ Genain quadruplets (Rosenthal, 1963) Hester, the one with the consistently worst school achievement and IQ, had the most severe schizophrenia, but Myra, with the best outcome, did not have the most intellectual competence among the four girls. Differences in ability in MZ pairs do not contribute to understanding discordance; similar conclusions can be reached for differences between discordant DZ pairs as well as for discordant nontwin siblings (Lane and Albee, 1968).

Dominance-Submissiveness

Most investigators of schizophrenic twins agree on the importance of childhood dominance, however defined, in accounting for discordance in identical twins. Table 8.1 shows that in the five studies giving relevant information for

TABLE 8.1

Submissiveness in Childhood and Schizophrenia in Discordant Twin Pairs

Investigator	More submissive twin is schizophrenic	
	MZ	DZ
Slater, 1953[a]	7/9	22/30
Inouye, 1963[a]	28/31	—
Tienari, 1963	14/15	—
Kringlen, 1967	23/29	16/20
Present study[a]	7/9	9/12
Total number of pairs:	79/93 (85%)	47/62 (76%)

[a]Includes pairs differing in severity as well as discordant pairs.

this variable, the future schizophrenic or more severely affected was the more submissive in 85% of 93 MZ pairs; in three studies with information on DZ pairs, it was true for 76% of 62 pairs. However, the observations about submissiveness are not specific to schizophrenia since they have also been made for twins with neurotic and affective disorders (Slater, 1953; Tienari, 1963). Furthermore, interpretations of the best meaning of such data are confounded by the fact that normal twin pairs or any dyadic relationship require such a division of roles for maximizing interpersonal effectiveness. Two individuals cannot go through a narrow door at the same time. Chance minor differences in conduct of the twins are often seized upon by parents and elaborated into "temperamental" differences, thereby misleading us about the degree of individuation.

Reliable ratings of the concept are difficult because of retrospective falsification (informants look for faults in the schizophrenic premorbidly and find them) and confusing cause with effect (submissiveness after the onset results from the disorder). In addition, unless the relatively independent traits subsumed under dominance, such as physical strength, psychic dominance, and initiative as a spokesman (Tienari, 1966), are distinguished, informants can mislead the investigator about the consistency of the behavior across various situations.

There is the further question of whether the relativity of intrapair dominance is a valid indicator of dominance in more absolute terms, e.g., both twins may have a low level of dominance compared to the general population, but one is still "more dominant" within the dyad. In our cases within-pair submissiveness was usually associated with lack of initiative in general.

Such criticisms of the construct as have been made do not permit it to be banished as irrelevant to which twin becomes schizophrenic. The division of roles may result in a relative failure of socialization (Rosenthal, 1963) in the more submissive twin and thus augment any existing predisposition to schizophrenia; it may be that the submissive one is less able to emancipate himself from parental emotional dependence and involvement.

Factors in Combination

So far we have considered birth weight, physical strength, intelligence, and dominance-submissiveness separately. In some pairs these variables were not independent; MZ 4 and 11 showed the future schizophrenic to be lower in birth weight, weaker, poorer in school, and the less dominant. These two pairs were the only ones to have the pattern reported in many of the selected sample of discordants studied by Pollin and Stabenau (1968). However it is important to note (Shields, 1968) that in a cohort of all male twins studied by Tienari (1966) there was no tendency for a twin who was simultaneously lighter, more submissive, and duller than his twin to be significantly greater at risk for mental illness of *any* kind.

A fuller understanding of the role of such factors, singly and in combination, will have to await a study that looks at them in pairs who are concordant and discordant for schizophrenia, and equal as well as different for the traits. The pairs will have to be distinguished with respect to the absolute differential and their position in the entire range as well (cf. Scarr, 1969).

Laterality

Slater (1953) noted that in three out of five MZ pairs discordant for schizophrenia one twin was left-handed (not necessarily the schizophrenic) and the other right-handed. This compared with one out of six concordant pairs where there was information. The suggested difference was confirmed in our series where we have information as to laterality in all but one of our MZ pairs. Two out of ten concordant pairs differed, a rate similar to that found in normal MZ twins (Shields, 1962); but eight of our eleven discordant pairs differed in laterality. There was no evidence of greater perinatal hazards in the opposite-handed pairs. The most plausible explanation of the tendency, if confirmed, seems to be along lines suggested by Satz *et al.* (1969). According to their hypothesis, opposite-handed twins are heterozygous for alleles determining cerebral dominance, such heterozygotes being susceptible to incomplete lateralization. If schizophrenia develops as the response of an abnormal CNS to stress, it might be experienced differently by MZ twins who differ in cerebral dominance and could then account for the discordance in schizophrenia. Although the pooled data on laterality are statistically significant, we may be dealing with nothing more than an interesting speculation.

REASONS FOR DISCORDANCE IN MONOZYGOTIC PAIRS

Having looked at given premorbid variables in the histories of discordant and partially discordant pairs, we can now approach the problem of environmental influence by starting from discordant pairs. By taking into account what we know about both twins, thus maximizing the information that can enlighten us on reasons for discordance, we are operating in the context of discovery rather than justification, of hypothesis formation rather than hypothesis testing. We pointed out in a previous review (Gottesman & Shields, 1966b) that different kinds of reasons, singly or in combination, might account for the fact that a given MZ pair was technically discordant for schizophrenia.

The first of these reasons was that the proband may have been misdiagnosed or have had a symptomatic schizophrenia. Two of the pairs, MZ 10 and 12, were accounted for in this way by the consensus and removed from the final sample. Of the remaining pairs, Slater diagnosed MZ 9*A* as having had an alcoholic hallucinosis and MZ 20*A* as having had a symptomatic schizophrenia resulting from thyroid disease. Most of the other judges were also impressed by the etiological role of alcohol and thyroid disease in these pairs. Neither of their co-twins was suspected of being schizoid. Given probands as schizophrenic, these conspicuous exogenous factors override other considerations and explain the discordances observed. Of course we can always ask why MZ 9*A* became addicted to alcohol in the first place while her sister did not. The reasons adduced could only have been very remotely contributory to schizophrenia itself compared to the toxic effects of alcohol. In MZ 20*A* the etiology may be more complex, but with the effects of thyroidectomy being the major contributor to variance. In MZ 3*A* and 5*A*, malnutritional brain damage and barbiturate addiction, respectively, appear to have been organic contributors, but symptomatic schizophrenias here are less easily ruled in.

Discordance can also be understood in some pairs as a quantitative difference on a dimension of schizophrenicity. Our previously (Gottesman & Shields, 1966a) clearest example, MZ 23, is now consensus concordant as S and ?S. MZ 14 and 16 can largely be accounted for by the *B*-twin being high on such a dimension. MZ 14*B* was personality disordered with marked inadequacy and hypochondriasis and had a schizophrenic-like MMPI profile. The difference in degree between 14*A* and *B* may be explained by the perinatal factors mentioned earlier. Whereas MZ 16*B* was a paranoid personality, his brother was a compensated pseudoneurotic schizophrenic; the quantitative difference here appears traceable to family psychodynamics.

A third kind of reason for discordance stems from pairs where the co-twin has not used up enough of the risk period to be out of the woods. At age 20, MZ 14*B* is by far the youngest discordant identical twin we have.

Five MZ pairs remain for whom the above reasons cannot be invoked. MZ 18 is among the most difficult to explain, but there may be similarities to MZ 14 (p. 121); we have been unable to identify the precipitant of her acute psychosis at age 30. The other four probands had relatively mild schizophrenias and contributed to our finding a strong positive association between severity in the proband and concordance in the co-twin. In MZ 8*A* it seems that conflict over sexuality precipitated a catatonic excitement which left him open to further decompensation. Rejection in a love affair could have initiated the paranoid illness in MZ 24*A*; she was the more contentious and had poorer family relationships than her sister. We have already suggested that in MZ 4*A* and 11*A* it may have been the accumulation of many factors, including submissiveness and distorted parental ties.

Although we have focused on what we saw as the main contributors to discordance, the difference in any one pair will of course be multiply determined. We find ourselves in the same position as any clinician who believes he understands a case. Like him we run the danger of fooling ourselves with post hoc answers, confusing chance causes with conclusions deserving generalization.

Suggestions generated by the strategy of studying discordant identical twins need to be tested prospectively in ordinary individuals.

PRECIPITATION OF ONSET OR RELAPSE

So far we have considered premorbid personality and predisposing factors in the environment. Turning now to precipitators of onset or relapse we shift our attention from intrapair comparisons to individual schizophrenics. Precipitating events are usually of a kind that occur to a majority of people, both healthy and sick, such as death of relatives, changing jobs, or in women, childbirth. The case of MZ 23 was rare in that he was actually being trained to "expand his consciousness." The kind of experience to which he was exposed before developing acute symptoms was not only a severe stressor, but one which seems to be related to his particular schizophrenic symptomatology.

In a careful study Birley and Brown (1970) confirmed the widely held view that crises and life changes can be significantly related to the onset or relapse of an acute schizophrenia. However, only 24% of consecutive admissions for schizophrenia had acute onsets preceded by a potentially disturbing event in the three weeks before onset. Reducing or stopping medication with phenothiazines was also shown to contribute as a precipitant. The symptomatology of acute schizophrenia was largely unrelated to its precipitants such as hearing of the sudden illness of a relative or a change of residence; the latter were not regarded as sufficient causes for the psychosis. Long-term tension in the home appeared to increase the chances that a patient would become disturbed after changes of the kind studied. The environmental contributors appeared to be additive.

The histories of our schizophrenic twins illustrate diverse precipitants, but as we said in the introduction, generalizations are impossible. Childbirth was a clear precipitant with recurring effects in DZ 24A. Concordant pair MZ 6 shows a connection between childbirth and outcome—the A-twin's first relapse followed delivery and then she deteriorated; her childless sister did not relapse after her one and only episode. In MZ 18 only A became psychotic, after her second child, while B had four pregnancies, including one abortion on psychiatric grounds, without becoming psychotic. In another pair, MZ 24A was schizophrenic and childless, whereas her sister had a nonschizophrenic postpartum depression treated with ECT. Obviously whether childbirth precipitates schizophrenia in someone with the genetic predisposition depends on other variables.

Objective stressors are sometimes more pronounced in the nonschizophrenic twin. MZ 5B experienced the combat death of her first husband, chronic marital conflict with her second husband, a metabolic disorder, and distress over her sick sister. These stresses led to neurosis but not schizophrenia. Stresses which an individual imposed upon himself, such as taking on increased responsibility, were more important than unavoidable stresses in a twin study of peptic ulcer (Eberhard, 1968).

Within a pair the same stressor may be more disrupting for one twin than the other. For example DZ 32*B* was better able to deal with the incestuous sex play of the father than the eventually schizophrenic *A*-twin. Even when twins have the same genotype, the same event may be experienced differently. Death of a mother precipitated a psychosis only in MZ 4*A* who was much closer to her than was her sister.

Everyday events can be elaborated in the mind to produce inappropriate responses. MZ 22*A* spoke convincingly of how ordinary contact with a young man at work reawakened sexual conflicts which then escalated into overt thought disorder and a relapse. We can hardly call a happening as subjectively apperceived and idiosyncratically elaborated as this an *objective* etiological stress. It would not be feasible to design environments so as to insulate people against these kinds of stresses.

THERAPEUTIC CONSIDERATIONS
IN SCHIZOPHRENIA

It is helpful to bear in mind the more positive aspects of the environment whether social or therapeutic. Attempts to remove MZ 19*B* from the influence of her sister met with qualified success. MZ 16*A* was probably wise to confine himself to a routine job rather than to compete with his Oxbridge brother, and to live at the opposite end of the country from his neurotic, overpossessive mother and rigid, pedantic father. Removal from the family environment may have helped MZ 7*B* too; in mental hospital, and not visited by her intrusive but now rejecting parents, she was no worse than her twin who lived at home and attended a Day Hospital. In other pairs more normal parents provided valuable therapeutic support by maintaining the patient at home (DZ 6*A*, DZ 9*A*), or by accepting less demanding or prestigious occupations than expected, permitting moderate stability in the community (DZ 18*A*).

The wife of MZ 15*B* was undoubtedly an asset that allowed him to remain in the community, working regularly, though obviously schizophrenic; his unmarried brother could not be weaned from a nearby mental hospital renowned for its rehabilitation program. Similar considerations play a part in MZ 2, whereas in MZ 6*A* marriage seems to have been more a source of stress than comfort. Divorce in the case of DZ 30*A* led to a relapse in a dependent male, while for MZ 20*A* divorce from her irresponsible husband seems to have led to long-term stability.

We are not in a position to offer any controlled experimental evidence on the efficacy of physical or psychological treatments. Our twins were all exposed to the practices current at the time of their hospitalization or outpatient care. Only when the twins' psychoses are similar and the effects of different treatments or placebo are systematically assessed, as happened to MZ 22 (Benaim, 1960), can firm conclusions be reached. Perhaps the closest we come to relevant anecdotal information is in MZ 2. The depressive symptoms of the first episodes for both twins responded to ECT. In their next episodes *A* refused insulin coma treatments and deteriorated, while *B* received 24 coma treatments and stayed out of hospital for 11 years. In MZ 22 the *B*-twin responded to physical treatments (insulin comas) in the first phase of her psychosis. The *A*-twin failed to respond to analytical psychotherapy; after 40 weeks she was removed to another hospital where she then began to improve with chemotherapy. Both twins were unusual in that they showed a favorable response on reserpine. The advent of the phenothiazines in the 1950's led to some remarkable changes in schizophrenics; for example, DZ 13*A* could be discharged after he had spent 6 years in hospital as a chronic patient. Successful treatment of schizophrenia is currently an empirical procedure; our findings from the study of twins is not compatible with therapeutic pessimism or nihilism.

SUMMARY AND CONCLUSIONS

So far it has not been possible to identify any ways in which the shared environment of pairs where both twins are schizophrenic is more abnormal than in those where only one twin is schizophrenic.

Neither in our sample nor in any other twin series was there a trend for the future schizophrenic to have lower birth weight or poorer school achievement than his co-twin. In all studies there was a marked trend for the schizophrenic to have been the more submissive in childhood. Although the observation is not specific to schizophrenia and is of questionable reliability and generalizability, it seems possible that small differences in intrapair dominance are important. It may augment any existing predisposition to schizophrenia in the more submissive twin through a relative failure in socialization. More difficult to interpret was the fact that discordant pairs were more often opposite-handed than concordant pairs. While reasons for discordance are undoubtedly multiple in the individual pair, there was no tendency for the twin who was simultaneously lighter at birth, less intellectually competent, and submissive to his partner to be at greater risk for mental disorder in a national sample of twins identified at birth.

We have grouped our eleven discordant pairs according to the most likely causes. Two probands probably had symptomatic schizophrenias which would not lead to an expectation of schizophrenia in a co-twin. In two others there was also a major organic contribution. Two pairs could be seen as having co-twins high on a dimension of schizophrenicity, one of whom, at age 20, is still at considerable risk. The remaining pairs were accounted for by various combinations of causes including interpersonal and psychodynamic processes.

Factors precipitating onset or relapse, when they could be found, ranged from the rare and specific through the common crises differing in their disrupting effects on the individual, to the subjective elaboration of everyday events. Some therapeutic aspects of the environment were mentioned.

We have emphasized the difficulty of generalizing about the effect of the environment more than once. Of course it is possible that critical objective environmental factors singly or in combination may yet be identified which are necessary before schizophrenia can develop. But, if we are correct in believing that the environmental factors are nonspecific and idiosyncratic, the genetic contribution to the interaction that results in schizophrenia would appear to be specific. Even when both the environment and the genes are necessary but insufficient, Meehl (personal communication) has suggested to us that the genes can have a "privileged status" and be the "uniformly most potent" contributor to the etiology.

After integrating and recapitulating the evidence from twins in Chapter 9, we shall examine the genetic theories of schizophrenia in the final chapter.

9
COMPARATIVE EVALUATION OF TWIN STUDIES IN SCHIZOPHRENIA

In this chapter we compare our study with other recent twin studies and attempt to evaluate the criticisms that have been made of the twin method as applied to schizophrenia.

CONCORDANCE RATES FOR OUR STUDY

Before doing so it will be necessary to report data deferred from Chapters 3 to 8. As we have pointed out before, there is no single *true* concordance rate for schizophrenia. What is valuable is to discover in what ways concordance may rise or fall as a function of population, sampling, and methods of data reporting. We have already reported what may seem like too many concordance rates. At the risk of devaluing the concept by overuse, we need to show how rates vary by sex and in other ways.

At the time of our follow-up, 42% of 24 MZ pairs and 9% of 33 DZ pairs were concordant for a chart diagnosis of schizophrenia; this was the Grade I of our previous report.

	MZ		DZ		
Grade I	10/24	42% ± 10	3/33	9% ± 5	(1)

As described in Chapter 5, a consensus diagnosis of S, ?S, O, or N was reached for each twin in the 57 pairs. This was based on the blind diagnosis of the histories by six experienced diagnosticians. As will be all too familiar to readers who have followed us so far, this resulted in the exclusion of two MZ pairs and a concordance rate for schizophrenia, including ?schizophrenia, of

	MZ		DZ		
$(S + ?S)/(S + ?S)$	11/22	50% ± 11	3/33	9% ± 5	(2)

In view of the sex differences in concordance rates in some of the earlier twin studies, highlighted by Rosenthal (1962b), we calculated these rates by sex.

	MZ		DZ		
Male	6/11	55% ± 15	1/17	6% ± 6	(3)
Female	5/11	45% ± 15	2/16	13% ± 8	(4)

The sexes were equally represented and concordance could not have been more similar.

For the reader who is interested in knowing our rate for schizophrenia without qualification—same consensus diagnosis for both twins—it was

	MZ		DZ		
S/S	8/20	40% ± 11	3/31	10% ± 5	(5)

This gives a range of concordance in MZ pairs of 40-50%. The criterion S + ?S gave better discrimination between MZ and DZ rates, not only with respect to the consensus but also with respect to the diagnoses of five out of six judges, and we think it a more reasonable estimate for our sample. The S criterion did not allow for the fact that some discordant co-twins had been hospitalized for psychotic disorders with schizophrenic-like features.

One of the criticisms of earlier twin studies was that investigators used a narrow concept of schizophrenia for probands but a wider one for co-twins (Rosenthal, 1962a, p. 129). However, the tendency is for recent studies to report rates for an even wider range of disorders among the relatives of strictly defined schizophrenics. Two of our consensus rates using broader criteria for co-twins than for probands were

	MZ		DZ		
$(S + ?S)/S$	11/21	52% ± 11	3/31	10% ± 5	(6)
$(S + ?S + 0)/(S + ?S)$	17/22	77% ± 9	12/33	36% ± 8	(7)

The first is hardly different from rate (2). The second reduces the discrimination between MZ and DZ rates, the MZ : DZ ratio being only 2.1. Example (7) is the consensus version of our previous (1966a) concordance through Grade III—any abnormality in the co-twin. They are lower than the Grade III rates of 79% and 45% by reason of the omission of twins who were abnormal on the MMPI only.

The "other" diagnoses cannot be regarded as part of a continuum of schizophrenic psychopathology. It was not intended that they should (Gottesman & Shields, 1966b, p. 67). Reasons for this opinion, given in Chapter 6 (p. 236) were supported by the psychometric evidence (p. 278-279). There was little or nothing schizophrenic-like even in the six hospitalized co-twins who comprised our original Grade II concordance (pp. 67, 237).

One further pairwise rate was deferred. We can now report the consensus S + ?S concordance for the subset of pairs in which one of the twins had been diagnosed schizophrenic at the Maudsley Hospital, omitting those additional probands diagnosed only subsequently.

	MZ		DZ		
Maudsley schizophrenics only	8/14	57% ± 13	2/23	9% ± 6	(8)

Discrimination between zygosities was good, but standard errors were larger.

Rates (1)-(8) are all simple, uncorrected, direct pairwise rates, reflecting a commonsense approach. A comparison of such rates carries most conviction for non-specialist readers. But the ingrained habit of thinking of twins as pairs and not as individuals has hazards not only for parents of twins but also for the most meaningful analysis of twin data on morbidity. Prevalence rates and other morbidity statistics in human genetics are expressed in terms of cases not pairs, whether these be cases found in the general population or among uncles and aunts, fathers, or adopted children of given probands. Similarly with twins it is the casewise rate, i.e., the proportion of probands with affected co-twins, that we should be interested in, rather than the pairwise rate or proportion of pairs in which both twins are affected. As was explained in Appendix B, such a casewise rate should be calculated only for the twins of independently ascertained probands. Only four of our concordant MZ pairs and one of our concordant DZ pairs are represented by two probands, so our probandwise concordance rate is

	MZ		DZ		
Proband method	15/26	58% ± 10	4/34	12% ± 6	(9)

The direct pairwise rate gives the most conservative estimate of the corrected casewise rate.[1]

In Chapter 6 we discussed the problem of allowing for the age of the co-twins, some of whom may yet fall ill with schizophrenia. If one more twin in each group were to become schizophrenic—a not unreasonable estimate—rate (9) would become

	MZ		DZ		
Proband method age-corrected	16/26	62% ± 10	5/34	15% ± 6	(10)

[1] Given (9) as the casewise rate for our data, corrected for ascertainment by the proband method, the corresponding pairwise rate is obtained indirectly (Allen *et al.*, 1967) by halving the number of concordant pairs to give indirect pairwise rates of MZ 7.5/18.5 (41%), and DZ 2/32 (6%). Although this is of less interest than the more conventional proband rate, it is still within the range of the pairwise rates reported above, and the MZ rate here is 6.5 times the DZ rate.

COMPARISON WITH OTHER RECENT STUDIES

In Table 2.7 we showed the range of pairwise concordance rates in recent twin studies. In comparing our investigation with them we need to discover which of the various criteria used correspond most closely to our preferred consensus S + ?S criterion [rate (2), above] and also to show probandwise rates.

In Kringlen's Norwegian study his "wide concept" of concordance which led to the MZ rate of 21/55 (38%) goes slightly beyond our consensus concept, but the rate of 17/55 (31%), which is that for schizophrenia or schizophreniform psychosis uses a concept narrower than ours. We cannot say how our consensus judges might have diagnosed his four borderline cases. To convert Kringlen's rates into probandwise rates we add the 14 MZ and 6 SS DZ pairs in which both pairs appeared on the Psychosis Register in their own right to obtain a new numerator and denominator. At the level of schizophrenia or schizophreniform psychosis the proband method rates are 31/69 (45%) and 14/96 (15%). With the inclusion of borderline cases they are 35/69 (51%) and 17/96 (18%). These MZ and DZ rates are similar to our corresponding rates [i.e., (9) above] of 58% and 12%. However, ours require more correction for age than Kringlen's, so that our MZ rate may be rather higher than his, whether one uses the proband method and age correction or not.

Although restricted to "process" schizophrenic probands, the MZ concordance rate in the Danish study is very similar to ours if, as seems likely from the case histories, the Grade II criterion of Fischer (1972) is close to our consensus. Her Grade II included schizophreniform, paranoid, and atypical psychotics who did not meet her criterion for schizophrenia. The probandwise MZ rate was 14/25 (56%). The Danish DZ rate of 12/45 (26%) is higher than any reported hitherto and is difficult to account for in the light of the typical ordinary sib rate of 10% in the same study. In practice the number of concordant SS DZ co-twins has been so small in all studies but Kallmann's that the addition or subtraction of one or two cases has a very large effect on the concordance rate. In nine studies the number of such pairs ranged from zero to eight.

In our judgment a reading of Tienari's (1971) histories makes it seem that, within his range of 6-36% MZ concordance, criteria similar to ours would probably lead to a probandwise rate of 7/20 (35%). It is impossible to say from the information so far provided how close to our criteria were those employed by Pollin et al. which resulted in a probandwise MZ rate for schizophrenia of 24% in Veteran's Administration twins (cf. p. 35) at first, and then 43%.

It is worth noting that the four recent European studies in aggregate yield a probandwise concordance approaching 50% in MZ twins, nearly three times the DZ rate. This is despite the inclusion of the low MZ Finnish and the high DZ Danish rates. Table 9.1 summarizes the recent studies, using the proband method and a concordance criterion of functional psychosis.

TABLE 9.1

Probandwise Concordance Rates in Recent Studies at the Level of Functional Psychoses

Investigation	MZ		SS DZ	
Finland (1971)	7/20	35%	3/23	13%
Norway (1967)	31/69	45%	14/96	15%
Denmark (1969)	14/25	56%	12/45	26%
Present study	15/26	58%	4/34	12%
United States (1972)	52/121	43%	12/131	9%

DIFFERENCES BETWEEN THE EARLIER
AND THE RECENT STUDIES

Sex and Concordance

One of the criticisms of the earlier twin studies centered on the observation that concordance tended to be higher in female pairs than in males—possibly for psychological reasons, such as a greater propensity toward *folie à deux* in females—and that the samples were overweighted with female pairs. As we pointed out earlier (Gottesman & Shields, 1966b) the sex difference in concordance was accounted for in the main by two studies only. We noted (Shields, 1968) that in studies investigating both sexes and reporting the results separately, the sex difference disappeared when analysis was restricted to samples based on consecutive admissions with systematic twin ascertainment. When borderline co-twins were included, concordance was 46% in 59 male pairs and 47% in 53 female pairs. This contrasted with 51% in 55 male pairs and 71% in 84 female pairs in other studies.

Table 9.2 analyzes the latest versions of the Norwegian, Danish, and Maudsley twin studies. (The Finnish and Veteran's Administration studies were restricted to males only.) When schizophrenia was defined more narrowly than in Table 9.1, both Kringlen and Fischer found numerically higher female rates in MZ pairs—Kringlen, 23% for males and 29% for females, both on the Psychosis Register; Fischer, only 1 out of 10 for males, but 4 out of 11 for females, both sexes consisting of process schizophrenics. The Danish male rate is the closest of any to the low rate found by Tienari in Finnish males. Strömgren's (1968) observation on data given him by Luxenburger in 1935 is relevant. At that time 5 out of 11 female MZ pairs were concordant for definite schizophrenia, compared with only 2 out of 9 male pairs. However, a further 5 male co-twins were "very probably schizophrenic." All three observations suggest a greater phenotypic variability in the male, possibly due to more environmental hetero-

TABLE 9.2

Concordance by Sex in MZ Pairs

Study	Level of concordance	Males		Females	
		Pairs	Concordance %	Pairs	Concordance %
Kringlen	Including borderline	31	42	24	33
Fischer *et al.*	Grade II	10	40	11	55
Present study	Consensus S + ?S	11	55	11	45
Total:		52	44	46	41

geneity, and may indicate part of the reason why some earlier investigators found a higher female concordance. Even should there be a genuinely higher concordance in males using wide standards and for females using a narrow standard, the difference is unlikely to be appreciable.

Zygosity Diagnosis

The recent studies made use of aids to zygosity determination, such as extensive blood grouping, which were not available formerly. It is unlikely they contain many misdiagnosed cases. However, the Danish and Veteran's Administration studies included, respectively, 11% and 20% pairs of uncertain zygosity. The point at issue is whether systematic errors of zygosity in the earlier studies account in part for their higher concordance rates. It is now known that the dangers of making an incorrect diagnosis on the evidence of physical appearance is not as hazardous as some critics supposed (Cederlöf *et al.*, 1961), so that this source of bias can be largely discounted. Diagnosis of zygosity continues to be more accurate than diagnosis of schizophrenia.

Preferential Reporting of Concordant MZ Pairs

Despite the efforts of earlier workers from Luxenburger onward to avoid such a bias, it is impossible to exclude it. The question does not arise so much in countries like Norway and Denmark with a good national twin register against which the twinship of hospitalized schizophrenics can be checked. In the Maudsley study twinship was determined systematically on admission (cf. p. 53). However, neither in Kallmann's nor in Slater's study, where there were fewer guarantees of adequate ascertainment, was an excess of MZ pairs found.

Contaminated Psychiatric Diagnosis

The additional information available on Kallmann's methods (cf. p. 28) and the results of our experiment with blind diagnosis in Chapter 5 suggest that experimenter bias in diagnosing the co-twin was less important than was sometimes suspected.

Severity and Heterogeneity

The earlier studies tended to be based on "typical," long-stay schizophrenics. The newer studies from consecutive admissions include a larger proportion of patients with a good premorbid personality. They may also include more cases where the schizophrenic-like illness could have a different etiology from those of the earlier probands, being symptomatic schizophrenias with primarily organic causes (Davison & Bagley, 1969) or misdiagnosed neurotics, of a kind that would be unlikely to have a schizophrenic co-twin.

Within most studies, old or new, some relationship between severity and concordance could be discerned. In the Finnish study the tendency could not be tested, since there was insufficient concordance. Restriction of range in the Finnish co-twins is paralleled by restriction of range in the Danish probands, all of whom were called process schizophrenics. Since all process schizophrenics on the register are already accounted for, milder schizophrenic probands, if they were to be included, could only have mild or unaffected co-twins. Restriction of range in either variable will attenuate the magnitude of correlation. It must be conceded, however, that the process schizophrenic probands of the Danish study showed variation in age at onset and outcome, but neither of these was strongly associated with concordance in the co-twin.

Not only were the Danish probands all process schizophrenics, but the Norwegian and Finnish probands included many that were severe cases by any standards. Severe probands from the Scandinavian studies, including Essen-Möller's from Sweden, did not have as high a proportion of affected or severely affected co-twins as ours.

Impressive though the association between severity and concordance was in the Maudsley study, we certainly cannot claim that it is always so marked. Nor do we intend to imply that differences in severity reflect differences in genotype in any rigid manner. Clearly environmental factors, including treatment or the lack of it, also play an important part, especially on such measures of severity as time spent in hospital or degree of rehabilitation achieved. Severity in the proband may be successful in predicting concordance in the co-twin largely through its association with premorbid personality. The hypothesis would predict a low concordance rate in twins selected for premorbid health in both. The prediction is well borne out in the Veteran's Administration study where both twins were well enough to be inducted and pairwise concordance was only 27%. It will be recalled (p. 251) that when the premorbid personality of our probands was assessed as normal, only two out of eight co-twins were schizophrenic.

When two of our mild probands with nonschizophrenic twins, MZ 10 and 12, were excluded by the consensus and diagnosed as personality disorders, the strength of the association between severity and concordance as measured earlier (Gottesman, 1968b) was slightly reduced. If we were to follow our inclination and also exclude MZ 9 and 20 as symptomatic schizophrenic reactions to alcohol and thyroid disease, respectively, the association would be reduced still further, since both probands were acute schizophrenics with short spells in hospital and unaffected twins. Were the above co-twins nonschizophrenic because the probands had mild/acute illnesses or because the probands were misdiagnosed? The answer depends on the personal cut-off points of the diagnostician. We have previously mentioned the difficulty of distinguishing between symptomatic schizophrenia and schizophrenia induced by "high value" environmental stress in an individual with a low genetic diathesis schizophrenia.

Gene Pool Differences

Other more speculative reasons for explaining the difference between the older and newer studies include population gene pool differences (Shields, 1968) between Scandinavian and other countries. The variation could have resulted from differential selection pressures in the remote past, perhaps related to climate and population density. Genetic variation would be reduced and concordance rates would be lowered (a) through selection pressure resulting in a disease resistant genotype and preserved by the lack of immigration, and (b) through a large amount of environmental variation masking what genotypic variation remained in a relatively homogeneous gene pool.

A further possibility is the rural character of the Scandinavian populations compared with Munich, New York, and London where Luxenburger, Kallmann, and Slater worked. There is some suggestion from our analysis of Kringlen's case histories that within Norway urban-reared twins tended to be concordant more often than rural-reared twins.

However, we do not wish to make too much of any of these ideas. To what then is the difference between the old and new rates mostly due? Kallmann's MZ rate of 86%, high both relatively and absolutely, is to a considerable extent due to an overcorrection for age (cf. p. 234). Apart from this, the differences are probably best accounted for by the severity of the older probands (insofar as it reflects a more highly predisposed genotype) and the variability of the schizophrenias tapped by the wider and more thorough sampling methods of the recent studies.

How Abrupt Is the Difference between the Old and New?

We are now in the fortunate position of knowing approximately what the outcome of our study would have been had Slater applied the same methods to our sample as he did to his in 1953. His diagnoses, unblindfolded (p. 244), result in the exclusion of MZ 9, 10, 12, 16, and 20, leaving 19 MZ pairs, 10 (53%) of which he would have counted as concordant. He would not have accepted MZ 23*B*, the messianic devotee of the occult, as a first choice diagnosis of schizophrenia, but retained him as a schizoid personality. When seeing the pair's case histories side by side, the fact that he now knew that the proband had a psychotic illness did not alter his subjective probability of psychosis in the *B*-twin. Using the proband method, Slater would have obtained an MZ rate of 61%. In his own study the corresponding rate was 68%. When diagnostician and method and location (London) are held constant, there is in this instance only a 7% difference between the old and the new.

The earlier and the more recent studies, given their standard errors, do not contradict one another as much as has been supposed. Within each study there is variation in the findings depending on the method used to calculate concordance rates. Between studies there is a wide range of results holding method constant —from Essen-Möller to Kallmann among the old, and from Tienari to ourselves among the new. We have already given our views about the relative importance of factors which might account for the undeniably lower rates in the newer studies. We feel justified in concluding that the twin studies of schizophrenia are replications of the same experiment, once provisions are made for differences in sampling and diagnostic practices.

THE NONSCHIZOPHRENIC MZ CO-TWINS

What about the co-twins who were not diagnosed as schizophrenic? What proportion are seen as normal across studies? What is the nature of the abnormalities in the remaining co-twins, and how far can they be regarded as related

TABLE 9.3

Pairwise MZ Rates for Normality, Other Psychiatric Conditions, and Schizophrenia in Some Older and Recent Twin Studies

Study	Normal		"Other"[a]		S + ?S	
	N	%	N		%	N	%
Luxenburger[b]	2	14	2	(2)	14	10	72
Rosanoff *et al.*	13	32	3	–	7	25	61
Kallmann	9	5	45	(36)	26	120	69
Slater	8	22	5	–	14	24	64
Kringlen	18	33	16	–	29	21	38
Fischer *et al.*	9	43	2	(1)	10	10	48
Present study	5	23	6	(2)	27	11	50

[a]Number schizoid in brackets.
[b]Only includes cases where probands were certain schizophrenics.

to schizophrenia? Table 9.3 shows what 7 of the investigations reported. The per cent of normals ranges from 5% in Kallmann (Shields *et al.*, 1967) to 43% in Fischer *et al.* In our study, as well as in others, normal co-twins could be found who were paired with severe schizophrenics. Such pairs would lend support to the importance of the concept of a threshold for understanding schizophrenia.

Not all investigators attempted to assess schizoid abnormalities in co-twins. Those who did so disagreed in their findings. The effects of contaminated diagnosis are especially dangerous here. Slater (1953) said one could not regard the nonschizophrenic MZ co-twin as consistently schizoid. Essen-Möller's and Inouye's results present difficulties to tabular presentation, but both reported a high proportion of schizoid or schizothymic co-twins; Tienari (1968) reported a noteworthy degree of introversion in 7 co-twins, many of whom were "healthy." The nonschizoid co-twins among the "other" in Table 9.3 were mostly neurotic and personality disordered. In Chapter 6 we gave our reasons for caution before concluding that "other" diagnoses in the families of schizophrenics were genetically related to schizophrenia. Some of the abnormalities can be explained largely on environmental grounds. Others may develop on a constitutional basis but not specifically connected to the schizophrenic diathesis—schizophrenics have many additional genetic potentialities besides the one for schizophrenia.

LIMITATIONS OF TWIN STUDIES

As a means of elucidating the relative importance of genetic factors in the etiology of schizophrenia the twin method has been quite successful. Its promise for elucidating the nature and role of life experience has not been realized. All investigators hoped that the discordant MZ pairs in each study would highlight the critical events. Instead, nothing out of the ordinary has been found to distinguish the sick from the healthy. When a factor emerged about which there was general agreement, namely submissiveness, the interpretation of its significance was confounded by the division of roles peculiar to twins. The range of environments to which twins are exposed is usually small, interfering with the efficient search for conspicuous environmental factors. Generalizing about identified environmental factors is also hazardous in other high risk groups, such as the children of schizophrenics, because they are atypical in their own ways. Most schizophrenics are neither twins, nor are they reared by schizophrenic parents.

Hypothesized environmental causes can be tested prospectively in individuals unselected for genetic relationship to a schizophrenic. We do not want to be too pessimistic about the prospects for advances in identifying such factors. However, the events may be too variable from person to person to permit any generalizations unless they can be reduced to a credible common denominator, such as anxiety-producing stresses. Up to the present no one has identified any experiential factor predicting an increased risk for schizophrenia, corresponding, say, to the increase in risk for lung cancer in heavy cigarette smokers.

SUMMARY AND CONCLUSIONS

The earlier twin studies from Luxenburger to Slater have been criticized for having incredibly high MZ concordance rates that led to a misleading impression about the importance of genetic factors in schizophrenia. The recent studies, including our own, permitted considered evaluation of the supposed biases. However, there are also ways in which a low level of concordance may give a misleading impression that genetics is unimportant (Shields, 1968, p. 105).

Neither contaminated diagnosis of zygosity nor of clinical state was a major source of error. The preferential reporting of concordant MZ pairs was overestimated. Sex differences in concordance disappeared in the newer studies. The possibility that differences in gene pools might in part account for the lower rates in the Scandinavian studies was considered. Higher raw rates in the earlier studies can to a considerable extent be understood as reflecting the higher proportion of chronic schizophrenics with poor premorbid personalities whose co-twin might be at a higher risk. Wider and more thorough sampling methods of the newer projects tap a wider range of schizophrenia, including more cases which a conservative clinician might have rejected.

It is easy to exaggerate the differences between the old and new. Our sample has been diagnosed by Slater in an attempt to replicate his 1953 methods. When diagnostician, method, and location were held constant, there was only a 7% difference between MZ rates in his sample and ours. Variability among the systematic twin studies can be considerably reduced by omitting the highest (Kallmann) and lowest (Tienari) MZ rates reported. Once provisions are made for differences in sampling and diagnostic practices, twin studies of schizophrenia can be seen as successful replications of the same experiment.

In order to appreciate the degree of risk to the identical twin of a schizophrenic, compared to that for a member of the general population, the pairwise rates of the new series had to be converted into probandwise rates. The four European samples, in aggregate, yielded a probandwise rate approaching 50% in MZ and 17% in DZ co-twins. Rates were calculated at the level of similarity for a functional schizophrenic-like psychosis in the co-twin. Identical twins of schizophrenics have a risk about 50 times greater and fraternal twins one about 17 times greater than a member of the general population comparable in age. The lower initial rates in the Veteran's Administration twin panel of 24% and 8% were in line with predictions for a sample selected on the basis of premorbid mental health and employing wider diagnostic standards for schizophrenia.

Investigators have different cut-off points for normality as well as for schizophrenia. The wide range of the proportion of co-twins who were called normal was most likely due to the imprecision surrounding the term schizoid. Pairing of typical schizophrenics with normal MZ co-twins was observed often and focuses attention on the threshold concept. Other psychiatric abnormalities in MZ co-twins could not be regarded as necessarily related to schizophrenia; generally they did not resemble schizophrenia clinically and their inclusion as "spectrum disorders" did not improve the discrimination between MZ and DZ co-twins in any of the studies.

Environmental stressors were too idiosyncratic to be easily summarized (cf. comments in Chapters 4 and 8). Unfortunately there are limitations to the resolving power of the twin method for highlighting environmental factors of general importance and specificity even with identical twins discordant for schizophrenia.

10
GENETIC THEORIZING AND SCHIZOPHRENIA

A basic postulate of contemporary human genetics is that all of a person's characteristics are the result of interaction between his genotype and his environment. Such a statement can be taken as an exhortation to get on with the task of searching for the nature and relative importance of the factors involved and how they interact for the characteristic with which we have been concerned—schizophrenia—or it can be interpreted as an end in itself permitting peaceful coexistence between formerly warring factions. The latter option would be a false peace not in the interest of mankind; such a *laissez faire* attitude would allow the continued practice or malpractice of one's "received" ideas about the origins and treatment of schizophrenia and thus perpetuate the notoriously poor record of the helping professions in regard to sufferers from this syndrome. Ask yourself how much schizophrenia has been prevented or cured on the basis of our current understanding of the condition; the obvious answer is in no way meant to detract from the great strides made in alleviating the anguish of patients and their relatives by humanitarian treatment and advances in chemotherapy.

It should be obvious that we opted for the exhortative meaning for the "G X E" postulate. From the vantage point of our twin study we could examine both genetic and environmental contributors to the schizophrenic phenotype. In our chapter on environment we pointed out that if the genes are necessary but not sufficient for the development of schizophrenia, it follows that the environment is also necessary but not sufficient. The evidence we have generated as well as that we have reviewed led us to conclude that *genetic factors specific to schizophrenia are conclusively involved in its etiology*. At no time did we imply that genetic factors were the only ones. Our evaluation of data suggested that the environmental factors were nonspecific and idiosyncratic but that the genetic contribution to the interaction resulting in schizophrenia appeared to have specificity. Even with both the genes and the environment being necessary but insufficient, the network of information above led us to the view that the genetic contribution had "privileged status" and was the "uniformly most potent" (Meehl) contributor to the etiology of schizophrenia. One of our main tasks throughout this book has been to contribute to a climate of opinion conducive to the continued, energetic application of *biological* techniques to the unsolved enigma of the etiology of schizophrenia, despite a missing *corpus delicti* so far.

We are reluctant to label schizophrenia as a disease and, even more so, as a reaction; most justice to our apprehension of the phenomenon is done by construing the disorder as an outcome of a genetically determined developmental predisposition. The word *developmental* is crucial in that it adds the dimension of time to our efforts at understanding schizophrenia; it is the ontogenetic unfolding of a particular phylogenetically given predisposition, buffeted by environmental influences with both graded and saltatory effects, that holds the attention of schizophreniologists. Some human pathologies, such as PKU and Huntington's chorea, clearly fall at one end of a continuum and are called inherited diseases because all persons with the genes develop the conditions. Other pathologies, such as cholera and plague, fall at the other end of the continuum and are called environmental diseases because virtually all persons sufficiently exposed to the vector develop the disease. Many other disorders, often common, fall in between the end points of the continuum; among these are schizophrenia, affective psychoses, diabetes, and some congenital malformations. For these latter conditions the unspecified genetic predisposition must share the spotlight with the unspecified factors in the environment which *cause* some and not others with the genotype to develop the disorder. Adding to the complexity of these middle of the continuum conditions is the probability (Shields, 1968) that the effects of environmental factors may be interaction effects only, operating on the relatively few genetically predisposed individuals to produce schizophrenia, but with no generally adverse effects on the population as a whole, unless the stressors become extreme ones.

MAIN GROUNDS FOR OUR EMPHASIS ON GENETIC FACTORS

By way of review, the following are some of the many points of evidence we have presented to support our position regarding genetic factors in theories about the etiology of schizophrenia.

1. Our species is extremely diverse genetically. It is logical to expect that this genetic variability will occasionally produce a combination of genes that results in a *phenodeviant* (Lerner, 1958) at the extreme of a distribution. The work of Lewontin (1967) on blood group antigens and of Harris (1970) on enzymes

suggests that about 30% of all human loci are polymorphic, i.e., two or more alleles at a given gene locus, each with frequencies greater than .01 (hence not explainable by mutation). The findings imply that 16% of the loci coding for the structure of proteins in any one person will be heterozygous; using conservative estimates for the number of such loci (50,000) each of us has about 8000 loci at which there are two different alleles, each locus resulting in a distinct protein. (Genes responsible for regulation and organization are excluded from consideration at this stage of our ignorance.) Harris calculated that the probability of two persons at random having the same type of enzymes at only eight loci was 1 in 200; the most commonly occurring types would be found in 1.8% of the population. He called the kind of diversity already demonstrated merely the tip of an iceberg.

2. Many morphological and physiological traits are known to be under some genetic control. Behavioral traits such as intelligence, social introversion, and anxiety have an appreciable genetic component, with data for some of these traits coming from animal strain difference and selection studies, as well as from work with twins and families. It would be surprising if schizophrenia were altogether exempted from analogous genetic influences.

3. No environmental causes have been found that will invariably or even with moderate probability produce schizophrenia in subjects unrelated to a schizophrenic. When cases of *folie à deux* are examined carefully, a high prevalence of schizophrenia is found among the genetic relatives of the induced (Scharfetter, 1970), thus shifting the focus from the role of inducer as a cause to one of precipitator, and a consequent refocusing on the predisposition of the induced.

4. Schizophrenia is present in all countries that have been studied extensively. In many the incidence is about the same despite great variations in ecologies such as child rearing practices. Such observations detract from assigning "culture" a major causal role in the etiology of schizophrenia.

5. Within modern urban communities there is a disproportionately higher incidence and prevalence of schizophrenia in the lowest social classes compared to the highest. On the face of it, such observations provide strong support for the role of social stressors as causes of schizophrenia. Our considered evaluation of the data (cf. Chapter 2) was that downward social drift of the patient was the major explanation for the excess of schizophrenia in the lower classes; however, we pointed out that some genetically predisposed individuals might have remained compensated had they been in a more sheltered class. Paradoxically, social stressors can be both predisposing and precipitating at different times. Kay and Roth (1961) in their study of late paraphrenia noted that social isolation was initially the effect of the preferences of schizoid people and secondarily a cause of their decompensation in that isolation removed various resources for adjustment in old age. Fuller and Collins (1970) provided a clear experimental model for such a phenomenon in mice susceptible to audiogenic seizures; sound

as a stressor precipitated seizures in certain predisposed genotypes (DBA) on first exposure, but not in others (C57BL); on a second trial "sensitization-induced seizure susceptibility" was observed in 60% of the C57BL mice, but it was seen in even more (81%) of the hybrids carrying half their genes from the DBA's. Even when the stressor predisposed the mice to seizures, it did so as a function of the genetic predisposition.

We would like to make it clear that we would not downgrade the part played by stress—it is, after all, half of the diathesis-stress model. We, as well as others, are unable to deal adequately with the concept of stress as an explanatory construct or as an intervening variable. Many of the difficulties plaguing the concept are confronted by Levine and Scotch (1970) and their colleagues and by Selye (1956). The simplistic flow chart

$$\text{Stressor} \longrightarrow \text{Stress} \longrightarrow \text{Disorder}$$

is acceptable as a starting point, but denies the important role we wish to assign to the "stressee."[1] Events which are apperceived as stressors depend on the genotypic *and* experiental uniqueness (e.g., intrauterine environment, perinatal hazards, learning history, exposure to CNS toxins, etc.) of the stressee; so do the kind and degree of stress responses and so do the various disordered outcomes of the stress responses. The problems of specificity are far from solved and we do not yet have the answers. But why might the outcomes vary from hypertension to ulcer to schizophrenia ?—Perhaps the answer depends upon the specific properties of the stressee.

6. There is an increasing risk of schizophrenia to the relatives of schizophrenics as a function of the degree of genetic relatedness. The familial distribution cannot entirely be due to environmental differences between families— the MZ concordance rate is higher than the DZ—or to gross differences within families such as sex or birth order. Although there is a need for more and more detailed studies of step-sibs, sibs, and half-sibs, Rüdin (1916) found a seven-fold increase in risk for schizophrenia in the full sibs compared to the half-sibs of dementia praecox probands, despite the high proportion of shared family environment.

7. The difference in identical vs. fraternal twin concordance rates is not due to aspects of the within-family environment that are more similar for MZ than DZ twins, although there are many such aspects. Studies of MZ twins reared apart as well as adoption and fostering studies show a markedly raised incidence of schizophrenia among relatives even when they were brought up in a different home by nonrelatives.

---

[1] M. Vartanyan (personal communication) in Moscow has combined stress, neuro-chemical, and genetic research strategies to yield interesting preliminary data which he interprets as evidence in support of a polygenic theory for schizophrenia.

8. Such implicitly causal constructs as schizophrenogenic mothers, double-binding, marital skew, and communication deviance, have been found wanting (by others as well as ourselves), although we would not categorically deny them a role as possible precipitators or exacerbators of schizophrenia. The offspring of male schizophrenics are as much at risk for the disorder as are the offspring of female schizophrenics. When both parents are schizophrenic the risk to their children is about 46%; it is difficult to account for the absence of schizophrenia in the rest of the children on environmental grounds given such a schizophrenogenic environment; in what might be perceived as an even worse environment, one where one parent is a schizophrenic and the other is psychopathic, the risk of schizophrenia in the offspring is only 15%. Both sets of data are, however, compatible with genetic theories of etiology.

We close this section with yet another reminder that paradoxically it is the data showing that identical twins are as often discordant as concordant for schizophrenia that provide the most impressive evidence for the important role of environmental factors in schizophrenia, whatever they may be.

GENETIC MODELS FOR THE MODE OF TRANSMISSION

Once the existence of genetic diathesis has been established, it becomes important to provide a theory for the mode of its transmission. In the first instance theories provide a scheme for systematizing diverse pieces in a jigsaw puzzle. In the second, they encourage the formation of testable and refutable hypotheses; ideally they should compete with each other in such a fashion that one theory is made more credible and another less so when subjected to a test. Different genetic models have different implications for the kinds of studies to be conducted, for the kind of molecular pathology involved and hence the rational treatment, for possibilities of detecting premorbid cases, and for recommendations about the prevention of schizophrenia, e.g., by genetic counseling.

Models for the genetic mode of transmission in schizophrenia can be roughly classed into three categories which can in turn be divided. The broad classes are monogenic or one major locus, genetic heterogeneity, and polygenic. Monogenic theories can be divided into recessive, requiring homozygosity or a double dose of a gene at one locus (one from each parent), and dominant, requiring only a single dose of some necessary gene (from one parent). Genes themselves are neither dominant nor recessive; the terms only have meaning with respect to a particular phenotypic characteristic. John and Lewis (1966) introduced the useful distinction between exophenotype (external phenotype) and endophenotype (internal), with the latter discernible only after aid to the naked eye, e.g., a biochemical test or a microscopic examination of chromosome morphology. As endophenotypes have become more available, the distinction between recessivity and dominance has become blurred; in a sense all genes are "dominant" (cf.

sickle-cell anemia vs. sickling trait) when we have a way of detecting gene action molecularly. Like most inborn errors of metabolism PKU is the result of an enzyme deficiency inherited in a recessive fashion (two doses of a gene), but the heterozygote (one dose of the gene) can usually be identified. Enzyme deficiencies can also be inherited in a dominant fashion, e.g., porphyria. The difference depends on how far the normal homozygous state produces an excess of the minimal level needed for health. To quote Harris (1970, p. 252), "Dominant inheritance of a disease due to an enzyme deficiency is most likely to occur where the enzyme in question happens to be rate limiting in the metabolic pathway in which it takes part, because the level of activity of such enzymes in the normal organism will in general be closer to the minimum required to maintain normal function."

Dominant gene theories of schizophrenia which provide for the modifying effects on the phenotype of genes at other loci or other alleles at the same locus (cf. the G6PD polymorphism) are in practice difficult to distinguish from polygenic models; Slater's particular model (Slater, 1958; Slater & Cowie, 1971) is discussed below. A simple monogenic theory for all schizophrenic psychoses where the gene is sufficient cause for the psychosis has no advocates.

Genetic heterogeneity, as we said in Chapter 1, means different things to many people. It can mean that schizophrenia, like low-grade mental deficiency, is comprised of many rare varieties of different recessive or dominant conditions with the mutation rate at each locus maintaining the genes in the population. One form of genetic heterogeneity we can agree with is that the model is like that of mental deficiency throughout its range; a very small percentage of schizophrenic cases are due to different dominant and recessive loci, a further group is due to symptomatic phenocopies (e.g., epilepsy, use of amphetamines, or psychic trauma), but the vast majority are segregants in a normal distribution of a liability toward schizophrenia.

Polygenic models can be divided into continuous phenotypic variation and quasi-continuous variation or threshold effect. Examples of the former are height and IQ scores where extremes of a distribution may be labeled as pathological (dwarf or retardate) at some arbitrary point in the distribution; individuals just to the other side of the point are not distinctively different. The most widely known polygenic trait models posit a large number of underlying genes all of whose effects are equal; with traits so determined we would expect the phenotypic correlation between relatives to be the same as the genetic correlation if the traits are completely heritable. We find for example that the parent-child and sib-sib correlations for height or fingerprint ridge count are very close to .50. A less well known polygenic model of importance to our thinking about a model for schizophrenia permits the gene effects to be unequal. Thoday (1961, 1967) has shown that although bristle number in *Drosophila* is under polygenic control, 87.5% of the genetic difference between the means of a high and a low line could be accounted for by only five of the many genetic loci involved. The implications of such a weighted gene model for schizophrenia are to encourage searching by the usual methods of segregation analysis and linkage for the few "handleable" genes which may prove to mediate a large part of the genetic variation in the liability to schizophrenia.

A polygenic model for handling discontinuous phenotypic variation, so-called threshold or quasi-continuous characters, also forms an important background to our thinking about schizophrenia. This model has made analysis of such traits as schizophrenia, cleft palate, diabetes, and seizure susceptibility feasible, provided one accepts the working hypothesis that the underlying liability is continuously and normally distributed. Falconer (1965, 1967), Edwards (1969), Morton *et al.* (1970), and Smith (1970, 1971) have illustrated the methods involved and we (Gottesman & Shields, 1967) were the first to study psychopathology with such methods. Data on the occurrence of cleft lip with or without cleft palate, CL(P), in the relatives of probands can be used to illustrate the threshold model (Carter, 1969a; Woolf, 1971). Schizophrenia is not present at birth like cleft lip so the analogy is wanting in this respect, but such elegant data for a disorder with a variable age of onset and with the capacity for remission are not available yet.

The population incidence (q_g) of CL(P) can be taken as .001 (Woolf, 1971). The risk in sibs is .04, a low absolute value but a 40-fold increase over the population risk; in second-degree relatives it is .0065 and in third-degree (first cousins), .0036. The sharp falling off of incidence as one moves to more remote relatives is one of the tests for polygenic theory; a dominant gene theory calls for the frequency of affected relatives to decrease by one-half in each step. An important parallel between CL(P) and schizophrenia is that the risk to parents is about one-half that in sibs although both are classes of first-degree relatives. In both disorders the reduced values probably represent the effects of social selection for who become parents; different values of q_g will be required to evaluate the significance to genetic theorizing of lower rates in parents when such selection is probable. Figure 10.1 shows a diagram for the hypothetical distribution of a genetic liability to CL(P) or other threshold character, for the general population as well as first- and third-degree relatives.

The X axis is for normal deviate values of the posited polygenically deter-mined predisposition or liability to the threshold trait. At a point on the X axis (not drawn to scale) corresponding to a value of .001 (q_g) of the general population we can erect a vertical line (T) to represent the threshold value of liability beyond which all persons are affected; such a line would cut off 4.0% of the sibs (q_r) and only .36% of first cousins. The distances x/2, and x/8 in Fig. 10.1 are the increased means of the liability distributions for first- and third-degree relatives and are predictable from our general knowledge about genetic correlation between relatives, once A and G, the mean liability of affected persons and of the general population, have been determined. A sharp threshold

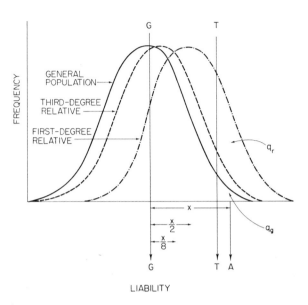

Fig. 10.1. Model for polygenic inheritance of threshold characters: Three distributions of the underlying liability in the general population, in first-degree relatives, and in third-degree relatives (see text for symbol definitions).

between the liability of affected and unaffected persons is artificial; the threshold model implies an increasing likelihood of being affected (i.e., a cumulative normal risk function) as the polygenic predisposition increases (Edwards, 1969; Smith, 1970, 1971).

Support for the threshold model arises from a demonstration of a relationship between the severity of the defect in the proband and the risk to his relatives; this is based on the assumption that the more genes involved, the more severe the condition, and the more genes involved, the more the relatives will have when the amount is halved, quartered, etc. For CL(P) unilateral and bilateral affectation form two levels of severity; in the sibs of unilateral cases the risk is 3.83%, in those of bilaterals, 6.71%; and the generalization holds for other degrees of relatives. Further support for the theory comes from the demonstration that the risk to probands' relatives, say sibs, increases with the number of other relatives affected, i.e., families with two patients are more "high risk" families than those with only one. In the case of CL(P), if no other relative is affected, the recurrence risk to a proband's sib is 2.24%; if an aunt or uncle is affected, the risk rises to 9.91%; finally, if a parent is affected, the risk to the sib rises to 15.55%. The malformation is too rare for there to have been extensive twin studies. From the available evidence Carter (1965, 1969a) estimates the risk to the identical twin of a proband to be about 40%.

From the above data, estimates of the heritability of the underlying liability to CL(P) can be made. Heritability (h^2) is defined as the proportion of the total variability of the trait in the population that is due to genetic differences, assuming the absence of dominance and interaction between genes (cf. Falconer, 1965). The risks to MZ twins, sibs, and first cousins yield h^2 estimates (Smith, 1970) of 88%, 92%, and 100%, respectively, which are reasonably consistent values.

COMPATIBILITY BETWEEN THEORY AND DATA

We shall deal with monogenic theories first. Recessive inheritance for schizophrenia is difficult to support, since sibs are not more often affected than children. Most monogenic theories invoke a dominant gene. Slater's final version of the general theory that he first proposed in 1958 fits the pooled family data best when the population lifetime risk for developing schizophrenia is taken as .85% and the gene frequency as .03. 90% of schizophrenics will then be heterozygotes, so the trait is basically a dominant one. Only 13% of heterozygotes manifest the psychosis; however, manifestation is complete in the 10% of schizophrenics who have inherited the gene in double dose (Slater & Cowie, 1971). Elston and Campbell (1970) proposed a similar theory, derived from the application of rigorous mathematical methods to the data of Kallmann. According to this theory, the manifestation rate in heterozygotes is only 6% or 7%, even lower than the 13% "penetrance" on Slater's theory. Clearly such theories are still viable and have the merit of simplicity. They suggest that the search for a simply inherited biological error underlying all cases may not be in vain. The problem of how the abnormal gene can maintain itself in the population in view of the low fertility of schizophrenics prompts a search for compensating selective advantage, such as an increased resistance to virus infections early in life (e.g., Carter & Watts, 1971). However, no mendelizing defect has so far been identified in schizophrenia—unless it is, the theory will remain implausible for many. Anderson (1972) has pointed out that "it is difficult to estimate the degree of penetrance unless the variations in phenotype can be identified unequivocally and unless there is independent information establishing the mode of inheritance." If the mode of inheritance is independently established as the result of a dominant gene, there is no objection in principle to invoking very low penetrances; Sewall Wright himself (1963, p. 178) cited a penetrance of 2% for a gene associated with a morphological character in hybrid guinea pigs.

To avoid invoking greatly reduced penetrance Meehl (1962, 1973) and Heston (1970) have concerned themselves with a phenotype broader than schizophrenia—the schizotype and schizoid disease, respectively. Heston considers most studies to have shown about 50% of the first-degree relatives of schizophrenics to have some kind of mental abnormality. The difficulty is that there is no reliable way of defining schizoid disease without reference to relatedness to a schizophrenic. If the concept is defined broadly enough to encompass abnormalities in some 50% of schizophrenics' parents, sibs, and children, and then generalized, the population base rate will be exaggerated and include many false positives. Nevertheless there is certainly merit in carrying out family investigations based on borderline schizophrenics, schizoid personalities and the like in order to test a Mendelian hypothesis.

Heterogeneity theories are less well defined, making it more difficult to say whether they are compatible with the data or not.

One class of heterogeneity theory claims that in principle schizophrenia can be divided etiologically, though not necessarily clinically, into two groups: (1) a high-risk genetic group comprising a large number of individually rare genetic disorders, each with a very high manifestation rate and inherited as a recessive (e.g., as in severe mental retardation, Dewey, Barrai, Morton, & Mi, 1965), or as a dominant trait (Erlenmeyer-Kimling, personal communication); and (2) a residual group of sporadic cases with a low risk of recurrence consisting on the one hand of fresh mutations and on the other of a group of cases of environmental or complex etiology. Deafness, blindness, low-grade retardation, and the muscular dystrophies are conditions that belong to this first class. The theory avoids the *ad hoc* assumption of low penetrance, though in practice there are few schizophrenic families in which the risk to sibs is as high as 25%. If there were many recessive loci for schizophrenia (as there are for deafness), dual mating parents would be unlikely to be of the same type, hence less than 100% of the children would be affected, as is the case; but the observed rate in children (13.9%) when only one parent is affected still remains unaccountable. The consequences of dominant gene heterogeneity are essentially the same as those of monogenic dominance as regards the risk of recurrence in the family; but the theory can account for the continuing prevalence of schizophrenia in the population without invoking either unrealistically high mutation rates (Erlenmeyer-Kimling & Paradowski, 1966) or speculative selective advantage for the heterozygote.

A second class of genetic heterogeneity comprises theories positing that clinically different types of schizophrenia have distinct etiologies. The most fully developed theory of this kind is probably Mitsuda's (1967); see also Rosenthal (1970) and Slater and Cowie (1971). A genetic distinction between typical and atypical schizophrenias is also proposed by Inouye (1963), another Japanese psychiatrist; by the Kleist-Leonhard school (e.g., von Trostorff, 1968); by many Scandinavian psychiatrists who differentiate between "true" and reactive or schizophreniform psychoses (e.g., Welner & Strömgren, 1958); and by those who hold that process and reactive schizophrenia (cf. Garmezy, 1968) are etiologically distinct and not zones on the same continuum. These theories differ in how far they claim dominance or recessive inheritance for the clinical groups, in how far the different groups are genetically determined (if at all), and in the specificity of disorders found in relatives.

Although the hypothesis of "the schizophrenias" rather than "schizophrenia" is a popular and potentially useful one, the genetic heterogeneity hypothesis has not so far met with as much success as in the affective psychoses. The hypothesis that bipolar manic-depressive psychosis is genetically distinct from unipolar depression has received support from the work of Angst (1966), Perris (1966), Winokur *et al.* (1969), and others. In schizophrenia some of the clinical distinctions are of uncertain reliability, some of the genetic work claiming heterogeneity is open to criticism, and findings, if any attempt is made to replicate them, are not confirmed at different centers (e.g., van Epen, 1969).

There is increasing evidence (Davison & Bagley, 1969) that some schizophrenic syndromes—the symptomatic schizophrenias—develop on the basis of an organic pathology, including Huntington's chorea, Wilson's disease, temporal lobe epilepsy (Slater *et al.*, 1963), and amphetamine intoxication (Connell, 1958). Slater believes the pathogenesis of such cases may give important clues to the pathogenesis of schizophrenia in general.

Multiple heterogeneity implies that the bulk of cases of schizophrenia can be accounted for by pooling a number of different rare or relatively rare causes, each of which is virtually sufficient to account for an instance of the disorder. To the extent that one of the causes, a major gene, is predominant, the theory merges with the monogenic. To the extent that the different causal factors are common and insufficient, etiology will depend on the combination of several elements. Here heterogeneity overlaps with polygenic and other multifactorial theories.

TESTS OF POLYGENIC THEORY

In this section we shall muster the various lines of evidence which can be brought to bear on the relative merit of polygenic theory in accounting for the body of data on schizophrenia. One indirect approach we favored in the past was to evaluate the compatibility and consistency of independent estimates of the heritability (h^2) of the liability to schizophrenia, after assuming a graded liability for a threshold trait (Gottesman & Shields, 1967). We now present a summary in Fig. 10.2 of an updated version of this approach which uses the improvements to Falconer's method for estimating the heritability of threshold disorders devised by Smith (1970, 1971, and personal communication), our consensus diagnosis pairwise concordance rates for MZ (50%) and DZ (9%) twins, and pooled risks for sibs (10.2%), offspring of dual mating schizophrenics (46%), and for second-degree relatives (3.3%) (Zerbin-Rüdin, 1967; Slater & Cowie, 1971; see Table 2.4). Six different values of q_p (the equivalent to q_g in Falconer's notation), the population risk, ranging from .85% to 3.0% were used so as to show the effects on estimations of heritability values. Probandwise twin rates might have been more technically correct here but they would not change the overall impression; the pooled probandwise rates for MZ twins in the recent

European studies approaches 50% (p. 306). The VA Hospital twin data from Chapter 2 (p. 35) lend themselves very well to an application of the technique since independent estimates of the MZ and DZ prevalence are available; the h^2 estimates for the probandwise concordance rates of 24.1% and 8.0% are 68% and 72%, respectively. For Fig. 10.2 we have taken the rates in relatives at the level including probable schizophrenia; again the overall impression would have been little affected had we used only "strictly" diagnosed rates. Earlier questions we and others raised about the suitability of the Falconer method for MZ twin data have been resolved by Smith's refinements; and since the child regression on midparent is the same as that for one MZ twin on his co-twin (1.0), we can calculate the h^2 from risks to dual mating schizophrenics' offspring.

It is the consistency of the estimates rather than their absolute values that is our main concern. Figure 10.2 shows the results of this procedure; the results are most consistent at values of population risk of about 1%, yielding heritability estimates close to 85%.[2] It can be seen that the MZ and dual mating data are least sensitive to changes in q_p, while the second-degree data are very much affected by changes in q_p exceeding 1%.

When pooled data on the risk to parents is subjected to the procedure, we must take account of the lower value of q_p in a sample selected for mental health (cf. Mednick *et al.* 1971); by halving a risk of 1% to .5% and entering Smith's nomograph with a risk q_R for parents of 5.5%, we obtain an estimate of heritability of 72%, not too much different from the values of unselected relatives at $q_p = 1$%. The consistency of h^2 estimates across relatives sharing different amounts of environmental communality provides one line of evidence in favor of polygenic theory.

As with the example of CL(P), we can show a sharp drop in the incidence of the condition as we move to more remote relatives rather than the reduction by one-half at each step predicted by simple dominant gene theory.

MZ co-twins	50% $\longrightarrow$	DZ co-twins	9% $\longrightarrow$	Second-degree	3.3%
		Sibs	10%	(pooled)	
		Children	14%		

The data on cousin risks for CL(P) gave good support for its polygenic inheritance. However, the incidence of schizophrenia in the general population is too high to allow the risks in third-degree relatives to be meaningful evidence for a choice between theories; the median risk to cousins in the literature is 2.6%, based on limited data.

---

[2] Nancy Mendell (personal communication) and Mendell and Elston (1971) kindly calculated the heritabilities for our data using tetrachoric functions as compared with Smith's solution by numerical integration of the normal curve; both approaches yielded very comparable results and neither implies high concordance rates in MZ twins until the heritability is virtually 100%. It should be noted that standard errors of heritability estimates are quite high since they depend on the number of affected relatives of probands.

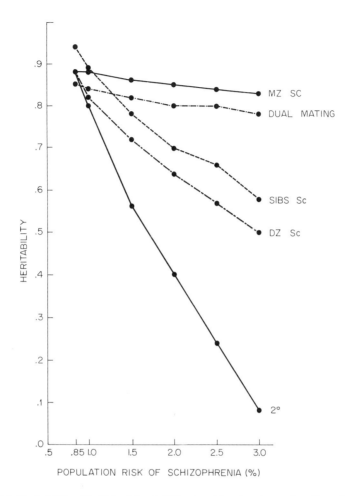

Fig. 10.2. Heritabilities (Smith) of the liability to schizophrenia as a function of varying population risks, estimated from risks in different classes of probands' relatives.

According to a polygenic hypothesis, affected antecedent relatives would occur less often on only *one* side of the family than under a monogenic hypothesis. Slater (1966) developed this idea to discriminate between the merits of monogenic and polygenic theories and concluded from an empirical test on family data that compatibility was somewhat better with dominant gene theory (Slater & Tsuang, 1968).

In Chapter 2 we reviewed some of the evidence which showed a higher schizophrenia risk to the children of hebephrenic and catatonic (i.e., severe) cases than to the children of paranoid and simple cases (Table 2.3). In Chapters 6 and

9 we showed that in most twin studies concordance was higher when the proband's illness was severe than when it was mild (pp. 225 and 309). The argument relating severity to risk can be extended (Kay, 1963; Shields, 1968) to disorders such as paranoia, late paraphrenia, recovered schizophrenia, and schizo-affective psychosis; in these the risk for schizophrenia in sibs was between 3% and 9%, that is, between the rate for sibs of classical schizophrenics and the population base rate. These kinds of data are analogous to the increased risk of CL(P) to relatives of bilateral over unilateral cases, and provide further support for polygenic theory. However, for a diathesis-stress disorder as opposed to a congenital malformation, considerations of the relationship between severity and risk of affectation may be confounded by environmental contributions to severity, for example in diabetes mellitus (Falconer, 1967).

Risk as a function of the number of relatives already affected with schizo-phrenia provides a further test of polygenic theory. Table 2.4 showed the increase in risk (a) to probands' sibs depending on whether a parent was schizophrenic or not and (b) to children depending on whether one or both parents were affected. Simple monogenic theory would predict no increase in the case of sibs and a rise from 50% to 75% for children. However, current modifications of monogenic theory predict a considerable rise under both (a) and (b). Slater and Cowie (1971) calculated the extent of these increases for Slater's theory. We compared the increased risks with those predicted by polygenic theory (Smith, 1971) taking q_p at 1% and h^2 at 80% as the tabled values closest to our interpretation of Fig. 10.2. The risks are given in Table 10.1. For sibs, both theories are equally successful at predicting the empirical risks. In the cases of children, when one parent is affected the observed risks are too high for both theoretical predictions; if Kallmann's data were omitted or if the median empirical risk (9.7%) were used as a criterion, the fit with both theoretical predictions would be much improved.

TABLE 10.1

Schizophrenia Risk as Function of Parent Status

Risk (%)	(a) To probands' sibs		(b) To probands' children	
Number of parents affected	0	1	1	2
Observed, Table 2.4	9.7	17.2	13.9	46.3
Predicted, polygenic	6.5	18.5	8.3	40.9
Predicted, monogenic	9.4	13.5	8.8	37.1

Kallmann (1938) found a higher rate of schizophrenia (25%, 16/64) in children of schizophrenics when an uncle or aunt was also affected than the 16.4% risk in the offspring of schizophrenics generally. Much more extensive data on risk to children or sibs, as a function of number of relatives affected, are available from Ødegaard's (1972) Norwegian study. Subdividing 1795 sibs of his probands, he found 8% were psychotic when there was no psychotic relative in the parental generation (parent, uncle, or aunt), 15% when there was one psychotic relative, and 21% when there were two or more.

Psychotic relatives in Ødegaard's study included all functional psychotics and not just schizophrenics. The presence of other abnormalities in the families of schizophrenics (e.g., Tables 2.2, 2.8, 2.10, 2.11, 2.12, 6.7, and 9.3) is taken as more consistent with polygenic than monogenic inheritance. If transmission were "simple" monogenic, we would usually expect to find (a) an excess of schizophrenia and no excess of any other abnormality and (b) an unambiguous bimodal distribution of affected and unaffected in the relatives of probands. When polygenic theories were adopted for the field of psychopathology (Slater & Slater, 1944, for psychoneurosis; and Ødegaard, 1950, 1963, for psychosis), it was on the assumption that they would account for the continuity observed fiom pathology to normality in both probands and relatives along more than one dimension. Such observations encouraged others to invoke a concept of schizophrenic spectrum to include some or all of the various subthreshold conditions found in the relatives of schizophrenics. Some recent proponents of a polygenic theory for schizophrenia have not explicitly rejected the equating of this view with the retrograde belief in a neuropathic taint or a unitary psychosis (*Einheitspsychose*); however, we concur with Ødegaard's (1972) rejection of the latter and with his conclusion "that there is a strong tendency for the psychotic relatives of schizophrenics to develop schizophrenia rather than other types of functional psychosis, and within the schizophrenic group, there is a tendency toward similarity in clinical picture ... [p. 268]."

We agree with Anderson (1972) that it is not too helpful to rely on evolutionary theory in deciding among genetic models; we simply do not know enough about how any human behavior evolved (cf. Gottesman & Heston, 1972). However, data on the fertility or Darwinian fitness of schizophrenics is interesting and important in its own right. The question of how a disadvantageous genetic condition can be maintained in the population over time despite the greatly reduced fitness of both male and female schizophrenics (e.g., Slater, Hare, & Price, 1971) can perhaps be answered more readily by polygenic than monogenic theory. The former would obviate the need to find a selective advantage in gene carriers hypothesized by the balanced polymorphism theory of Huxley *et al.* (1964). Response to natural selection against a polygenic trait associated with lowered marriage and fertility rates would be very slow. Genes in the system would only be eliminated from the gene pool when they were present in the rare individual at the tail end of the distribution, while those below the threshold would not be subject to negative selection. Schizophrenics could be thought of as part of the genetic load, the price paid for conserving genetic diversity. In passing, we may note that high heritabilities suggest that the traits concerned may not have been objects of directional selection pressures and so may be irrelevant to the evolution of our species.

The evidence we have adduced in favor of polygenic threshold inheritance shows that it is an equal contender with current monogenic theories. On general grounds polygenic inheritance appears more likely to us; the commonest disorder for which single-gene inheritance has been established, cystic fibrosis of the pancreas, is about 20 times rarer than schizophrenia. However, there is considerable overlap between the two principal models, and the tests proposed for differentiating between them are far from efficient. As Slater and Cowie (1971) state, "Two genetical models are available, either of which provides an adequate framework for the observations, so that the worker is entitled to choose the model which suits his purposes best." To these we would add heterogeneity theories. Our own preference for a polygenic framework leads us to look for specific and important contributing factors on both the diathesis and the stressor sides of the model. Refutation of a polygenic theory would come about by the discovery of an endophenotype which segregated in a monogenic way in all schizophrenics.

FACTORS CONTRIBUTING TO LIABILITY

It would be both defeatist and incorrect to assume that because a trait such as the liability to schizophrenia is inherited polygenically, the search for cause has ended and relevant specific genetic loci are undiscoverable in principle. The genes underlying continuous variation are not qualitatively different from those associated with discontinuous traits at the molecular level—both are subject to the same rules of inheritance because they are chromosomal and thus segregate, show dominance, epistasis, linkage, and genotype-environment interactions (cf. Penrose, 1938). From the beginning of this century (see Mather, 1949, for references) geneticists have succeeded in identifying specific loci in polygenic systems and in locating them on specific chromosomes by linkage with major genes[3]; however, the feats were accomplished with genetically tractable organisms such as wheat and *Drosophila* (Thoday, 1961, 1967). It is heuristically important to us to learn that whenever polygenic variation has been studied under *laboratory conditions* (e.g., inbreeding, backcrossing, availability of chromosome markers), "a few handleable genes have proved to mediate a large part of the genetic variance under study" (Thoday, 1967). One locus in wheat

---

[3] The difficulties and complexities of establishing linkage groups, let alone locating them on particular chromosomes, should not be underestimated (Mohr, 1964; Jayakar, 1970), but the growing list of polymorphic markers and the impressive progress in mapping human autosomes (Renwick, 1969) suggest that the optimism of Thoday (1967) and Elston (personal communication) may be warranted.

accounted for 83% of the variance in ripening date, with three others jointly accounting for 14%. Such "weighted genes" for bristle number were mentioned on pp. 320-321, but we must add the striking fact that separable components of the complex character permitted their study as more or less discontinuous variables. Wright (1934a,b) concluded that three or four major factors (genes) controlled the threshold character polydactyly in inbred lines of guinea pigs.

Encouraged by the demonstration that the genes in polygenic inheritance need not be and often are not roughly equal in their effects on the phenotype, and bolstered by our clinical observations on what appears to be "excess" similarity between pairs of affected relatives on the simpler equal effects assumption, we hazard the speculation that there are a few genes of large effect in the polygenic system underlying many schizophrenias. In other words, we view the etiology of schizophrenia as being due to a weighted kind of polygenic system with a threshold effect.[4] Some of the heuristic implications of our speculations about high-value genes in the polygenic system underlying schizophrenia include a focusing on partitionable facets of the syndrome such as catatonia, paranoid features, protein polymorphisms in brain and blood, and neurophysiology, on the chance that family studies will reveal one or more of the high-value genes. Perhaps the credibility of this possibility can be buttressed with a quotation from Thoday (1967).

> ... the different genes of a polygenic system may have quite specific and quite different effects on the character and if we will but make serious attempts to isolate relevant genes these specific effects and their modes of interaction are open to study and might thereafter be used as diagnostic criteria of the genes, permitting them to be studied by the normal techniques used in the study of discontinuous variation. [Furthermore] the technique of genetic analysis of continuous variables may be used as a powerful tool for the developmental analysis of complex character differences with the aim of determining the unitary genetic components which may thus be opened to physiological and biochemical studies. I need hardly stress how valuable this approach might be in behavior genetics, whether of mice or man [p. 344].

There are already suggestions in the human genetics literature that a gene associated with a biochemically different insulin in juvenile diabetics may be identified as one of the polygenes causing early-onset diabetes (Falconer, 1967), and one facet of congenital dislocation of the hip, joint laxity, may segregate as a mendelian trait (Wynne-Davies, 1970).

The contribution of specific genetic factors to the genetic liability to schizophrenia analogous to the specific contributors in the diabetes and hip examples

---

[4] It is reasonable to raise questions about the number of genes in the system, the number of alleles at each locus and their frequency in the population, and whether there are e.g., 2 or 3 genes of large effect. Such questions have been entertained by Lush (1945), Wright (1968), and Hammond and James (1970) for animals and plants, but they must be deferred to the future for human polygenic traits.

above forms only part of the picture in respect of the *total* liability to schizophrenia. General genetic contributors which serve as modifiers or potentiators,
together with general environmental contributors which serve as modifiers or
potentiators, each define dimensions of liability which combine with the specific
genetic liability to determine the net liability and the position of an individual
vis à vis the threshold at a particular time. The concepts involved are illustrated
in Fig. 10.3. Additional axes to provide for dimensions of genetic and environmental *assets* are also shown; they are intended to illustrate the role of factors
which may decrease the total liability to schizophrenia, permitting an individual
to stay compensated or to go into a remission (e.g., via phenothiazines and

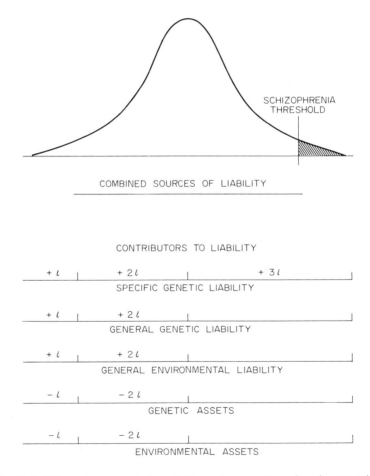

Fig. 10.3. Schematic representation of the various genetic and environmental contributors to the liability to schizophrenia.

rational psychotherapy) despite high genetic and high environmental liability. The five diagrammatic dimensions are marked off in liability units (1*l*, 2*l*, etc.) of varying length to indicate the differential weighting of both gene effects and environmental effects. At this rough stage of model building we may take the contribution of each dimension to the combined liability as additive, but with the reservation that important interaction effects must be provided for in the future. The curve in the figure shows the distribution of combined liability units in the general population together with the point defining the threshold value of liability determining the presence of schizophrenia. If the three liability values plus the two asset values sum to a suprathreshold value for an individual, he is a diagnosable schizophrenic at that point in time. The concepts illustrated in Figure 10.3 are admittedly oversimplified and overschematic. There may well turn out to be specific environmental liability to schizophrenia in addition to the general one we have provided for. The number, weight, and population frequency of the factors contributing to the five (three positive and two negative) liability axes are presently debatable as is their specificity to schizophrenia. A major advantage of the scheme is that it does show clearly how two individuals matched in their specific genetic liability or in their general environmental liability can have different combined liabilities and hence have different phenotypes with respect to schizophrenia.

DIATHESIS-STRESS AND THE UNFOLDING OF SCHIZOPHRENIA

The static depiction of schizophrenia in the figure above is not very satisfying when it comes to communicating knowledge about the changes over time which add to or subtract from the combined genetic predisposition to schizophrenia. A more dynamic picture of an individual's trajectory through life is needed to do justice to the concept of a genetic diathesis interacting with stress to produce the varieties of schizophrenic phenotypes (cf. Bleuler, 1968). Figure 10.4 presents a trial scheme that incorporates the ideas of changes in the combined liability over time from environmental sources and from the environmental triggering of genes that had not been switched on at birth. The time axis starts with zero time at the moment of conception. Both chance (random) factors and ontogenetic constitutional changes will influence the trajectories, leading to both upward and downward inflections. Environmental stressors coming close together in time would be expected to exert a cascade effect and have more effect than the same stressors spread out in time. Our comments on the case histories about vicious circles, aversive drift and escalation, and feedback effects may be more intelligible in the light of Fig. 10.4.

G_1 is intended to indicate the trajectory of a person with a low (for schizophrenics generally) combined *genetic* liability to schizophrenia; over time environmental contributors to liability, say first the death of a spouse and then

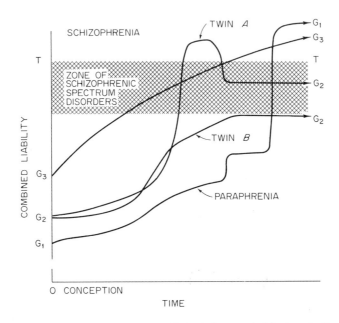

Fig. 10.4. Schematic proposal for how diathesis interacts with stress in the ontogenesis of schizophrenia.

the onset of deafness, cause upward deflections of his trajectory, culminating in a late-onset paraphrenia. The dashed line at the bottom of the zone of schizophrenic spectrum disorders is intended to convey the idea of a possible need for a second threshold in our model; Wright (1934b) invoked a second threshold to account for the imperfectly formed fourth digit seen in crosses between a high and a moderate line of guinea pigs with liabilities to polydactyly.

G_2 could be the divergent trajectories of a pair of MZ twins; only the A-twin encounters the sufficient factors over time leading to schizophrenia for a person with his genotype. The B-twin at the time of observation is discordant for schizophrenia, but close to the threshold of schizophrenic spectrum disorders. Subthreshold values of combined liability make it clear why so many of our first-degree relatives (Chapter 7) could have had normal MMPI's and why two phenotypically normal parents are typical for the vast majority of schizophrenics. The A-twin is shown to have an acute onset with an undistinguishable premorbid personality, and a remission from schizophrenia into a chronic schizoid state.

G_3 is the posited trajectory of a person with a high genetic loading needing very little in the way of environmental contributors to make him schizoid; he is shown as having a poor premorbid personality, an insidious onset, and a deteriorating course. Many other life trajectories could have been drawn to illustrate the unfolding of our twins' schizophrenias. It is easy to see how the hospitalization data in Fig. 6.1 and the fascinating histories of the Genain quadruplets (Rosenthal, 1963) would augment the total perspective about the pathogenesis of schizophrenia.

BREADTH OF THE SCHIZOPHRENIC SPECTRUM

As long ago as 1911, Bleuler said "We should know which diseases, particularly which psychoses, in a family have any connection with the schizophrenia of one of its members." That question has not yet been resolved; there are almost as many opinions as investigators. Mitsuda (1967) holds that certain forms of schizophrenia are related to epilepsy. Heston's (1970) schizoid disease includes not only schizoid psychopathy but perhaps some neurotic and affective disorders and mental retardation. The schizophrenia spectrum disorders as defined by Kety *et al.* (1968) include borderline schizophrenia and inadequate personality and their "extended spectrum" (Table 2.12), character disorder, psychopathy, criminality, and suicide; at one time it was thought it might even include manic-depressive psychosis. Some proponents of polygenic inheritance believe that inheritance must be weak and nonspecific because neuroses and personality disorders outnumber schizophrenias in the twins and other relatives of schizophrenics. We have already argued (pp. 32-33, 40, 329) that the diagnostic similarity between relatives does not support this opinion, and in Chapters 6 and 7 we examined the nonschizophrenic abnormalities found in our probands' relatives. It is the recurrence of the same rare disorder in families that supports the existence of specific factors in inheritance. An excess of schizophrenia is found in co-twins and children of schizophrenics whether reared with them or not. The observation of neurosis in relatives must take into account that the prevalence of neurosis in the general population, while difficult to assess, may be 25 to 50 times greater than that of schizophrenia itself.

With regard to the different etiologies of the psychoses, Elsässer's (1952) study of dual matings illustrates the apparent genetic distinction not only of schizophrenia and manic-depression but also that of some of the hard-to-classify "atypical" psychoses. He analyzed the 134 dual mating families collected by Schulz, Kahn, and himself, rediagnosing the cases according to a uniform standard. The total risk for psychosis in the children did not vary greatly from one kind of mating to another, but their diagnostic distribution did. The psychotic offspring of two schizophrenics (S X S matings) were all schizophrenic, and those of two manic-depressives (M X M) nearly all M. In 19 S X M matings, eight of the psychotic children were S, eight M, one atypical (A), and two of uncertain diagnosis; the parental psychoses "segregated" rather than blended. Among the offspring of 17 A X A matings there were 5 S, 6 M, 11 A and two uncertain, suggesting that while some of the atypical parents were genetically S or M, several of them had psychoses which were genetically distinct from either S or M. Of course, as in most family studies, Elsässer's diagnoses were not made blindly.

Rosenthal (1970) concluded that the combined weight of the evidence was that schizophrenia and manic-depressive psychosis were different mental illnesses; "Each seems to represent the endpoint of a gradation or spectrum of clinical-behavioral manifestations that are genetically linked" [p. 276]. The clinical features of each spectrum were considered to be distinct. But he urged the adoption of strict methodological precautions in attempts to decide what are the basic characteristics of the schizophrenic and manic-depressive diatheses. We suggested (Shields, 1971) that views as to what belongs to the schizophrenic spectrum can be tested by means of family studies of schizoid, paranoid, and inadequate personalities, criminals, and so on. If they were all basically of the same genotype as schizophrenics, or were at least in the upper reaches of the genetic liability to the disorder, one would expect to find a raised incidence of typical schizophrenia in the children of two such parents. Family studies of psychopaths[5] and neurotics do not show an excess of schizophrenia. It would be misleading to regard "spectrum disorders" in the population as regularly reflecting a schizophrenic genotype. It is now commonplace to expect one gene to affect many traits and for each trait to be affected by many genes.

Enlarging the sample of adoptees in the Danish study summarized in Table 2.11 from 86 to 143 (Rosenthal, 1971) confirmed the strong view we are taking about the specificity of schizophrenia; while schizophrenia and borderline schizophrenia discriminated the index adoptees from controls well, conditions lower in the spectrum did not. Our multiple judge diagnostic study reported in Chapter Five indicated that at our present stage of knowledge a middle-of-the-road diagnosis of schizophrenia still has value as a phenotype for genetic analysis. We are optimistically hopeful that the current mass of research on families of schizophrenics will discover an endophenotype, either biological or behavioral (psychometric pattern), which will not only discriminate schizophrenics from other psychotics, but will also be found in all the identical co-twins of schizophrenics whether concordant or discordant. All genetic theorizing would benefit from the development of such an indicator.

ENVOI

Prevention is the ultimate goal of research into etiology. Paradoxically, advances on the genetic front lead to focusing on the environment and to the

---

[5] The intriguing possibility exists that some combinations of psychopathology may cancel each other's effects. Mednick (personal communication) has preliminary data sug-

mechanisms underlying the interaction with heredity. Harris (1970), a leading human biochemical geneticist, concluded that "one of the most important social and medical applications of genetical research will lie in the control of the environment, since the more it becomes possible to characterize the genetical constitution of an individual precisely, the more likely are we to see how to modify or tailor the environment according to his needs" [p. 265]. Environment in this context refers to the inputs from the external world of the individual as well as those from the internal environment of organs, tissues, and cells.

It is important to understand the implications of finding that a trait such as the liability to schizophrenia has a high heritability. In the samples so far studied, it means that environmental factors were unimportant as causative agents of the schizophrenias. However, and this cannot be emphasized too strongly, these data do not permit the conclusion that curative or preventive measures will be ineffective. As Falconer (1965, p. 69) has pointed out, "The environmental factors proved to be unimportant are those operating in the population sampled and these do not include special treatments or preventive measures. *No prediction can be made from a knowledge of the degree of genetic determination about the efficacy of curative or preventive treatments. All that could be said in such a case is that one will have to look outside the range of normal environments experienced by the untreated population (italics added)."*

The beauty of a diathesis-stressor theory, or philosophy if you will, is that it fills the chasm between geneticism and environmentalism. Our preferred model for construing the syndrome of schizophrenia permits the clearer separation of etiological and phenomenological considerations. It comprises a network of events connected by sequential causal arrows. A chain of consequences is set into motion by a genetically caused predisposition and culminates in a set of symptoms recognizable as schizophrenia. Feedback loops and chance have important roles in the total picture. Our construction clarifies how psychotherapy or chemotherapies or a good mother may each contribute to symptom amelioration without necessarily casting light on etiological questions. It is our hope that a heuristic genetic theory about the etiology of schizophrenia will hasten the day that brings it under man's control.

---

gesting that the GSR characteristics of the children of schizophrenics are less schizophrenic-like and more normal when the other parent was psychopathic, than children whose other parent was normal.

APPENDIX

A. CUE SHEET FOR TWIN INTERVIEW

1. Introduction of self and project.

2. (Affect) How are you feeling? How are your spirits? Appetite? Weight changes? What are you doing now in the way of work? How do you feel about the future? Do you show your feelings by crying?

3. (Anxiety) What worries you most right now? Are you afraid of anything? Do you have any trouble sleeping? Are you taking any kind of medicine now?

4. (Intrafamily and self-concepts) What kind of a person is (was) your mother? . . . your father? What kind of person are you? Do you take more after your mother or your father in your personality? In what ways do you differ from your twin?

5. (Anhedonia) How do you pass the time? What interests you in the newspapers or on television? What do you think about the present world situation? Who are your close friends? What kinds of nice things do you do for yourself? What gives you the most pleasure?

6. (Thinking, Delusions) How is your mind working? How does your head feel? How is your concentration? Do you have any aches or pains? What do you think people feel about you? How do you compare with your twin and other people?

7. (Past, childhood) What led up to your illness? What important things happened to you as a child that influenced you over the course of time? How did you get on with your twin? Were you ever mistreated by your family? What would you like to change about the past if you had the power to? How did you feel about being a twin?

8. (Symptoms) Have you any habits which seem a little silly to you? Do any ideas keep repeating themselves to you? Do you ever see or feel or hear anything unusual? Did you in the past? Does the world seem strange to you in any way?

9. (For co-twins) Where were you when your twin first had a breakdown? How did you feel about it? Did you worry that the same thing might happen to you? How do you account for the fact that your twin got ill and you did not? How do you think his breakdown could have been prevented? Were you satisfied with the care he received?

10. Do you have any questions you'd like to ask me?

B. THE CALCULATION OF TWIN CONCORDANCE RATES

Different Kinds of Concordance

The concept of concordance has led to a certain amount of confusion. There are a number of different ways in which the term may legitimately be used, and these need to be clearly distinguished (Allen, Harvald, & Shields, 1967). We shall illustrate the main sources of confusion by means of our earlier data on hospitalization and diagnoses of schizophrenia in MZ twins, reported on page 67.

Concordance: Complete and Partial, Narrow and Broad

The concept of concordance implies dichotomy, and dichotomies can be drawn arbitrarily. Therefore concordance, though it is usually taken to denote the occurrence of the *same* condition, e.g., schizophrenia, in both twins, as in our Grade I, can also usefully be reported in such a way that it includes pairs where the second twin has a lesser, or less certain, degree of abnormality, than the proband along some dimension not easily quantifiable. For instance, pairs may be reported as concordant, or *partially concordant,* in which one twin is schizophrenic and the other is doubtfully schizophrenic, "borderline" or even schizoid.

It may also be relevant to ask whether twins are similar in respect of a less specific category. It is sometimes claimed that it is not so much schizophrenia that is inherited as a more general and nonspecific tendency to psychiatric abnormality. A *broader type of concordance* may therefore be reported so as to include abnormalities in the second twin of a kind apparently different from that by which the original case was selected. For instance, pairs in which the first twin was schizophrenic and the second twin psychopathic, neurotic or even mentally retarded, may be regarded as concordant, both being abnormal on the broad dimension of *psychopathology.*

Our Grade II concordance includes pairs where both twins were *hospitalized,* not necessarily for the same or even related conditions: they are concordant in respect of the category of psychiatric hospitalization. As stated in Chapter 2, the purpose remains to contrast one rate with another, especially MZ:DZ, so as to permit an inference about the role of genetic factors.

Casewise and Pairwise

Concordance may refer to the proportion of cases with an affected partner (casewise rate) or the proportion of pairs in which both twins are affected (pairwise rate). For a given set of data, casewise and pairwise rates are related to one another in a constant manner, since *each concordant pair consists of two cases.* If the casewise rate is defined as $C/(C + D)$, where C = the number of cases with affected partners and D = the number with unaffected partners, the corresponding pairwise rate will = $\frac{1}{2} C/(\frac{1}{2} C + D)$. Our MZ Grade I pairwise rate of 10/24 (42%) has as its corresponding casewise rate 20/34 (59%); the 24 pairs included 34 schizophrenics, 20 from concordant and 14 from discordant pairs. However, all schizophrenics did not come to notice independently.

The Proband Method

Probands are the index cases of a genetic investigation, the cases having been independently ascertained in a defined manner. Confusion arises in some studies over whether concordance has or has not been calculated according to the proband method.

The kind of problem which the proband method of Weinberg (1912, 1928, and Crow, 1965) was originally designed to tackle was how to estimate the incidence of a disorder in a sibship in order to discover how closely it approximated to the 25% expected in the offspring of two heterozygotes under a recessive hypothesis. Some early attempts to estimate parameters of such a kind had fallen into the naive error of pooling sibships and calculating the proportion of affected to unaffected members without first excluding the cases by which the sibships came to be ascertained; inflated rates were thereby produced. The proband method omits probands from the calculation, but if there is more than one proband in the sibship the family will enter more than once. The method need not, of course, imply the testing of recessivity or any other monogenic hypothesis, but is applicable to all data where independent ascertainments of affected individuals had been recorded. In twin studies, concordance calculated by the proband method is intended to give an unbiased estimate of the ratio of affected to normal in the twins of given cases: twin pairs are treated as sibships of two.

The proband method is not a device thought up by wicked geneticists to make a condition appear to be "more genetic" than it really is by the unwarranted double counting of pairs in which both twins are probands. The proband method rate should not be thought of as the inflation of a pairwise rate, but as the reduction of a casewise rate. The method does not, strictly speaking, count pairs twice at all, but co-twins. Probands are omitted from the calculation unless they also happen to be co-twins in their own right. The obtained concordance is a rate in co-twins, and as such it may be compared with rates in other relatives or in unrelated controls. Provided the probands in a study are representative of cases in a population, the proband method rate will approximate the casewise rate for twins in the population.

It will be clear that concordance calculated according to the proband method expresses the proportion of probands who have affected partners. In other words it is a *casewise rate, corrected for mode of ascertainment.* We ascertained via our twin registration system not 34 but 28 schizophrenics from MZ pairs. Of these, 14 had affected co-twins and 14 had unaffected co-twins. The proband method rate is therefore 50%. If all the 34 schizophrenics observed by us had been ascertained as probands, the proband rate would have been the same as the directly observed casewise rate. If no pair had included two probands, it would have been the same as the directly observed pairwise rate. In fact, as we have already reported, four out of the ten concordant MZ pairs consisted of twins both of whom were independently ascertained as schizophrenic probands on our register. The proband method rate lies in between the two direct rates of 42% and 59%.

The pairwise rate which corresponds to the (casewise) proband method rate cannot be directly observed, but is estimated indirectly by halving the number of probands (cases) with affected co-twins, as in the formula shown in the previous section (Allen, 1955). The usefulness of the indirect pairwise rate, which is the estimate of a population parameter, appears to have limited value for our purpose.

The relationship between casewise and pairwise on the one hand and directly observed and proband-based on the other hand can be seen in Table B.1. The nomenclature of the four kinds of concordance generated is also shown.

TABLE B.1

Four Different MZ Grade I Concordance Rates Uncorrected for Age

		Calculations based on:	
		All schizophrenics observed	Ascertained schizophrenic probands
Casewise	$\dfrac{C}{C + D}$	20/34 = 58.8% raw casewise	14/28 = 50.0% proband method
Pairwise	$\dfrac{\frac{1}{2}C}{\frac{1}{2}C + D}$	10/24 = 41.7% direct pairwise	7/21 = 33.3% indirect pairwise

In our own study we shall normally report the direct pairwise rate. It has the merit of simplicity and it is the only rate based on the raw number of pairs observed, from which the number of degrees of freedom is calculated. It may be regarded as a conservative estimate of the proband method rate. However, when it is appropriate, some analyses will be given in terms of co-twins of Maudsley Hospital probands, i.e., probandwise. Slater and Cowie (1971, Appendix C) have shown how the proband method has the advantage of taking into account pairs where neither twin is affected, and V. E. Anderson (personal communication) has shown that this conclusion holds independently of the probability of ascertainment (π) (cf. Crow, 1965).

Age-Corrected Concordance

Estimates of the morbid risk for co-twins have to take account of their age because the disorder has a variable age of onset and the sample is a current one. Various methods, none of them entirely satisfactory, have been suggested for age correcting. They involve reducing the denominator in such a way as to allow for the proportion of the risk period for the condition not yet lived through by the co-twins (see Chapter 6 and Appendixes D and E in Slater and Cowie, 1971).

C. INDIVIDUAL JUDGE AND CONSENSUS DIAGNOSES

Twin	Consensus	P M	E S	K A	J P	J B	L M
MZ							
1A	S	S	S	S	S	O	S
1B	S	S	S	S	?S	O	S
2A	S	S	S	S	S	S	S
2B	S	S	S	S	S	S	S
3A	S	S	S	S	S	S	S
3B	N	N	N	N	N	N	N
4A	S	S	S	S	S	S	O
4B	N	N	N	N	N	N	N
5A	S	S	S	S	S	O	S
5B	O	O	O	O	O	O	O
6A	S	S	S	S	?S	S	S
6B	?S	S	O	S	?S	O	?S
7A	S	S	S	S	S	S	S
7B	S	S	S	S	S	S	S
8A	S	S	S	S	S	S	S
8B	N	N	N	N	N	N	N
9A	?S	S	O	O	S	?S	O
9B	O	O	O	O	O	O	O
10A	O	S	O	O	S	O	O
10B	O	O	O	O	O	O	O
11A	S	S	S	S	S	O	S
11B	N	N	N	N	N	N	N
12A	O	S	O	O	O	O	O
12B	N	N	N	N	N	N	N
13A	?S	S	O	O	S	O	?S
13B	S	S	S	S	S	?S	S

C. Continued

Twin	Consensus	P M	E S	K A	J P	J B	L M
MZ							
14*A*	S	S	S	S	S	S	S
14*B*	O	S	O	N	O	O	O
15*A*	S	S	?S	S	S	S	S
15*B*	S	S	S	S	S	?S	S
16*A*	S	S	O	O	S	O	S
16*B*	O	?S	O	O	O	O	N
17*A*	S	S	S	S	S	S	S
17*B*	S	S	S	S	S	S	S
18*A*	S	S	S	O	S	O	S
18*B*	O	S	O	O	O	O	O
19*A*	S	S	S	S	S	S	S
19*B*	S	S	S	S	S	?S	S
20*A*	S	S	O	O	S	S	?S
20*B*	N	N	N	N	N	N	N
21*A*	S	S	S	?S	S	O	S
21*B*	S	S	S	S	S	S	S
22*A*	S	S	S	S	S	O	S
22*B*	S	S	S	S	S	S	S
23*A*	S	S	S	S	S	S	?S
23*B*	?S	S	?S	N	?S	O	O
24*A*	S	S	S	O	S	S	S
24*B*	O	O	O	O	O	O	O

Twin	Consensus	P M	E S	K A	J P	J B	L M
DZ							
1*A*	S	S	S	S	S	S	S
1*B*	N	N	N	N	N	N	N
2*A*	S	S	S	O	S	S	?S
2*B*	N	N	N	N	O	N	O
3*A*	S	S	S	S	S	O	S
3*B*	N	N	N	N	N	N	N
4*A*	S	S	S	S	S	S	S
4*B*	S	S	S	S	S	S	S
5*A*	S	S	S	S	S	S	?S
5*B*	O	O	O	O	O	O	O
6*A*	S	S	S	S	S	S	S
6*B*	N	N	N	N	N	O	O
7*A*	S	S	S	S	S	S	S
7*B*	S	S	S	S	S	S	S
8*A*	S	S	S	S	S	S	S
8*B*	N	N	N	N	N	N	N
9*A*	S	S	S	S	S	S	S
9*B*	N	N	N	N	N	N	N
10*A*	S	S	S	O	S	S	O
10*B*	S	S	S	S	S	S	S

C. Continued

Twin	Consensus	P M	E S	K A	J P	J B	L M
DZ							
11*A*	S	S	S	O	?S	O	S
11*B*	O	O	O	N	O	O	O
12*A*	S	S	S	S	S	S	S
12*B*	N	N	N	N	N	N	N
13*A*	S	S	S	S	S	S	S
13*B*	N	N	N	N	N	N	N
14*A*	S	S	S	S	S	S	S
14*B*	O	S	N	N	O	N	O
15*A*	S	S	S	S	S	S	S
15*B*	N	N	N	N	N	N	O
16*A*	S	S	S	S	S	S	S
16*B*	N	N	N	N	N	N	N
17*A*	S	S	S	S	S	?S	O
17*B*	O	O	O	O	O	O	O
18*A*	S	S	S	S	S	S	S
18*B*	N	N	N	N	N	N	N
19*A*	S	S	S	S	S	S	S
19*B*	N	N	N	N	N	N	N
20*A*	S	S	S	S	S	?S	S
20*B*	O	N	O	N	O	O	N
21*A*	S	S	S	S	S	S	S
21*B*	O	O	O	O	O	O	O
22*A*	S	S	S	S	S	S	S
22*B*	N	N	N	N	O	N	N
23*A*	S	S	S	S	S	S	S
23*B*	N	N	N	O	N	N	O
24*A*	S	S	S	O	S	S	S
24*B*	O	S	O	O	O	O	O
25*A*	S	S	S	S	?S	O	O
25*B*	O	N	O	O	N	N	O
26*A*	S	S	S	S	S	O	?S
26*B*	N	N	N	N	N	N	N
27*A*	?S	S	O	O	S	O	?S
27*B*	O	S	O	O	O	O	O
28*A*	S	S	?S	S	S	O	S
28*B*	N	N	N	N	N	O	N
29*A*	S	S	S	S	S	S	?S
29*B*	N	N	N	N	N	N	O
30*A*	S	S	S	S	S	O	S
30*B*	N	N	N	N	N	N	N
31*A*	S	S	S	S	S	S	S
31*B*	N	S	N	N	N	N	O
32*A*	?S	S	O	O	S	O	?S
32*B*	N	?S	N	N	N	N	?S
33*A*	S	S	S	S	S	?S	O
33*B*	N	N	N	N	N	N	N

D. ESSEN-MÖLLER'S CLASSIFICATION OF THE TWINS

	Consensus-concordant			Consensus-discordant	
Pair[b]	Classification group[a]		Pair	Classification group[a]	
MZ	A	B	MZ	A	B
1	4	2	3	1	4
2	1	3	4	2	N
6	2	4	5	5	5
7*	1	1	8	1	?4
13	3	2	9	5	5
15	3	1	11	1	4
17	1	1	14	4	?4
19*	1	1	16	4	4
21*	1	1	18	3	5
22*	3	1	20	5	N
23	3	4	24	2(4)	5

[a] The twins were classified into six diagnostic groups (see text, p. 219 for fuller details). Essen-Möller classified the two consensus-discarded MZ pairs as follows: MZ 10–3, 5; MZ 12–5, N.

The following DZ *A* twins were classified as 2: 2, 8, 10, 28, 30, 33; as 3: 4, 5, 21, 24; as 4: 17, 25, 26, 32; as 5: 27; the remainder as 1. The DZ *B* twins were classified as–1: 7*, 10; 2: 4; ?4: 5, 14; 5: 6, 11, 17, 20, 21, 24, 25, 27, 29, 31, 32; the remainder as N.

[b] Asterisk indicates that both twins are probands.

E. INDEX OF PAIRED CONSENSUS DIAGNOSES[a]

	Consensus diagnosis		Case numbers	
	Twin A	Twin B	MZ	DZ
Concordant for schizophrenia	Schizophrenia	Schizophrenia	1, 2, 7*, 15, 17, 19*, 21*, 22*	4, 7*, 10
	Schizophrenia	?Schizophrenia	6, 23	
	?Schizophrenia	Schizophrenia	13	
Discordant for schizophrenia	Schizophrenia	Other	5, 14, 16, 18, 24	5, 11, 14, 17, 20, 21, 24, 25
	?Schizophrenia	Other	9	27
	Schizophrenia	Normal	3, 4, 8, 11, 20	1, 2, 3, 6, 8, 9, 12, 13, 15, 16, 18, 19, 22, 23, 26, 28, 29, 30, 31, 33
	?Schizophrenia	Normal		32
Neither twin schizophrenic	Other	Other	10	
	Other	Normal	12	

[a] Asterisk indicates that both twins are probands.

F. STANDARD DEVIATIONS FOR MEAN MMPI PROFILES IN CHAPTER 7[a]

		L	F	K	1	2	3	4	5	6	7	8	9	0
Figure 7.1														
MZ Probands	$\bar{Y}$:	54	73	51	59	71	58	68	61	71	72	78	62	63
(N = 18)	±S.D.:	8	17	10	15	19	10	17	15	16	15	17	14	12
MZ Co-twins	$\bar{Y}$:	55	64	58	61	65	63	66	56	62	65	70	61	57
(N = 16)	±S.D.:	12	13	11	17	14	12	13	9	10	12	17	11	10
1. Page														
MZ Probands	$\bar{Y}$:	53	73	51	59	73	59	69	60	70	72	78	61	64
(N = 21)	±S.D.:	8	16	10	15	19	10	17	15	15	15	17	14	12
MZ Co-twins	$\bar{Y}$:	56	66	57	62	66	63	67	57	63	67	73	62	59
(N = 20)	±S.D.:	11	15	12	19	14	11	14	11	11	12	17	10	10
2. Page														
MZ Consensus Sc	$\bar{Y}$:	54	72	52	59	70	59	68	61	70	71	77	62	62
(N = 26)	±S.D.:	10	15	11	14	18	10	16	14	14	14	17	14	12
Normal and "other"	$\bar{Y}$:	55	57	61	64	62	65	63	52	55	62	63	60	53
MZ Co-twins (N = 8)	±S.D.:	10	10	10	23	11	14	14	8	7	14	16	9	7
Figure 7.2														
DZ Probands	$\bar{Y}$:	56	71	58	67	74	69	71	61	70	70	78	58	61
(N = 19)	±S.D.:	9	19	11	14	14	10	8	13	16	16	20	12	11
DZ Co-twins	$\bar{Y}$:	54	56	57	53	61	57	60	54	51	54	58	56	57
(N = 25)	±S.D.:	9	9	8	11	12	9	10	8	8	8	9	10	10
3. Page														
MZ Co-twins (N = 20)	±S.D.:	11	15	12	19	14	14	11	11	11	12	17	10	10
DZ Co-twins (N = 26)	±S.D.:	9	14	9	12	12	10	10	8	12	11	13	11	11
Figure 7.8														
Nonworking Sc (N = 14)	$\bar{Y}$:	53	81	51	64	81	65	71	66	74	83	89	65	63
Recent hospital discharge (N = 10)	$\bar{Y}$:	54	73	52	67	77	65	70	64	70	72	80	64	63
Working Sc (N = 22)	$\bar{Y}$:	57	65	58	58	65	60	68	56	66	62	70	56	59
Table 7.6														
1. True Sc (N = 21)	$\bar{Y}$:	54	75	53	63	77	63	70	65	69	74	82	63	64
2. Sc-form Psychosis (N = 10)	$\bar{Y}$:	54	66	56	69	69	68	64	55	67	66	74	57	62
3. Possible Sc (N = 6)	$\bar{Y}$:	52	72	50	50	58	50	61	60	66	66	71	64	54
4. Schizoid personality (N = 10)	$\bar{Y}$:	59	65	60	67	74	68	72	60	67	71	73	56	60
5. Other Dx (N = 15)	$\bar{Y}$:	58	59	60	57	63	62	68	51	60	60	64	58	58
6. Normal (N = 16)	$\bar{Y}$:	51	54	57	50	57	55	56	54	49	51	54	56	55

[a] $\bar{Y}$ = Mean; S.D. = standard deviation, rounded to whole numbers to reflect the stage of development of personality testing.

BIBLIOGRAPHY

Abe, K. Patterns of relapse in remitting psychotics. *Psychiatria et Neurologia (Basel)*, 1965, **150**, 129-140.

Abe, K. Susceptibility to psychosis and precipitating factor: A study of families with two or more psychotic members. *Psychiatria et Neurologia (Basel)*, 1966, **151**, 276-290.

Abe, K. The morbidity rate and environmental influence in monozygotic co-twins of schizophrenics. *British Journal of Psychiatry*, 1969, **115**, 519-531.

Alanen, Y. O. The family in the pathogenesis of schizophrenic and neurotic disorders. *Acta Psychiatrica Scandinavica*, 1966, Suppl. 189.

Alanen, Y. O. From the mothers of schizophrenic patients to interactional family dynamics. In D. Rosenthal and S. S. Kety (Eds.), *The transmission of schizophrenia*. Oxford: Pergamon, 1968. Pp. 201-212.

Allen, G. Comments on the analysis of twin samples. *Acta Geneticae Medicae et Gemellologiae*, 1955, **4**, 143-160.

Allen, G. Twin research: Problems and prospects. In A. G. Steinberg and A. G. Bearn (Eds.), *Progress in medical genetics*. Vol. IV. New York: Grune & Stratton, 1965. Pp. 242-269.

Allen, G., Harvald, B., & Shields, J. Measures of twin concordance. *Acta Genetica et Statistica Medica (Basel)*, 1967, **17**, 475-481.

Allen, M. G., Cohen, S., & Pollin, W. Schizophrenia in veteran twins: A diagnostic review. *American Journal of Psychiatry*, 1972, **128**, 939-945.

American Psychiatric Association. *Diagnostic and Statistical Manual of Mental Disorders.* (2nd ed.) Washington: Committee on Nomenclature and Statistics of the American Psychiatric Association, 1968.

Anderson, V. E. Genetic hypotheses in schizophrenia. In A. R. Kaplan (Ed.), *Genetic factors in "schizophrenia."* Springfield, Illinois: Thomas, 1972. Pp. 490-494.

Angst, J. Zur Ätiologie und Nosologie endogener depressiver Psychosen. *Monographien aus dem Gesamtgebiete der Neurologie und Psychiatrie*, 1966, No. 112. Berlin and New York: Springer-Verlag.

Ash, P. The reliability of psychiatric diagnosis. *Journal of Abnormal and Social Psychology*, 1949, **44**, 272-276.

Astrup, C., & Noreik, K. *Functional psychoses: Diagnostic and prognostic models*. Springfield, Illinois: Thomas, 1966.

Bell, R. Q. A reinterpretation of the direction of effects in studies of socialization. *Psychological Review*, 1968, **75**, 81-95.

349

Benaim, S. The specificity of reserpine in the treatment of schizophrenia in identical twins. *Journal of Neurology, Neurosurgery and Psychiatry*, 1960, **23**, 170-175.

Birley, J. L. T., & Brown, G. W. Crises and life changes preceding the onset or relapse of acute schizophrenia: Clinical aspects. *British Journal of Psychiatry*, 1970, **116**, 327-333.

Bleuler, E. *Dementia praecox or the group of schizophrenias.* Leipzig: Deuticke, 1911. Republished: Translated by Joseph Zinkin. New York: International Universities Press, 1950.

Bleuler, M. Vererbungsprobleme bei Schizophrenen. *Zeitschrift für die gesamte Neurologie und Psychiatrie*, 1930, **127**, 321-388.

Bleuler, M. A 23-year longitudinal study of 208 schizophrenics and impressions in regard to the nature of schizophrenia. In D. Rosenthal and S. S. Kety (Eds.), *The transmission of schizophrenia.* Oxford: Pergamon, 1968. Pp. 3-12.

Bleuler M. *Die schizophrenen Geistesstörungen im Lichte langjähriger Kranken- und Familiengeschichten.* Stuttgart: Thieme, 1972.

Block, J. *The challenge of response sets.* New York: Appleton, 1965.

Böök, J. A. A genetic and neuropsychiatric investigation of a north-Swedish population. *Acta Genetica et Statistica Medica (Basel)*, 1953, **4**, 1-100.

Böök, J. A. Genetical aspects of schizophrenic psychoses. In D. D. Jackson (Ed.), *The etiology of schizophrenia.* New York: Basic Books, 1960. Pp. 23-36.

Bradford, N. H. Comparative perceptions of mothers and maternal roles by schizophrenic patients and their normal siblings. Unpublished doctoral dissertation, Univ. of Minnesota, 1965.

Bratfos, O., Eitinger, L., & Tau, T. Mental illness and crime in adopted children and adoptive parents. *Acta Psychiatrica Scandinavica*, 1968, **44**, 376-384.

Briggs, P. F. Eight item clusters for use with the M-B History Record. *Journal of Clinical Psychology*, 1959, **15**, 22-28.

Briggs, P. F., Taylor, M., & Tellegen, A. A study of the Marks and Seeman MMPI profile types as applied to a sample of 2,875 psychiatric patients. Report No. PR-66-5, Reports from the Research Laboratories of the Department of Psychiatry, University of Minnesota, 1966.

Broen, W. E. *Schizophrenia, research and theory.* New York: Academic Press, 1968.

Brown, G. W. The family of the schizophrenic patient. In A. Coppen and A. Walk (Eds.), *Recent developments in schizophrenia.* British Journal of Psychiatry Special Publication No. 1. Ashford, Kent: Headley, 1967. Pp. 43-59.

Bruetsch, W. L. Neurosyphilitic conditions. In S. Arieti (Ed.), *American handbook of psychiatry.* Vol. II. New York: Basic Books, 1959. Pp. 1003-1020.

Butcher, J. N. (Ed.) *MMPI: Research developments and clinical application.* New York: McGraw-Hill, 1969.

Cabot, R. C. Diagnostic pitfalls identified during a study of three thousand autopsies. *Journal of the American Medical Association*, 1912, **59**, 2295-2298.

Cameron, N. Reasoning, regression and communication in schizophrenics. *Psychological Monographs*, 1938, **50** (1, Whole No. 221), 1-33.

Cancro, R., & Pruyser, P. W. A historical review of the development of the concept of schizophrenia. In R. Cancro (Ed.), *The schizophrenic reactions: A critique of the concept, hospital treatment and current research.* New York: Brunner/Mazel, 1970. Pp. 3-12.

Carter, C. O. The inheritance of common congenital malformations. In A. G. Steinberg and A. G. Bearn (Eds.), *Progress in medical genetics.* Vol. IV. New York: Grune & Stratton, 1965. Pp. 59-84.

Carter, C. O. Genetics of common disorders. *British Medical Bulletin*, 1969, **25**, 52-57. (a)

Carter, C. O. An ABC of medical genetics. VI-Polygenic inheritance and common diseases. *Lancet*, 1969, **1**, 1252-1256. (b)

Carter, M., & Watts, C. A. H. Possible biological advantages among schizophrenics' relatives. *British Journal of Psychiatry*, 1971, **118**, 453-460.

Cederlöf, R., Friberg, L., Jonsson, E., & Kaij, L. Studies on similarity diagnosis in twins with the aid of mailed questionnaires. *Acta Genetica et Statistica Medica (Basel)*, 1961, **11**, 338-362.

Cerasi, E., & Luft, R. Insulin response to glucose infusion in diabetic and non-diabetic monozygotic twin pairs. Genetic control of insulin response? *Acta Endocrinologica*, 1967, **55**, 330-345.

Clark, J., & Mallett, B. A follow-up study of schizophrenia and depression in young adults. *British Journal of Psychiatry*, 1963, **109**, 491-499.

Clausen, J. A. Interpersonal factors in the transmission of schizophrenia. In D. Rosenthal and S. S. Kety (Eds.), *The transmission of schizophrenia*. Oxford: Pergamon, 1968. Pp. 251-263.

Connell, P. H. *Amphetamine psychosis*. Maudsley Monograph No. 5. London: Chapman & Hall, 1958.

Cooper, J. E., Kendell, R. E., Gurland, B. J., Sartorius, N., & Farkas, T. Cross-national study of diagnosis of the mental disorders: Some results from the first comparative investigation. *American Journal of Psychiatry*, 1969, **125** (10, April Suppl.) Pp. 21-29.

Coppen, A. J., Cowie, V. A., & Slater, E. Familial aspects of "neuroticism" and "extraversion". *British Journal of Psychiatry*, 1965, **111**, 70-83.

Craike, W. H., & Slater, E. (with the assistance of G. Burden). Folie a deux in uniovular twins reared apart. *Brain*, 1945, **68**, 213-221.

Crittenden, L. B. An interpretation of familial aggregation based on multiple genetic and environmental factors. *Annals of the New York Academy of Science*, 1961, **91**, 769-780.

Cronbach, L. J. *Essentials of psychological testing*. (3rd ed.) New York: Harper, 1970.

Cronbach, L. J., & Meehl, P. E. Construct validity in psychological tests. *Psychological Bulletin*, 1955, **52**, 281-302.

Crow, J. F. Problems of ascertainment in the analysis of family data. In J. V. Neel, M. W. Shaw, and W. J. Schull (Eds.), *Genetics and the epidemiology of chronic diseases*. Washington, D.C., U.S. Department of Health, Education and Welfare, 1965. Pp. 23-44.

Dahlstrom, W. G. Recurrent issues in the development of the MMPI. In J. N. Butcher (Ed.), *MMPI: Research developments and clinical applications*. New York: McGraw-Hill, 1969. Pp. 1-40.

Dahlstrom, W. G., Welsh, G. S., & Dahlstrom, L. E. *An MMPI handbook*: Vol. 1. *Clinical interpretation*. (2nd ed.) Minneapolis: Univ. of Minnesota Press, 1972.

Davison, K., & Bagley, C. R. Schizophrenia-like psychoses associated with organic disorders of the central nervous system: A review of the literature. In R. N Herrington (Ed.), *Current problems in neuropsychiatry*. British Journal of Psychiatry Special Publication No. 4. Ashford, Kent: Headley, 1969. Pp. 113-184.

Deming, W. E. A recursion formula for the proportion of persons having a first admission as schizophrenic. *Behavioral Science*, 1968, **13**, 467-476.

Dewey, W. J., Barrai, I., Morton, N. E., & Mi, M. P. Recessive genes in severe mental defect. *American Journal of Human Genetics*, 1965, **17**, 237-256.

Dobzhansky, T. On some fundamental concepts of Darwinian biology. In T. Dobzhansky, M. K. Hecht, & W. M. C. Steere (Eds.), *Evolutionary biology*, Vol. 2. New York: Appleton, 1968. Pp. 1-34.

Dunham, H. W. *Community and schizophrenia*. Detroit: Wayne State Univ. Press, 1965.

Dunn, L. C. Cross currents in the history of human genetics. *American Journal of Human Genetics*, 1962, **14**, 1-13.

Eberhard, G. Peptic ulcer in twins. A study in personality, heredity and environment. *Acta Psychiatrica Scandinavica*, Suppl. 205, 1968.

Edwards, J. H. The simulation of Mendelism. *Acta Genetica et Statistica Medica (Basel)*, 1960, **10**, 63-70.

Edwards, J. H. The genetic basis of common disease. *American Journal of Medicine*, 1963, **34**, 627-638.

Edwards, J. H. Familial predisposition in man. *British Medical Bulletin*, 1969, **25**, 58-64.

Eisenberg, L. The interaction of biological and experiential factors in schizophrenia. In D. Rosenthal and S. S. Kety (Eds.); *The transmission of schizophrenia*. Oxford: Pergamon, 1968. Pp. 403-409.

Elsässer, G. *Die Nachkommen geisteskranker Elternpaare*. Stuttgart: Thieme, 1952.

Elston, R. C., & Campbell, M. A. Schizophrenia: Evidence for the major gene hypothesis. *Behavior Genetics*, 1970, **1**, 3-10.

vanEpen, J. H. Defect schizophrenic states (residual schizophrenia). *Psychiatria, Neurologia, Neurochirurgia*, 1969, **72**, 371-394.

Erlenmeyer-Kimling, L. Studies on the offspring of two schizophrenic parents. In D. Rosenthal and S. S. Kety (Eds.), *The transmission of schizophrenia*. Oxford: Pergamon, 1968. Pp. 65-83.

Erlenmeyer-Kimling, L., & Paradowski, W. Selection and schizophrenia. *American Naturalist*, 1966, **100**, 651-665.

Essen-Möller, E. Empirische Ähnlichkeitsdiagnose bei Zwillingen. *Hereditas, Genetiskt Arkiv*, 1941, **27**, 1-50. (a)

Essen-Möller, E. Psychiatrische Untersuchungen an einer Serie von Zwillingen. *Acta Psychiatrica et Neurologica Scandinavica*, 1941, Suppl. 23. (b)

Essen-Möller, E. The concept of schizoidia. *Monatsschrift für Psychiatrie und Neurologie*, 1946, **112**, 258-271.

Essen-Möller, E. The calculation of morbid risk in parents of index cases as applied to a family sample of schizophrenics. *Acta Genetica et Statistica Medica (Basel)*, 1955, **5**, 334-342.

Essen-Möller, E. Twenty-one psychiatric cases and their MZ co-twins: A thirty years' follow-up. *Acta Geneticae Medicae et Gemellologiae*, 1970, **19**, 315-317.

Evans, C., & McConnell, T. R. A new measure of introversion-extraversion. *Journal of Psychology*, 1941, **12**, 111-124.

Falconer, D. S. *Introduction to quantitative genetics*. Edinburgh: Oliver & Boyd, 1960.

Falconer, D. S. The inheritance of liability to certain diseases, estimated from the incidence among relatives. *Annals of Human Genetics*, 1965, **29**, 51-76.

Falconer, D. S. The inheritance of liability to diseases with variable age of onset, with particular reference to diabetes mellitus. *Annals of Human Genetics*, 1967, **31**, 1-20.

Fischer, M. Psychoses in the offspring of schizophrenic monozygotic twins and their normal co-twins. *British Journal of Psychiatry*, 1971, **118**, 43-52.

Fischer, M. A Danish twin study of schizophrenia. *Acta Psychiatrica Scandinavica*, Suppl., 1972, in press.

Fischer, M., Harvald, B., & Hauge, M. A Danish twin study of schizophrenia. *British Journal of Psychiatry*, 1969, **115**, 981-990.

Fish, F. *Clinical psychopathology*. Bristol: Wright, 1967.

daFonseca, A. F. Affective equivalents. *British Journal of Psychiatry*, 1963, **109**, 464-469.

Fuller, J. L., & Collins, R. L. Genetics of audiogenic seizures in mice: A parable for psychiatrists. *Seminars in Psychiatry*, 1970, **2**, 75-88.

Fuller, J. L., & Thompson, W. R. *Behavior genetics*. New York: Wiley, 1960.

Gajdusek, C. Physiological and psychological characteristics of stone age man. *Engineering and Science*, 1970, **33**, 6, 26-33; 56-62.

Gajdusek, C., Gibbs, C. J., & Alpers, M. Transmission and passage of experimental 'Kuru' to chimpanzees. *Science*, 1967, **155**, 212-214.

Galton, F. The history of twins, as a criterion of the relative powers of nature and nurture. *Fraser's Magazine*, 1875, **12**, 566-576.

Galton, F. Family likeness in stature. *Proceedings of the Royal Society (London)*, 1886, **40**, 42-63.

Garmezy, N. Process and reactive schizophrenia: Some conceptions and issues. In M. M. Katz, J. O. Cole, and W. E. Barton (Eds.), *The role and methodology of classification in psychiatry and psychopathology*. Washington, D.C.: Superintendent of Documents, U.S. Government Printing Office, 1968. Pp. 419-466.

Garrone, G. Étude statistique et génétique de la schizophrénie à Genève de 1901 à 1950. *Journal de Génétique Humaine*, 1962, **11**, 89-219.

Gilberstadt, H., & Duker, J. *A handbook for clinical and actuarial MMPI interpretation*. Philadelphia: Saunders, 1965.

Goldberg, E. M., & Morrison, S. L. Schizophrenia and social class. *British Journal of Psychiatry*, 1963, **109**, 785-802.

Gottesman, I. I. The efficiency of several combinations of discrete and continuous variables for the diagnosis of zygosity. *Proceedings of the Second International Congress of Human Genetics*. Rome: Istituto G. Mendel, 1963. Pp. 346-347. (a)

Gottesman, I. I. Heritability of personality: A demonstration. *Psychological Monographs*, 1963, **77** (9, Whole No. 572). (b)

Gottesman, I. I. Personality and natural selection. In S. G. Vandenberg (Ed.), *Methods and goals in human behavior genetics*. New York: Academic Press, 1965. Pp. 63-74.

Gottesman, I. I. A sampler of human behavioral genetics. In T. Dobzhansky, M. K. Hecht, and W. C. Steere (Eds.), *Evolutionary biology*. Vol. 2. New York: Appleton, 1968. Pp. 276-320. (a)

Gottesman, I. I. Severity/concordance and diagnostic refinement in the Maudsley-Bethlem schizophrenic twin study. In D. Rosenthal and S. S. Kety (Eds.), *The transmission of schizophrenia*. Oxford: Pergamon, 1968. Pp. 37-48. (b)

Gottesman, I. I. Biogenetics of race and class. In M. Deutsch, I. Katz, and A. R. Jensen (Eds.), *Social class, race, and psychological development*. New York: Holt, 1968. Pp. 11-51. (c)

Gottesman, I. I., & Heston, L. L. Human behavioral adaptations—speculations on their genesis. In L. Ehrman and G. Omenn (Eds.), *Genetic endowment and environment in the determination of behavior*. New York: Academic Press, 1972, in press.

Gottesman, I. I., & Shields, J. Schizophrenia in twins: 16 years' consecutive admissions to a psychiatric clinic. *British Journal of Psychiatry*, 1966, **112**, 809-818. (a)

Gottesman, I. I., & Shields, J. Contributions of twin studies to perspectives on schizophrenia. In B. A. Maher (Ed.), *Progress in experimental personality research*. Vol. 3. New York: Academic Press, 1966. Pp. 1-84. (b)

Gottesman, I. I., & Shields, J. A polygenic theory of schizophrenia. *Proceedings of the National Academy of Sciences*, 1967, **58**, 199-205.

Gottesman, I. I., & Shields, J. In pursuit of the schizophrenic genotype. In S. G. Vandenberg (Ed.), *Progress in human behavior genetics*. Baltimore, Maryland: Johns Hopkins Press, 1968. Pp. 67-103.

Gruenberg, E. M. Epidemiology and medical care statistics. In M. M. Katz, J. O. Cole, and W. E. Barton (Eds.), *The role and methodology of classification in psychiatry and psychopathology*. Public Health Service Publication No. 1584. Washington, D.C.: U.S. Government Printing Office, 1968. Pp. 76-99.

Gruenberg, E. M. How can the new diagnostic manual help? *International Journal of Psychiatry*, 1969, 7, 368-374.

Grüneberg, H. *Animal genetics and medicine.* London: Hamilton; New York; Hoeber, 1947.

Grüneberg, H. Genetical studies on the skeleton of the mouse. IV. Quasicontinuous variations. *Journal of Genetics*, 1952, 51, 95-114.

Guilford, J. P. *Fundamental statistics in psychology and education.* (3rd ed.) New York: McGraw-Hill, 1956.

Halevy, A., Moos, R. H., & Solomon, G. S. A relationship between blood serotonin concentrations and behavior in psychiatric patients. *Journal of Psychiatric Research*, 1965, 3, 1-10.

Haller, M. H. *Eugenics: Hereditarian attitudes in American thought.* Brunswick: Rutgers Univ. Press, 1963.

Hallgren, B., & Sjögren, T. A clinical and genetico-statistical study of schizophrenia and low-grade mental deficiency in a large Swedish rural population. *Acta Psychiatrica et Neurologica Scandinavica*, 1959, Suppl. 140.

Hammond, K., & James, J. W. Genes of large effect and the shape of the distribution of a quantitative character. *Australian Journal of Biological Science*, 1970, 23, 867-876.

Hare, E. H. The epidemiology of schizophrenia. In A. Coppen and A. Walk (Eds.), *Recent developments in schizophrenia.* British Journal of Psychiatry Special Publication No. 1. Ashford, Kent: Headley, 1967. Pp. 9-24.

Harris, H. *The principles of human biochemical genetics.* Amsterdam: North-Holland Publ., 1970.

Harvald, B. Heredity in epilepsy, an electroencephalographic study of the relatives of epileptics. *Opera ex Domo Biologiae Hereditariae Humanae Universitatis Hafniensis.* Vol. 35. Copenhagen: Munksgaard, 1954.

Hathaway, S. R., & McKinley, J. C. *The Minnesota Multiphasic Personality Inventory manual.* (Rev. ed.) New York: The Psychological Corporation, 1951.

Hathaway, S. R., Monachesi, E., & Salasin, S. A follow-up study of MMPI high 8, schizoid, children. In M. Roff and D. Ricks (Eds.), *Life history research in psychopathology.* Minneapolis: Univ. of Minnesota Press, 1970. Pp. 171-188.

Heath, R. G., & Krupp, I. M. Schizophrenia as an immunologic disorder. *Archives of General Psychiatry*, 1967, 16, 1-33.

Heston, L. L. Psychiatric disorders in foster home reared children of schizophrenic mothers. *British Journal of Psychiatry*, 1966, 112, 819-825.

Heston, L. L. The genetics of schizophrenic and schizoid disease. *Science*, 1970, 167, 249-256.

Heston, L. L., & Denney, D. Interactions between early life experience and biological factors in schizophrenia. In D. Rosenthal and S. S. Kety (Eds.), *The transmission of schizophrenia.* Oxford: Pergamon, 1968. Pp. 363-376.

Heston, L. L., Denney, D., & Pauly, I.B. The adult adjustment of persons institutionalized as children. *British Journal of Psychiatry*, 1966, 112, 1103-1110.

Heston, L., & Shields, J. Homosexuality in twins. *Archives of General Psychiatry*, 1968, 18, 149-160.

Higgins, J. Effects of child rearing by schizophrenic mothers. *Journal of Psychiatric Research*, 1966, 4, 153-167.

Hirsch, S. R., & Leff, J. P. Parental abnormalities of verbal communication in the transmission of schizophrenia. *Psychological Medicine*, 1971, 1, 118-127.

Hoch, P. H., & Polatin, P. Pseudoneurotic forms of schizophrenia. *Psychiatric Quarterly*, 1949, 23, 248-276.

Hoffer, A., & Pollin, W. Schizophrenia in the NAS-NRC panel of 15,909 veteran twin pairs. *Archives of General Psychiatry*, 1970, **23**, 469-477.

Hollingshead, A. B., & Redlich, F. C. *Social class and mental illness*. New York: Wiley, 1958.

Hunt, W. A., Wittson, C. L., & Hunt, E. B. A theoretical and practical analysis of the diagnostic process. In P. H. Hoch and J. Zubin (Eds.), *Current problems in psychiatric diagnosis*. New York: Grune & Stratton, 1953. Pp. 53-65.

Huxley, J., Mayr, E., Osmond, H., & Hoffer, A. Schizophrenia as a genetic morphism. *Nature (London)*, 1964, **204**, 220-221.

Inouye, E. Similarity and dissimilarity of schizophrenia in twins. *Proceedings Third International Congress of Psychiatry*, 1961, **1**, 524-530 (Montreal: Univ. of Toronto Press, 1963).

Inouye, E. Monozygotic twins with schizophrenia reared apart in infancy. *Excerpta Medica International Congress Series*, 1971, No. 233, 93.

Jablon, S., Neel, J. V., Gershowitz, H., & Atkinson, G. F. The NAS-NRC twin panel: Methods of construction of the panel, zygosity diagnosis, and proposed use. *American Journal of Human Genetics*, 1967, **19**, 133-161.

Jackson, D. D. A critique of the literature on the genetics of schizophrenia. In D. D. Jackson (Ed.), *The etiology of schizophrenia*. New York: Basic Books, 1960. Pp. 37-87.

Jacobs, E. G., & Mesnikoff, A. M. Alternating psychoses in twins: Report of 4 cases. *American Journal of Psychiatry*, 1961, **117**, 791-797.

Jayakar, S. D. On the detection and estimation of linkage between a locus influencing a quantitative character and a marker locus. *Biometrics*, 1970, **26**, 451-464.

Jinks, J. L., & Fulker, D. W. A comparison of the biometrical genetical, MAVA, and classical approaches to the analysis of human behavior. *Psychological Bulletin*, 1970, **73**, 311-349.

John, B., & Lewis, K. R. Chromosome variability and geographic distribution in insects. *Science*, 1966, **152**, 711-721.

Kahn, E. Studien uber Vererbung und Entstehung geistiger Storungen. IV. Schizoid und Schizophrenie im Erbgang. *Monographien aus dem Gesamtgebiete der Neurologie und Psychiatrie*, 1923, **36**.

Kaij, L. *Alcoholism in twins*. Stockholm: Almqvist & Wiksell, 1960.

Kallmann, F. J. *The genetics of schizophrenia*. New York: Augustin, 1938.

Kallmann, F. J. The genetic theory of schizophrenia: An analysis of 691 schizophrenic twin index families. *American Journal of Psychiatry*, 1946, **103**, 309-322.

Kallmann, F. J., & Roth, B. Genetic aspects of pre-adolescent schizophrenia. *American Journal of Psychiatry*, 1956, **112**, 599-606.

Karlsson, J. L. *The biologic basis of schizophrenia*. Springfield, Illinois: Thomas, 1966.

Kasanin, J. S. (Ed.) *Language and thought in schizophrenia*. Berkeley: Univ. of California Press, 1946.

Katz, M. M., Cole, J. O., & Barton, W. E. (Eds.) *The role and methodology of classification in psychiatry and psychopathology*. Public Health Service Publ. No. 1584. Washington, D.C.: U.S. Government Printing Office, 1968.

Kay, D. W. K. Late paraphrenia and its bearing on the aetiology of schizophrenia. *Acta Psychiatrica Scandinavica*, 1963, **39**, 159-169.

Kay, D. W., & Roth, M. Environmental and hereditary factors in the schizophrenias of old age ('late paraphrenia') and their bearing on the general problem of causation in schizophrenia. *Journal of Mental Science*, 1961, **107**, 649-686.

Kety, S. S. Biochemical theories of schizophrenia. *Science*, 1959, **129**, 1528-1532; 1590-1596.

Kety, S. S., Rosenthal, D., Wender, P. H., & Schulsinger, F. The types and prevalence of mental illness in the biological and adoptive families of adopted schizophrenics. In D. Rosenthal and S. S. Kety (Eds.), *The transmission of schizophrenia*. Oxford: Pergamon, 1968. Pp. 345-362.

Kirk, R. L. Population genetic studies of the indigenous peoples of Australia and New Guinea. In A. G. Steinberg and A. G. Bearn (Eds.), *Progress in medical genetics*. Vol. IV. New York: Grune & Stratton, 1965. Pp. 202-241.

Klein, R. G., & Klein, D. F. Premorbid asocial adjustment and prognosis in schizophrenia. *Journal of Psychiatric Research*, 1969, 7, 35-53.

Kohn, M. L. Social class and schizophrenia: A critical review. In D. Rosenthal and S. S. Kety (Eds.), *The transmission of schizophrenia*. Oxford: Pergamon, 1968. Pp. 155-173.

Kreitman, N. The reliability of psychiatric diagnosis. *Journal of Mental Science*, 1961, 7, 876-886.

Kreitman, N., & Smythies, J. R. Schizophrenia: Genetic and psychosocial factors. In J. R. Smythies, *Biological psychiatry*. London: Heinemann, 1968. Pp. 1-24.

Kringlen, E. Schizophrenia in male monozygotic twins. *Acta Psychiatrica Scandinavica*, 1964, Suppl. 178.

Kringlen, E. Schizophrenia in twins, an epidemiological-clinical study. *Psychiatry*, 1966, **29**, 172-184.

Kringlen, E. *Heredity and environment in the functional psychoses*. London: Heinemann, 1967.

Kringlen, E. An epidemiological-clinical twin study on schizophrenia. In D. Rosenthal, and S. S. Kety (Eds.), *The transmission of schizophrenia*. Oxford: Pergamon, 1968. Pp. 49-63.

Laing, R. D. Is schizophrenia a disease? *International Journal of Social Psychiatry*, 1964, **10**, 184-193.

Lane, E., & Albee, G. W. Childhood intellectual differences between schizophrenic adults and their siblings. *American Journal of Orthopsychiatry*, 1965, **35**, 747-753.

Lane, E., & Albee, G. W. Comparative birth weights of schizophrenics and their siblings. *Journal of Psychology*, 1966, **64**, 227-231.

Lane, E., & Albee, G. W. On childhood intellectual decline of adult schizophrenics: A reassessment of an earlier study. *Journal of Abnormal Psychology*, 1968, **73**, 174-177.

Lerner, I. M. *The genetic basis of selection*. New York: Wiley, 1958.

Levine, S. & Scotch, N. A. (Eds.) *Social stress*. Chicago, Illinois: Aldine, 1970.

Lewis, A. J. The offspring of parents both mentally ill. *Acta Genetica et Statistica Medica (Basel)*, 1957, 7, 349-365.

Lewontin, R. C. An estimate of average heterozygosity in man. *American Journal of Human Genetics*, 1967, **19**, 681-685.

Lidz, T., Fleck, S., & Cornelison, A. R. *Schizophrenia and the family*. New York: International Universities Press, 1965.

Lidz, T., Wild, C., Schafer, S., Rosman, B., & Fleck, S. Thought disorders in the parents of schizophrenic patients: A study utilizing the object sorting test. *Journal of Psychiatric Research*, 1962, 1, 193-200.

Lovibond, S. H. The object sorting test and conceptual thinking in schizophrenia. *Australian Journal of Psychology*, 1954, 6, 52-70.

Luborsky, L. Clinicians' judgments of mental health: A proposed scale. *Archives of General Psychiatry*, 1962, 7, 407-417.

Lush, J. L. *Animal breeding plans*. (3rd ed.) Ames: Iowa State Univ. Press, 1945.

Luxenburger, H. Vorläufiger Bericht über psychiatrische Serienuntersuchungen an Zwillingen. *Zeitschrift für die gesamte Neurologie und Psychiatrie*, 1928, **116**, 297-326.

Luxenburger, H. Psychiatrisch-neurologische Zwillingspathologie. *Zeitblatt für die gesamte Neurologie und Psychiatrie*, 1930, **56**, 145-180.

Luxenburger, H. Die Manifestations wahrscheinlichkeit der Schizophrenie im Lichte der Zwillingsforschung. *Zeitschrift für Psychische Hygiene*, 1934, 7, 174-184.

Luxenburger, H. Untersuchungen an schizophrenen Zwillingen und ihren Geschwistern zur Prüfung der Realität von Manifestationsschwankungen. *Zeitschrift für die gesamte Neurologie und Psychiatrie*, 1936, **154**, 351-394.

Maher, B. A. *Principles of psychopathology*. New York: McGraw-Hill, 1966.

Mandell, A. J., & Mandell, M. P. (Eds.) *Psychochemical research in man*. New York: Academic Press, 1969.

Marks, P. A., & Seeman, W. *The actuarial description of abnormal personality: An atlas for use with the MMPI*. Baltimore, Maryland: Williams & Wilkins, 1963.

Mather, K. *Biometrical genetics*. New York: Dover; London: Methuen, 1949.

Mather, K. *Human diversity*. Edinburgh: Oliver & Boyd, 1964.

Mayr, E. From molecules to organic diversity. *Fed. Proc. Fed. Amer. Soc. Exp. Biol.*, 1964, **23**, 1231-1235.

McConaghy, N. The use of an object sorting test in elucidating the hereditary factor in schizophrenia. *Journal of Neurology, Neurosurgery and Psychiatry*, 1959, **22**, 243-246.

McConaghy, N., & Clancy, M. Familial relationships of allusive thinking in university students and their parents. *British Journal of Psychiatry*, 1968, **114**, 1079-1087.

McKinley, J. C., & Hathaway, S. R. A multiphasic personality schedule (Minnesota): IV. Psychasthenia. *Journal of Applied Psychology*, 1942, **26**, 614-624.

McKusick, V. A. *Mendelian inheritance in man*. (3rd ed.) Baltimore, Maryland: Johns Hopkins Press, 1971.

Mednick, S. A. A learning theory approach to research in schizophrenia. *Psychological Bulletin*, 1958, **55**, 316-327.

Mednick, S. A., Mura, E., Schulsinger, F., & Mednick, B. Prenatal conditions and infant development in children with schizophrenic parents. In I. I. Gottesman and L. Erlenmeyer-Kimling (Eds.), *Differential reproduction in individuals with mental and physical disorders. Social Biology*, 1971, **18**, S103-S113.

Mednick, S. A., & Schulsinger, F. Some premorbid characteristics related to breakdown in children with schizophrenic mothers. In D. Rosenthal & S. S. Kety (Eds.), *The transmission of schizophrenia*. Oxford: Pergamon, 1968. Pp. 267-291.

Meehl, P. E. The dynamics of "structured" personality tests. *Journal of Clinical Psychology*, 1945, 1, 296-303.

Meehl, P. E. Configural scoring. *Journal of Consulting Psychology*, 1950, **14**, 165-171.

Meehl, P. E. *Clinical versus statistical prediction: A theoretical analysis and a review of the evidence*. Minneapolis: Univ. of Minnesota Press, 1954.

Meehl, P. E. Schizotaxia, schizotypy, schizophrenia. *American Psychologist*, 1962, **17**, 827-838.

Meehl, P. E. Manual for use with checklist of schizotypic signs. Unpublished manuscript, Univ. of Minnesota Medical School, Minneapolis, 1964.

Meehl, P. E. Specific genetic etiology, psychodynamics, and therapeutic nihilism. *International Journal of Mental Health*, 1972, **1**, 10-27.

Meehl, P. E. MAXCOV–HITMAX: A taxonomic search method for loose genetic syndromes. In P. E. Meehl, *Psychodiagnosis: Selected papers*. Minneapolis: Univ. of Minnesota Press, 1973, in press.

Mendell, N. R., & Elston, R. C. Analyses of quasicontinuous traits. *Excerpta Medica International Congress Series*, 1971, No. 233, 120.

Menninger, K. A. *The vital balance*. New York: Viking Press, 1963.

Menninger, K. A. Syndrome, yes; disease entity, no. In R. Cancro (Ed.), *The schizophrenic reactions: A critique of the concept, hospital treatment, and current research*. New York: Brunner/Mazel, 1970. Pp. 71-78.

Mishler, E. G., & Scotch, N. A. Socio-cultural factors in the epidemiology of schizophrenia: A review. *Psychiatry*, 1963, **26**, 315-351.

Mishler, E. G., & Waxler, N. E. Family interaction and schizophrenia: Alternative frameworks of interpretation. In D. Rosenthal and S. S. Kety (Eds.), *The transmission of schizophrenia*. Oxford: Pergamon, 1968. Pp. 213-222. (a)

Mishler, E. G., & Waxler, N. E. *Interaction in families: An experimental study of family processes and schizophrenia*. New York: Wiley, 1968. (b)

Mitsuda, H. *Clinical genetics in psychiatry*. Tokyo: Igaku Shoin, 1967.

Mohr, J. Practical possibilities for detection of linkage in man. *Acta Genetica et Statistica Medica*, 1964, **14**, 125-132.

Morton, N. E. The detection of major genes under additive continuous variation. *American Journal of Human Genetics*, 1967, **19**, 23-34.

Morton, N. E., Yee, S., Elston, R. C., & Lew, R. Discontinuity and quasi-continuity: Alternative hypotheses of multifactorial inheritance. *Clinical Genetics*, 1970, **1**, 81-94.

Mosher, L. R., Pollin, W., & Stabenau, J. R. Families with identical twins discordant for schizophrenia: Some relationships between identification, thinking styles, psychopathology and dominance-submissiveness. *British Journal of Psychiatry*, 1971, **118**, 29-42.

Murphy, H. B. M. Cultural factors in the genesis of schizophrenia. In D. Rosenthal and S. S. Kety (Eds.), *The transmission of schizophrenia*. Oxford: Pergamon, 1968. Pp. 137-153.

Neel, J. V., Fajans, S. S., Conn, J. W., & Davidson, R. T. Diabetes mellitus. In J. V. Neel, M. W. Shaw, & W. J. Schull (Eds.), *Genetics and the epidemiology of chronic diseases*. Washington, D.C.: U.S. Department of Health, Education and Welfare, 1965. Pp. 105-132.

Neel, J. V., & Schull, W. J. *Human heredity*. Chicago, Illinois: Univ. of Chicago Press, 1954.

Norris, V. *Mental illness in London*. Maudsley Monograph, No. 6. London: Chapman & Hall, 1959.

Ødegaard, Ø. La génétique dans la psychiatrie. *Proceedings First World Congress of Psychiatry*, Paris, 1950. Paris: Hermann, 1952. *Comptes rendus*, VI. Pp. 84-90.

Ødegaard, Ø. The psychiatric disease entities in the light of a genetic investigation. *Acta Psychiatrica Scandinavica*, 1963, Supplement 169, 94-104.

Ødegaard, Ø. The multifactorial theory of inheritance in predisposition to schizophrenia. In A. R. Kaplan (Ed.), *Genetic factors in "schizophrenia."* Springfield, Illinois: Thomas, 1972. Pp. 256-275.

Pap, A. *Semantics and necessary truth*. New Haven, Connecticut: Yale Univ. Press, 1958.

Partanen, J., Bruun, K., & Markkanen, T. *Inheritance of drinking behavior*. Helsinki: The Finnish Foundation for Alcohol Studies, 1966.

Payne, R. W. The measurement and significance of overinclusive thinking and retardation in schizophrenic patients. In P. H. Hoch & J. Zubin (Eds.), *Psychopathology and schizophrenia*. New York: Grune & Stratton, 1966. Pp. 77-97.

Payne, R. W. Disorders of thinking. In C. G. Costello (Ed.), *Symptoms of psychopathology: A handbook*. New York: Wiley, 1970. Pp. 49-94.

Penrose, L. S. Genetic linkage in graded human characters. *Annals of Human Genetics*, 1938, **8**, 233-237.

Penrose, L. S. The genetical background of common diseases. *Acta Genetica et Statistica Medica (Basel)*, 1953, **4**, 257-265.

Penrose, L. S. *The biology of mental defect*. (2nd ed.) London: Sidgwick & Jackson, 1963.

Perris, C. A study of bipolar (manic-depressive) and unipolar recurrent depressive psychoses. *Acta Psychiatrica Scandinavica*, 1966, Suppl. 194.

Peterson, D. R. The diagnosis of subclinical schizophrenia. *Journal of Consulting Psychology*, 1954, **18**, 198-200.

Phillips, J. E., Jacobson, N., & Turner, W. J. Conceptual thinking in schizophrenics and their relatives. *British Journal of Psychiatry*, 1965, **111**, 823-839.

Phillips, L. Case history data and prognosis in schizophrenia. *Journal of Nervous and Mental Disease*, 1953, **117**, 515-525.

Pollin, W., Allen, M. G., Hoffer, A., Stabenau, J. R., & Hrubec, Z. Psychopathology in 15,909 pairs of veteran twins: Evidence for a genetic factor in the pathogenesis of schizophrenia and its relative absence in psychoneurosis. *The American Journal of Psychiatry*, 1969, **126**, 597-610.

Pollin, W., & Stabenau, J. R. Biological, psychological and historical differences in a series of monozygotic twins discordant for schizophrenia. In D. Rosenthal & S. S. Kety (Eds.), *The transmission of schizophrenia*. Oxford: Pergamon, 1968. Pp. 317-332.

Pollin, W., Stabenau, J. R., Mosher, L., & Tupin, J. Life history differences in identical twins discordant for schizophrenia. *American Journal of Orthopsychiatry*, 1966, **36**, 492-509.

Price, J. S. Personality differences within families: Comparison of adult brothers and sisters. *Journal of Biosocial Science*, 1969, **1**, 177-205.

Pyke, D. A., Cassar, J., Todd, J., & Taylor, K. W. Glucose tolerance and serum insulin in identical twins of diabetics. *British Medical Journal*, 1970, **4**, 649-651.

Race, R. R., & Sanger, R. *Blood groups in man*. (5th ed.) Oxford: Blackwell, 1968.

Rado, S. Theory and therapy: The theory of schizotypal organization and its application to the treatment of decompensated schizotypal behavior. In S. Rado (Ed.), *Psychoanalysis of behavior*. Vol. II. New York: Grune & Stratton, 1962. Pp. 127-140.

Rapaport, D. *Diagnostic psychological testing*. Chicago, Illinois: Yearbook Publ., 1946.

Reed, S. C., Hartley, C., Anderson, V. E., Phillips, V. P., & Johnson, N. A. *The psychoses: Family studies*. Philadelphia, Pennsylvania: Saunders, 1972, in press.

Reichenbach, H. *The rise of scientific philosophy*. Berkeley: Univ. of California Press, 1951.

Reisby, N. Psychoses in children of schizophrenic mothers. *Acta Psychiatrica Scandinavica*, 1967, **43**, 8-20.

Renwick, J. H. Progress in mapping human autosomes. *British Medical Bulletin*, 1969, **25**, 65-73.

Reznikoff, M., & Honeyman, M. S. MMPI profiles of monozygotic and dizygotic twin pairs. *Journal of Consulting Psychology*, 1967, **31**, 100.

Roberts, J. A. F. Multifactorial inheritance and human disease. *Progress in Medical Genetics*, 1964, **3**, 178-216.

Roe, A., Burks, B. S., & Mittelmann, B. Adult adjustment of foster-children of alcoholic and psychotic parentage and the influence of the foster home. *Quarterly Journal for Studies of Alcohol*, 1945, No. 3.

Romney, D. Psychometrically assessed thought disorder in schizophrenic and control patients and in their parents and siblings. *British Journal of Psychiatry*, 1969, **115**, 999-1002.

Rosanoff, A. J., Handy, L. M., Plesset, I. R., & Brush, S. The etiology of so-called schizophrenic psychoses with special reference to their occurrence in twins. *American Journal of Psychiatry*, 1934, **91**, 247-286.

Rosanoff, A. J., & Orr, F. I. A study of heredity in insanity in the light of the Mendelian theory. *American Journal of Insanity*, 1911, **68**, 221-261.

Rosen, A. Development of MMPI scales based on a reference group of psychiatric patients. *Psychological Monographs*, 1962, **76**, (8, Whole No. 527).

Rosenthal, D. Some factors associated with concordance and discordance with respect to schizophrenia in monozygotic twins. *Journal of Nervous and Mental Disease*, 1959, **129**, 1-10.

Rosenthal, D. Confusion of identity and the frequency of schizophrenia in twins. *Archives of General Psychiatry*, 1960, **3**, 297-304.

Rosenthal, D. Sex distribution and the severity of illness among samples of schizophrenic twins. *Journal of Psychiatric Research*, 1961, **1**, 26-36.

Rosenthal, D. Problems of sampling and diagnosis in the major twin studies of schizophrenia. *Journal of Psychiatric Research*, 1962, **1**, 116-134. (a)

Rosenthal, D. Familial concordance by sex with respect to schizophrenia. *Psychological Bulletin*, 1962, **59**, 401-421. (b)

Rosenthal, D. (Ed.), and Colleagues. *The Genain quadruplets*. New York: Basic Books, 1963.

Rosenthal, D. The offspring of schizophrenic couples. *Journal of Psychiatric Research*, 1966, **4**, 169-188.

Rosenthal, D. The heredity-environment issue in schizophrenia: Summary of the conference and present status of our knowledge. In D. Rosenthal and S. S. Kety (Eds.), *The transmission of schizophrenia*. Oxford: Pergamon, 1968. Pp. 413-427.

Rosenthal, D. *Genetic theory and abnormal behavior*. New York: McGraw-Hill, 1970.

Rosenthal, D. Two adoption studies of heredity in the schizophrenic disorders. In M. Bleuler and J. Angst (Eds.), *The origin of schizophrenia*. Bern: Huber, 1971. Pp. 21-34.

Rosenthal, D., & Kety, S. S. (Eds.) *The transmission of schizophrenia*. Oxford: Pergamon, 1968.

Rosenthal, D., & VanDyke, J. The use of monozygotic twins discordant as to schizophrenia in the search for an inherited characterological defect. *Acta Psychiatrica Scandinavica*, 1970, Suppl. 219. Pp. 183-189.

Rosenthal, D., Wender, P. H., Kety, S. S., Schulsinger, F., Welner, J., & Østergaard, L. Schizophrenics' offspring reared in adoptive homes. In D. Rosenthal and S. S. Kety (Eds.), *The transmission of schizophrenia*. Oxford: Pergamon, 1968. Pp. 377-391.

Rosenthal, R. *Experimenter effects in behavioral research*. New York: Appleton, 1966.

Rosman, B., Wild, C., Ricci, J., Fleck, S., & Lidz, T. Thought disorders in the parents of schizophrenic patients: A further study utilizing the object sorting test. *Journal of Psychiatric Research*, 1964, **2**, 211-221.

Rüdin, E. *Zur Vererbung und Neuentstehung der Dementia Praecox*. Berlin and New York: Springer-Verlag, 1916.

Sargant, W. *Battle for the mind*. London: Heinemann, 1957.

Satz, P., Fennell, E., & Jones, M. B. Comments on: A model of the inheritance of handedness and cerebral dominance. *Neuropsychologia*, 1969, **7**, 101-103.

Scarr, S. Effects of birth weight on later intelligence. *Social Biology*, 1969, **16**, 249-256.

Scharfetter, C. On the hereditary aspects of symbiontic psychoses—a contribution towards the understanding of the schizophrenia-like psychoses. *Psychiatria Clinica (Basel)* 1970, **3**, 145-152.

Schmidt, H. O., & Fonda, C. P. The reliability of psychiatric diagnosis: A new look. *Journal of Abnormal and Social Psychology*, 1956, **52**, 262-267.

Schopler, E., & Loftin, J. Thought disorders in parents of psychotic children (A function of test anxiety). *Archives of General Psychiatry*, 1969, **20**, 174-181.

Schuham, A. I. The double-bind hypothesis a decade later. *Psychological Bulletin*, 1967, **68**, 409-416.

Schulz, B. Zur Erbpathologie der Schizophrenie. *Zeitschrift für die gesamte Neurologie und Psychiatrie*, 1932, **143**, 175-293.

Schulz, B. Kinder schizophrener Elternpaare. *Zeitschrift für die gesamte Neurologie und Psychiatrie*, 1940, **168**, 332-381.

Scott, J. P., & Fuller, J. L. *Genetics and the social behavior of the dog.* Chicago, Illinois: Univ. of Chicago Press, 1965.

Searles, H. F. *Collected papers on schizophrenia and related subjects.* New York: International Universities Press, 1965.

Selye, H. *The stress of life.* New York: McGraw-Hill, 1956.

Shakow, D. The role of classification in the development of the science of psychopathology with particular reference to research. In M. M. Katz, J. O. Cole, and W. E. Barton (Eds.), *The role and methodology of classification in psychiatry and psychopathology.* Public Health Service Publication No. 1584. Washington, D.C.: U.S. Government Printing Office, 1968. Pp. 116-143.

Shakow, D. On doing research in schizophrenia. *Archives of General Psychiatry*, 1969, **20**, 618-642.

Shakow, D. Some observations on the psychology (and some fewer on the biology) of schizophrenia. *Journal of Nervous and Mental Diseases*, 1971, **153**, 300-316.

Shepherd, M., Brooke, E. M., Cooper, J. E., & Lin, T. An experimental approach to psychiatric diagnosis. *Acta Psychiatrica Scandinavica*, 1968, Suppl. 201, 1-89.

Shields, J. Personality differences and neurotic traits in normal twin schoolchildren. *Eugenics Review*, 1954, **45**, 213-246.

Shields, J. *Monozygotic twins brought up apart and brought up together.* London and New York: Oxford Univ. Press, 1962.

Shields, J. The genetics of schizophrenia in historical context. In A. Coppen and A. Walk (Eds.), *Recent developments in schizophrenia.* British Journal of Psychiatry Special Publication No. 1. Ashford, Kent: Headley, 1967. Pp. 25-41.

Shields, J. Summary of the genetic evidence. In D. Rosenthal and S. S. Kety (Eds.), *The transmission of schizophrenia.* Oxford: Pergamon, 1968. Pp. 95-126.

Shields, J. Concepts of heredity for schizophrenia. In M. Bleuler and J. Angst (Eds.), *The origin of schizophrenia.* Bern: Huber, 1971. Pp. 59-75.

Shields, J. Heredity and psychological abnormality. In H. J. Eysenck (Ed.), *Handbook of abnormal psychology.* (2nd ed.) London: Pitman, 1972. Pp. 540-603.

Shields, J., Gottesman, I. I., & Slater, E. Kallmann's 1946 schizophrenic twin study in the light of new information. *Acta Psychiatrica Scandinavica*, 1967, **43**, 385-396.

Shields, J., & Slater, E. La similarité du diagnostic chez les jumeaux et le problème de la spécificité biologique dans les névroses et les troubles de la personnalité. *L'Evolution Psychiatrique*, 1966, **31**, 441-451. [Original English version in J. Shields and I. I. Gottesman (Eds.), *Man, mind, and heredity. Selected papers of Eliot Slater on psychiatry and genetics.* Baltimore, Maryland: Johns Hopkins Press, 1971. Pp. 552-557.

Simpson, G. G. *This view of life.* New York: Harcourt, 1964.

Sines, J. O. Actuarial methods in personality assessment. In B. A. Maher (Ed.), *Progress in experimental personality research.* Vol. 3. New York: Academic Press, 1966. Pp. 133-193.

Sines, J. O. Actuarial versus clinical prediction in psychopathology. *British Journal of Psychiatry*, 1970, **116**, 129-144.

Singer, M. T., & Wynne, L. C. Thought disorder and family relations of schizophrenics: IV. Results and implications. *Archives of General Psychiatry*, 1965, **12**, 201-212.

Singer, M. T., & Wynne, L. C. Principles for scoring communication defects and deviances in parents of schizophrenics: Rorschach and TAT scoring manuals. *Psychiatry*, 1966, **29**, 260-288.

Sjöbring, H. *La personnalité. Structure et développement*. Paris: Doin, 1963.

Slater, E. (with the assistance of Shields, J.) Psychotic and neurotic illnesses in twins. *Medical Research Council Special Report Series No. 278*. London: Her Majesty's Stationery Office, 1953.

Slater, E. The twin-study method in wider perspective: Discussion. First International Conference on Human Genetics, Copenhagen, 1957. *Acta Genetica et Statistica Medica (Basel)*, 1957, 7, 20.

Slater, E. The monogenic theory of schizophrenia. *Acta Genetica et Statistica Medica (Basel)*, 1958, 8, 50-56.

Slater, E. Thirty-fifth Maudsley lecture: 'Hysteria 311.' *Journal of Mental Science*, 1961, **107**, 359-381.

Slater, E. Diagnosis of zygosity by finger prints. *Acta Psychiatrica Scandinavica*, 1963, **39**, 78-84.

Slater, E. Expectation of abnormality on paternal and maternal sides: A computational model. *Journal of Medical Genetics*, 1966, 3, 159-161.

Slater, E. A review of earlier evidence on genetic factors in schizophrenia. In D. Rosenthal and S. S. Kety (Eds.), *The transmission of schizophrenia*. Oxford: Pergamon, 1968, Pp. 15-26.

Slater, E., Beard, A. W., & Glithero, E. The schizophrenia-like psychoses of epilepsy. *British Journal of Psychiatry*, 1963, **109**, 95-150.

Slater, E., & Cowie, V. A. *The genetics of mental disorders*. London and New York: Oxford Univ. Press, 1971.

Slater, E., Hare, E. H., & Price, J. S. Marriage and fertility of psychiatric patients compared with national data. In I. I. Gottesman and L. Erlenmeyer-Kimling (Eds.), *Differential reproduction in individuals with mental and physical disorders. Social Biology*, 1971, 18, S60-S73.

Slater, E., & Roth, M. *Mayer-Gross, Slater and Roth Clinical psychiatry*. (3rd ed.) London: Baillière, 1969.

Slater, E., & Slater, P. A heuristic theory of neurosis. *Journal of Neurology and Psychiatry*, 1944, 7, 49-55.

Slater, E., & Tsuang, M-t. Abnormality on paternal and maternal sides: Observations in schizophrenia and manic-depression. *Journal of Medical Genetics*, 1968, 5, 197-199.

Smith, C. Heritability of liability and concordance in monozygous twins. *Annals of Human Genetics*, 1970, **34**, 85-91.

Smith, C. Recurrence risks for multifactorial inheritance. *American Journal of Human Genetics*, 1971, **23**, 578-588.

Smith, S., & Penrose, L. S. Monozygotic and dizygotic twin diagnosis. *Annals of Human Genetics*, 1955, **19**, 273-289.

Smythies, J. R. *Biological psychiatry*. London: Heinemann, 1968.

Stabenau, J., Tupin, J., Werner, M., & Pollin, W. A comparative study of families of schizophrenics, delinquents and normals. *Psychiatry*, 1965, **28**, 45-59.

Stengel, E. A study on some clinical aspects of the relationship between obsessional neurosis and psychotic reaction types. *Journal of Mental Science*, 1945, **91**, 166-187.

Stengel, E. Recent developments in classification. In A. Coppen and A. Walk (Eds.), *Recent developments in schizophrenia*. British Journal of Psychiatry Spec. Publ. No. 1. Ashford, Kent: Headley, 1967. Pp. 1-7.

Storrs, E. E., & Williams, R. J. A study of monozygous quadruplet armadillos in relation to mammalian inheritance. *Proceedings of the National Academy of Sciences*, 1968, **60**, 910-914.

Strömgren, E. Zum Ersatz des Weinbergschen "abgekürzten Verfahrens", zugleich ein Beitrag zur Frage von der Erblichkeit der Erkrankungsalters bei der Schizophrenie. *Zeitschrift für die gesamte Neurologie und Psychiatrie*, 1935, **153**, 784-797.

Strömgren, E. "Schizophreniform psychoses". *Acta Psychiatrica Scandinavica*, 1965, **41**, 483-489. (a)

Strömgren, E. Psychiatrische Genetik. In H. W. Gruhle, R. Jung, W. Mayer-Gross, and M. Müller (Eds.), *Psychiatrie der Gegenwart*. Vol. 1/1a, 1-69. Berlin and New York: Springer-Verlag, 1965. (b)

Strömgren, E. Etiology of schizophrenia in the light of family and twin studies. *Acta Jutlandica* (Medical Series), 1968, **40**, 59-71.

Sullivan, H. S. *Schizophrenia as a human process*. New York: Norton, 1962.

Sullivan, P. L., & Welsh, G. S. A technique for objective configural analysis of MMPI profiles. *Journal of Consulting Psychology*, 1952, **16**, 383-388.

Szasz, T. S. The myth of mental illness. *American Psychologist*, 1960, **15**, 113-118.

Then-Bergh, H. Die Erbbiologie des Diabetes Mellitus. Vorläufiges Ergebnis der Zwillingsuntersuchungen. *Archiv für Rassen-und Gesellschaftsbiologie*, 1938, **32**, 289-340.

Thoday, J. M. Location of polygenes. *Nature (London)*, 1961, **191**, 368-370.

Thoday, J. M. New insights into continuous variation. In J. F. Crow & J. V. Neel (Eds.), *Proceedings of the Third International Congress of Human Genetics*. Baltimore, Maryland: Johns Hopkins Press, 1967. Pp. 339-350.

Tienari, P. Psychiatric illnesses in identical twins. *Acta Psychiatrica Scandinavica*, Suppl. 171, 1963.

Tienari, P. On intrapair differences in male twins with special reference to dominance-submissiveness. *Acta Psychiatrica Scandinavica*, Suppl. 188, 1966.

Tienari, P. Schizophrenia in monozygotic male twins. In D. Rosenthal & S. S. Kety (Eds.), *The transmission of schizophrenia*. Oxford: Pergamon, 1968. Pp. 27-36.

Tienari, P. Schizophrenia and monozygotic twins. In K. A. Achté (Ed.), *Psychiatria Fennica 1971*, Helsinki: Helsinki University Central Hospital, 1971. Pp. 97-104.

Toms, E. C. Personality characteristics of mothers of schizophrenic veterans. Unpublished doctoral dissertation, Univ. of Minnesota, 1955.

Trostorff, S. von Über hereditäre Belastung bei den zykloiden Psychosen, den unsystematischen und systematischen Schizophrenien. *Psychiatrie, Neurologie und Medizinische Psychologie*, 1968, **20**, 98-106.

Tuke, D. H. *A dictionary of psychological medicine*. New York: McGraw-Hill (Blakiston), 1892.

Turner, R. J., & Wagenfeld, M. O. Occupational mobility and schizophrenia: An assessment of the social causation and social selection hypotheses. *American Sociological Review*, 1967, **32**, 104-113.

Tutko, T. A., & Spence, J. T. The performance of process and reactive schizophrenics and brain injured subjects on a conceptual task. *Journal of Abnormal and Social Psychology*, 1962, **65**, 387-394.

Vandenberg, S. G. Hereditary factors in normal personality traits (as measured by inventories). In J. Wortis (Ed.), *Recent advances in biological psychiatry*. Vol. 9. New York: Plenum, 1967. Pp. 65-104.

Venables, P. H. Experimental psychological studies of chronic schizophrenia. In M. Shepherd and D. L. Davies (Eds.), *Studies in Psychiatry*. London and New York: Oxford University Press, 1968. Pp. 83-105.

Vigotsky, I. S. Thought in schizophrenia. *Archives of Neurology and Psychiatry*, 1934, **31**, 1063-1077.

Waddington, C. H. *The strategy of the genes*. London: Allen & Unwin, 1957.

Waring, M., & Ricks, D. Family patterns of children who became adult schizophrenics. *Journal of Nervous and Mental Disease*, 1965, **140**, 351-363.

Waterhouse, J. A. H. Twinning in twin pedigrees. *British Journal of Social Medicine*, 1950, **4**, 197-216.

Watson, C. G. Interrelationships of six overinclusion measures. *Journal of Consulting Psychology*, 1967, **31**, 517-520.

Weinberg, W. Methode und Fehlerquellen der Untersuchung auf mendelschen Zahlen beim Menschen. *Archiv für Rassen- und Gesellschaftsbiologie*, 1912, **9**, 165-174.

Weinberg, W. Mathematische Grundlage der Probandenmethode. *Zeitschrift für induktive Abstammungs- und Vererbungslehre*, 1928, **48**, 179-338.

Welner, J., & Strömgren, E. Clinical and genetic studies on benign schizophreniform psychoses based on a follow-up. *Acta Psychiatrica Scandinavica*, 1958, **33**, 377-399.

Welsh, G. S. An extension of Hathaway's MMPI profile coding system. *Journal of Consulting Psychology*, 1948, **12**, 343-344.

Wender, P. H., Rosenthal, D., & Kety, S. S. A psychiatric assessment of the adoptive parents of schizophrenics. In D. Rosenthal and S. S. Kety (Eds.), *The transmission of schizophrenia*. Oxford: Pergamon, 1968, Pp. 235-250.

Wiggins, J. S. Content dimensions in the MMPI. In J. N. Butcher (Ed.), *MMPI: Research developments and clinical applications*. New York: McGraw-Hill, 1969. Pp. 127-180.

Wild, C., Singer, M., Rosman, B., Ricci, J., & Lidz, T. Measuring disordered styles of thinking in the parents of schizophrenic patients on the object sorting test, part I. In T. Lidz, S. Fleck, and A. R. Cornelison (Eds.), *Schizophrenia and the family*. New York: International Universities Press, 1965. Pp. 400-422.

Williams, G. R., Fischer, A., Fischer, J. L., & Kurland, L. T. An evaluation of the Kuru genetic hypothesis. *Journal de Génétique Humaine*, 1964, **13**, 11-21.

Wilson, E. B., & Deming, J. Statistical comparison of psychiatric diagnosis of some Massachusetts State Hospitals during 1925 and 1926. *Quarterly Bulletin of the Massachusetts Department of Mental Disease*, 1927, **11**, 1-15.

Wing, J. K., Birley, J. L. T., Cooper, J. E., Graham, P., & Isaacs, A. D. Reliability of a procedure for measuring and classifying "present psychiatric state." *British Journal of Psychiatry*, 1967, **113**, 499-515.

Winokur, G., Clayton, P. J., & Reich, T. *Manic depressive illness*. St. Louis, Missouri: Mosby, 1969.

Wittermans, A. W., & Schulz, B. Genealogischer Beitrag zur Frage der geheilten Schizophrenien. *Archiv für Psychiatrie und Nervenkrankheiten*, 1950, **185**, 211-232.

Woolf, C. M. Congenital cleft lip: A genetic study of 496 propositi. *Journal of Medical Genetics*, 1971, **8**, 65-83.

Wright, S. An analysis of variability in number of digits in an inbred strain of guinea pigs. *Genetics*, 1934, **19**, 506-536. (a)

Wright, S. The results of crosses between inbred strains of guinea pigs, differing in number of digits. *Genetics*, 1934, **19**, 537-551. (b)

Wright, S. Genetic interaction. In W. J. Burdette (Ed.), *Methodology in mammalian genetics*. San Francisco, California: Holden-Day, 1963. Pp. 159-192.

Wright, S. *Evolution and the genetics of populations*. Vol. 1. *Genetic and biometric foundations*. Chicago, Illinois: Univ. of Chicago Press, 1968.

Wynne, L. C. Methodologic and conceptual issues in the study of schizophrenics and their families. In D. Rosenthal & S. S. Kety (Eds.), *The transmission of schizophrenia.* Oxford: Pergamon, 1968. Pp. 185-199.

Wynne, L. C. Schizophrenics and their families. *British Journal of Psychiatry*, 1972, in press.

Wynne-Davies, R. A family study of neonatal and late-diagnosis congenital dislocation of the hip. *Journal of Medical Genetics*, 1970, 7, 315-333.

Yolles, S. F., & Kramer, M. Vital statistics. In L. Bellak & L. Loeb (Eds.), *The schizophrenic syndrome.* New York: Grune & Stratton, 1969. Pp. 66-113.

Zerbin-Rüdin, E. Endogene Psychosen. In P. E. Becker (Ed.), *Humangenetik, ein kurzes Handbuch.* Vol. V/2. Stuttgart: Thieme, 1967. Pp. 446-577.

Zigler, E., & Phillips, L. Psychiatric diagnosis: A critique. *Journal of Abnormal and Social Psychology*, 1961, 63, 69-75.

A CRITICAL AFTERWORD

PAUL E. MEEHL

Here we have a definitive work, a magnum opus, a beautiful book indeed. It is an absolute *must* for theoreticians and investigators in clinical psychology, psychiatry, sociology, personology, and behavior genetics. It is also, I venture to suggest, required reading for clinical practitioners who feel the need to know something scientific about their patients, although I am aware that in some circles such an impulse is an aberration, viewed almost as a perversion.

This fine book shows what *can* be accomplished by a couple of first-rate intellects when they apply brains, drive, efficiency, thoroughness, and methodological sophistication to a problem, even so fuzzy and intractable a problem as the genetics of schizophrenia. I am pleased to have been invited, along with such a distinguished European psychiatrist as Dr. Eliot Slater, of whose early work at Maudsley the authors' project is a continuation, to comment on this book. And I am proud to think that when coauthor Irving Gottesman was studying for his doctorate at Minnesota in the late 1950's, I perhaps had some small part, directly or indirectly, in shaping and stimulating his mind. But then, he began his career as a physics student, a fact I suspect is relevant. (I am unfamiliar with James Shields' early vocational choice, but I'll take an even bet it wasn't clinical psychology.) There are certain "Minnesota habits of thought"–including some that I am not unwilling to label "prejudices," as I believe them to be good ones!–that are easily discerned herein. It is a pleasure to see a work product like this one emerging partly from one's academic subculture.

I take it to be obvious that the publication of this book "settles" a long-debated question, to wit: Is there something–a fairly "big" something–genetic about schizophrenia? Some of us may have become a trifle bored with the persistence of that debate over recent years, given the sizable and largely consistent research literature summarized and criticized so ably by Gottesman and Shields in Chapter 2, and I would like to urge, in the light of the present volume, that we now bring that phase of the debate to its long-deserved close. Nothing but American social-science prejudice plus "establishment psychiatry" brainwashing (given the usual dash of plain muddleheadedness) can, I think, lead anyone who has read this book to persist in strong doubt as to the prime importance of genetic factors in schizophrenia. Hence I should now be inclined to adopt Otto Neurath's maxim in the Vienna Circle, when (quoted to me by my friend and philosophy colleague Herbert Feigl) he opined, "Feigl, you know, one can seriously discuss only with those who are in the club." For my part, I am no longer interested in a debate with anyone who continues to believe that we are all born with equal talents for developing schizophrenia, and who is prepared to exercise unlimited theoretical ingenuity (and counter-Bayesian defiance of the reasonable prior probabilities!) in "explaining away" all genetic evidence by this or that *ad hoc* hypothesis (see *infra*). Let us henceforth consider this psychiatric variant of the *tabula rasa* doctrine, which, by the way, never did have anything impressive going for it, to have been now refuted; and let us concentrate our efforts on devising methods for answering the next big questions, namely: (1) What behavioral dispositions (tending to schizophrenia) are inherited?, (2) How (polygenic? one "big gene?" what penetrance?) are they inherited?, (3) How are these inherited dispositions *acted upon* by life experiences to yield schizophrenia versus compensated schizotypy?, and (4) What is the biochemical or neurophysiological *substrate* of those (molar) behavior dispositions?

The second of these, being a "statistical" question in population genetics, may perhaps be researchable without solving any of the other three, although one may be permitted to suspect that for a "geneticist's nightmare" (Neel's characterization of diabetes mellitus), like schizophrenia, this atheoretical development could turn out to be unusually difficult. For example, if, as the authors' MMPI data on discordant MZ twins discouragingly suggest, a molar-level, social-learning-mediated phenotypic indicator such as the schizotype's verbal response to a personality inventory item is just too far removed from the schizogene(s), too many links getting into the causal chain between DNA and laryngeal twitches, then we must "move further back," probably to *non*-behavioral indicators (belonging to the endophenotype) in order to diagnose the subclinical condition with sufficient accuracy to test competing genetic models against one another. I hope this is not the case, as it would (almost) put the "molar-level" behavior geneticist out of the schizophrenia business. But it might be so. We psychologists may have to say to the neurochemists, brain-stimulators, etc., "Look, Gottesman and Shields have now gone as far as we can go while relying on 'psychological' indicators of the schizoid genotype. It appears that no psychometric refinements or souped-up objectification-*cum*-quantification of the Mental Status Examination (or life history) will *ever* be 'valid' enough to do appreciably better. We have skimmed the cream; we are near the research asymptote. Until you can chemically identify the chromatographic Schizotypal

Purple Spot or the schizotaxic single-neurone's hypokrisia, we are at an impasse. Choice between genetic models is, until further notice, not feasible." I consider this molar-level impasse to be a possibility, but in what follows I shall set it aside in favor of more optimistic outcomes.

Even genetically oriented readers lacking the usual (American) social-science bias may have been shocked by my remark *supra* that the "pure environment-alist" theory never did have much of anything going for it. Since this sounds rather high-handed, and since the Gottesman-Shields data (despite their valiant efforts to squeeze blood out of the environmentalist turnip) lend essentially zero support to some of the received theories (e.g., that mother's mental health is so much more important than father's; that having a twin is, per se, a schizophren-ogenic factor because of the alleged identity-mixups; that lower social class is primarily a *cause*, rather than an *effect*, of schizoid personality), I should say a few brief words by way of defense of so strong a statement.

1. The initial impetus for a family-dynamics emphasis in schizophrenia the-ory was largely *theoretical*, to wit, conceptual extrapolation from the Freudian and post-Freudian psychodynamics. (Among American psychologists the en-vironmentalist bias of Watson and his behaviorist successors potentiated this influence.) There is, of course, nothing wrong with relying upon theory to begin one's understanding of a puzzling domain of phenomena, provided the theory is well-corroborated and that it really speaks, without undue ambiguity, to the facts. Further, if—even given those two conditions—the going theory is used to *exclude* a class of factors (as the alleged psychodynamics have been used in preempting the etiological field so as to rule out or down-play genetics) there must be some plausible showing that the received theory is "complete" and that its logical form, content, and *quantitative (parameter) assignments* leave no room for the set of competing causal factors thereby excluded. It should be obvious that these conditions for "exclusionary use" of sociopsychological theory have never been met, or even approached at any time during the entire history of clinical and scientific research on schizophrenia. I am somewhat embarrassed to be writing down these elementary methodological truths, but the intellectual habits of my profession regrettably necessitate it.

As to the Freudian and learning theory influences on extreme environment-alist views of schizophrenia, two comments will suffice. (a) While I myself accept a rather large portion of it, honesty requires admission that the Freudian theoretical corpus remains, three-fourths of a century after its inception, largely uncorroborated (and, on its *clinical* side, as a helping mode, increasingly in disrepute); and (b) The question of "who falls ill" is a learning-function *param-eter* question, not a question of the psychosocial *contents* of what is learned—a methodological point on which Freud himself was clear (and, hence, a lifelong "constitutionalist" even as regards the psychoneuroses). The received "laws" of, e.g., object-cathexis or of secondary reinforcement can, of course, be accepted without thereby settling anything whatever as to the relative potency of genes and early life history on the parameters that play crucial roles in those psychodynamisms or learning functions. It really should not be necessary to explain this but, alas, it is.

2. A second powerful influence was the (impressionistic) experience of psychotherapists in working intensively with their schizophrenic patients. I myself can testify to this influence, as a practitioner who has spent thousands of hours trying to help (sometimes successfully, I believe) schizoid and schizophrenic patients by psychotherapeutic means. I do not dispute the "raw data" that so impress clinicians of environmentalist persuasion, although I believe them to be somewhat overdone at times. My schizophrenic patients seem to talk, feel, and think in psychotherapy pretty much the same as those treated by antigenetic helpers. But these "data" are *almost completely ambiguous* as to etiological interpretation. I find it puzzling that so many American psychotherapists fail to understand this methodological point. *Consider*: A schizophrenic patient tells me (and/or acts it out repetitively in the transference) that "Mother—the old bitch—never liked me, she always preferred my brother Seydlitz." If I encourage him, he then "documents" this generalization from his childhood by narrating several pro-Seydlitz, antipatient episodes of maternal behavior. *What does this tell us about schizophrenic etiology?* For a critical mind, next to nothing. The relative importance of contributions by the following causal relations to such intratherapeutic narratives is, of course, simply impossible to assess on the basis of therapy-session evidence.

a. Mother preferred Seydlitz, rejected patient, thereby made him subsequently schizophrenic.

b. Patient got the schizogene from mother (a compensated schizotype of the battle-ax variety) and she preferred Seydlitz (because he didn't)—although this maternal preference played no causal role in the patient's subsequent illness.

c. However mother really behaved to Seydlitz versus patient, the latter (due to his genetic schizotypy) overreacted to all "aversive social inputs" and selectively misperceived things as a child.

d. As an adult, having decompensated, the patient does not now recall accurately what he may have perceived accurately at the time.

e. The patient learns to narrate battle-ax mother episodes because such therapy behavior is what his environmentalist therapist reinforces in the sessions.

f. The therapist selectively remembers these kinds of episodes because of his theoretical prejudices.

Only the most egregious naiveté (not to say arrogance about one's clinical discernment and "factor-unscrambling" powers) could enable a psychotherapist to deny the possibility of these factors operating to produce the received impression of how bad mothers create schizophrenics. In the aggregate, they reduce the *evidentiary* value of one's therapeutic experience to almost nil. *Point:*

The main empirical source of strong environmentalist views of schizophreno-genesis was the experience of psychotherapists, and this experience is methodologically next to worthless as to the disputed issue.

3. A third influence, stemming from more sophisticated environmentalists' recognizing the shaky evidentiary status of therapist impressions, consists of systematic, experimental and quantitative file-data research on schizophrenic perception and learning, social class, community isolation, intrafamily dynamics, etc., in an effort to objectify the therapist etiological impressions. Unfortunately, despite many ingenious efforts to smuggle real life into the laboratory, this research suffers from the same kind of causal-arrow ambiguity present in therapist impressions. It is, for example, a relatively pointless enterprise to investigate whether the mothers of schizophrenics "act differently" toward them than toward their nonschizophrenic sibs; or to show that future schizophrenics often do badly in intelligence tests, or flunk fractions; or that schizophrenics tend to be socially isolated or to live in neighborhoods of such-and-such kinds. Such correlations are predictable, of course, from a *wholly* genetic theory, or a *wholly* environmental theory, or *any intermediate mixture* of genetic/environmental emphasis. What then is the point in conducting such investigations?

If the preceding destructive analysis is substantially correct, it is inappropriate for behavior geneticists to adopt a defensive posture *vis-à-vis* the received psychosociodynamic doctrine—to stand on the scientific streetcorner asking timidly, "Who will buy my violets?" The received doctrine stood on very shaky foundations from the beginning. To repeat my challenging remark *supra*, it never did have much going for it—except the zeitgeist. In fact, if one were going to rely on "general background knowledge" and "extrapolated theory" in his personalistic assignment of prior probabilities to etiological notions about mental disorder, the *prima facie* case goes rather the other way—given the massive body of evidence, in both humans and animals (not to mention the folklore of animal breeders, surely admissible in evidence if we admit psychotherapist folklore?) that genes play a role in determining a variety of the individual differences found in any species of animal. One could, of course, have said this with reasonable assurance from the extrapolative armchair. Can it plausibly be supposed that while genes largely determine species characteristics, and mice show pronounced strain differences in dominance behavior, and the Basenji dog breed forms a weak super-ego compared with the spaniel, and the determination of human intelligence is (conservatively) at least 50% genetic—still, differences in potential for major mental breakdown have nothing importantly genetic about them? Why on earth would anyone have bet on such a proposition in the first place?

But enough of polemics. I repeat, we should take the Gottesman-Shields collaboration as a dispositive closing to a discussion that has gone on quite long enough. The burden of proof has been sustained, and from here on it should be shifted to the opposition. He who alleges a particular class of life-history determiners now has the burden; and he who denies that heredity plays a major etiological role bears a heavy burden indeed. Meanwhile I attempt to dissuade my Minnesota colleague Gottesman from lecturing on the theme, "Genes have a lot to do with schizophrenia." It is too late in the day for that lecture topic. Better he should now lecture on, e.g., "Why Gottesman, Shields, Bleuler, and Co., are betting on a polygenic theory of schizophrenia," and "What evidence shows that Slater, Heston, Meehl and Co., are wrong in pushing a monogenic theory." That is the kind of question worth our serious attention henceforth!

My colleague Professor Gottesman did not ask me to "review the book" in this afterword, but rather to express my overall evaluation, to raise questions for future research, to "needle" the authors wherever I had methodological stomachaches about their design or argument, and—he seductively added—"put in whatever of Meehl's current ideas you think appropriate." Having got some antienvironmentalist polemics out of my system, these things I shall now (briefly and, as a result, perforce a bit dogmatically) proceed to do.

The methodological merits of this work are so clear as to require nothing more than mention here. Among the most important are the care in ascertainment of cases; the clear definition of a population; the accurate determination of zygosity; the tracking down of officially "normal" cotwins; the taperecording of interview sessions; the use of an objective personality test (having respectable evidence in the research literature as to its construct validity); the reliance on blind diagnoses by several expert clinicians; the fact that these clinicians varied widely in their conception of schizophrenia, their etiological persuasions, and their national origins; and the combination of these clinical judgments to form a "best available criterion" for concordance analysis. This on the empirical front, to which one must add the authors' thorough knowledge of the work of their great (and not-so-great) predecessors, their (so rarely found!) fusion of criticality and openness, and their ingenuity in "slicing the pie" of data with various competing hypotheses in mind. I shall make no further remarks on these admirable features, more detailed praise being surely a presumptuous gilding the lily.

At the risk of obscuring the forest with some rather small trees, I present comments and questions in the order they are raised by a reading of the text, expanding with some discussion of my own views here and there. I trust that some of the minor "criticisms"—I simply *have* no major ones—will not appear nit-picking. They are meant as gentle prods to alert the reader that a *caveat* is perhaps in order.

In Chapter 1 a niggling comment: I could wish the authors had not described this fine investigation as an "experiment," a terminology I think better employed, as in the physical and biological sciences, to mean a more controlled situation in which the investigator imposes fixed values of certain variables upon a (relatively) isolated system, and "at his will" manipulates the independent variables of interest.

In their reference to Seymour Kety's review of the failures to date in the search for a specific biochemical lesion, the authors assert that "the frequency of methodological errors and technical limitations prevented finding the needle in the haystack." I would put it otherwise. While these errors and limitations were responsible for investigators' mistakenly supposing they had found the lesion, what presumably "prevented" genuinely finding it—assuming *arguendo* that it exists—would be the very fact of its *being* "a needle in the haystack," i.e., no one knows what kind of biochemical aberration he should be *looking for* in the CNS. One of the functions of an integrated theory of schizophrenia—including a highly speculative one like my own—is to suggest to workers knowledgeable in neurochemistry what general *kind* of substance ought to be involved in order for it to produce the kind of neuro-integrative ("schizotaxic") defect that such theories postulate. A theory that does not attempt, however conjecturally, to sketch in the causal chains leading from a mutated gene to the many-steps-removed clinical phenomenology of schizophrenic disorders just does not say enough about what sort of thing a Seymour Kety might profitably look for. Ever since Kety's important review (and I hasten to add that Dr. Kety himself is not the least bit responsible for this) "sociotropes" have had a tendency to treat the biochemical search problem as though there were some kind of huge statistical population of biochemical culprits, such that the misleadingly positive studies properly criticized by Kety can somehow be viewed as samples from the "biochemical etiology domain." On this strange view, the more negatives you pile up, the less probable it becomes that a positive exists! But we are of course not estimating some hypothetical parameter from a universe of biochemical agents. Rather, we are in the sad state of wondering whether there *is* one schizospecific aberrated neurohumor, an "existential" question that surely has no appreciable light shed upon it by discovering that schizophrenics do not really have an alleged thyroid deficiency or that a reported difference in blood chemistry is an artifact of bad institutional hygiene or peculiar dietary habits. *Point:* If there were some specific neurohumoral substance X whose presence, absence or molecularly aberrant form constituted the first link in the endo-phenotypic causal chain underlying schizoidia (and, hence, predisposing to schizophrenia), it would not be *expected* that anything other than this substance would turn out to be a specific diagnostic differentiator, given adequate controls.

When they point out that schizophrenia is, by usual geneticists' standards, a remarkably common genetic disorder (compared, say, with the incidence of the many Mendelizing varieties of mental deficiency), being more frequent than these by several orders of magnitude, I wish they had hinted that this extremely high incidence of the clinical syndrome itself ought to alert the geneticist to the possibility that it might be rather "special" or "different" from other main-gene disorders in important respects. For instance, one is repeatedly told that it is methodologically sinful to postulate extremely low penetrance (I don't like that word in this connection anyway, *vide infra*, but I will use it here roughly to mean "penetrance of the diagnosable clinical syndrome schizophrenia") suggesting that less than one-fourth or one-fifth or even as few as one-tenth of schizotypes decompensate to the point of being properly labelled "schizophrenia." If you combine the remarkably high incidence of the clinical entity with its persistence in all cultures studied despite its well-documented reproductive disadvantage, you already have reason to suspect that it may have something "special" about it as genetic disorders go, and that the extrapolation of higher-order inductions concerning reasonable penetrance values and the like from the more familiar and well-understood entities should be viewed with caution.

I am unhappy with their flat statement that "It is obvious from the discussion above that the classical methods of genetics for discrete entities cannot be applied to the study of schizophrenia in order to prove a genetic etiology; the approach must be biometrical and data on frequencies must be converted to measures of similarity between relatives (e.g., Falconer, 1965)," but this involves a perennial hassle between Gottesman and Meehl that cannot be developed here. Suffice it to say that where we differ is in my "faith" that sufficiently powerful statistical taxonomic methods, such as I myself have been working on recently, should theoretically permit the assignment of probability numbers to an individual's being or not being a schizotype, despite the fact that the phenotypic indicators are dimensional variables and that they are all, singly *and collectively*, highly fallible. It remains to be settled whether such a taxonomic procedure generates numerical predictions for, e.g., a dominant monogene of low penetrance (affected by numerous modifiers or, as I prefer to say, for reasons of semantic clarity, "potentiators") differing from a Gottesman-Shields polygenic model enough to permit a definitive empirical test. I surely don't wish to prejudge the issue by dogmatizing that the answer to this question is affirmative; but I am mildly distressed when our authors state so flatly that the "classical methods of genetics for discrete entities *cannot* be applied to the study of schizophrenia in order to prove a genetic etiology." I am still reasonably hopeful that they can, provided that statistics of loose syndromes can be made sufficiently powerful, and especially made to embody enough "internal consistency tests" as to the correctness of the postulated latent model, so that the clear identification of individuals in pedigrees relied on in traditional genetics can be adequately substituted for by trustworthy numerical probability statements. But this is a large and difficult topic beyond the scope of these remarks.

I don't think I am willing to agree unreservedly with their statement that "No one or two clinical symptoms are pathognomonic" unless they mean here "Two-way pathognomonic," i.e., infallible both as an inclusion and an exclusion test. I think—and I rather suspect that Gottesman and Shields would agree if pressed—that you can find at least a one symptom-pair that is *one*-way pathognomonic (the meaning in Dorland's Dictionary) that is, quasi-infallible used as an inclusion test. I proposed one myself a decade ago, to wit, the clear presence of the characteristic and unique schizoid thought disorder in the absence of clouded sensorium. A patient who is well oriented for time, place, and person and is listening attentively and communicating cooperatively and who says, "Of course, Doctor, you realize that naturally I am growing my father's hair," or who justifies plugging the alarm clock insert into the sidewalk on the Healy Picture Completion II (the frame where the boy is spilling his books) by telling the examiner "That is so he will have time to get to school," is in my view unquestionably schizophrenic. I doubt that any clinically experienced reader would think otherwise. I urge that there is a kind of thought disorder in the calm, oriented and cooperative patient that is a more powerful inclusion test than a positive Wassermann for lues (remember yaws!), and almost as good as the cherry-red retinal spot for Tay-Sachs' Disease. Of course if the authors meant two-way pathognomonicity, I agree with them entirely. It is regrettably as true today, as it was when the great Eugen Bleuler wrote it in 1911, that we have no trustworthy method for excluding schizophrenia.

I should like to have seen, toward the end of this chapter a fuller exposition of what they intend to claim by alleging a "specific genetic etiology" for schizophrenia. This is one of the few places where I might be inclined to fault them in other than a minor way. It is clear that they have a strong theoretical investment in defending the diagnostic identification of schizophrenia as an entity, of denying the notion of "one (functional) psychosis," and that they want to press for a considerable genetic homogeneity over almost all correctly diagnosed schizophrenics despite the well known variations in momentary symptom picture (the subtypes) and the wide range of clinical status (Rado's compensation, semicompensation, decompensation, disintegration, and deterioration). I rather think they owed it to us to deliver the conceptual goods in a little more detail at this place, even though the empirical evidence is not in. I do not complain, let it be emphasized, of the authors' "pushing" their polygenic theory against a monogenic one. I am a noninductivist neo-Popperian and believe that everybody should "do his own (theoretical) thing" and bet on what he thinks—guesses or intuits!—is the best horse. Rather, I express dissatisfaction with what seems to me an unclarity in *meaning* of their combined doctrine of specific etiology and polygenicity. They do not, if I read them right, want to say that already-named polygenic systems determinative of such non-schizospecific variables as general intelligence, "garden-variety (nonpathological) social introversion," energy-level, or anxiety-readiness are the components of the schizogenotype. That is, I take them to suppose that you don't get to be a schizophrenic just by having an unfortunate combination of high anxiety, low intelligence, low dominance, low energy, low mesomorphic toughness, and high social introversion. There is a *something else* allegedly specific to schizophrenia (better, the disposition thereto) that is, however, not a "big-effect monogene," as postulated by Slater, Heston, and myself. (This is not, of course, an empty claim on their part, since it has at least a quantitative formal meaning within the already existing general framework of genetics, permitting one to conjecture about a "whatever it is, a not further specified something" that (a) provides the disposition for a recognizable clinical disease, but (b) is nevertheless polygenic.) Do we have a set of genes that all tend to affect the same first-order phenotypic disposition, whatever it is, that is specifically a disposition toward schizophrenia? One understands fairly clearly what it means to conjecture that a "big-effect monogene" is the specific etiology of a disease, and can write a general (albeit useless) mathematical expression for it simply by putting down the function that would characterize all of the genome-*cum*-environment main and interaction effects, and then plugging in as a multiplier on this function a variable that takes on only the two values: 1 [= monogene present] and 0 [= monogene absent]. But once we have excluded that simple situation, the *very meaning* of the phrase "specific etiology" begins to "fuzz up," and I wish the authors had said more about how they conceive it. Furthermore, I cannot suppose that they believe the polygenic systems related to nonspecific variables (like anxiety, social introversion, low energy level, low aggressiveness, or weak social dominance) play no part in contributing to clinical decompensation and, if the patient does decompensate and becomes diagnosably schizophrenic, what form, direction, duration and outlook he has. (I know from conversations that Dr. Gottesman inclines to share my opinions about these potentiators.) But if that is so, we should have some explanation of why these important potentiators of the occurrence of clinical illness, and even of its form and content, its severity and duration, are not considered part of "what is specific" for schizophrenia? I do not assert, nor do I believe, that they would have no interesting and perhaps quite satisfactory answers to these questions. But I wish they had included them in their discussion of specificity.

In Chapter 2 it is not clear to me why they lump alcoholism and general paresis together as examples where cross-cultural stability of rates earlier led to the "false implication of genetic etiology." The two seem to me sufficiently different not to be used for this same purpose. Surely no one, however simpleminded a hereditarian of the old "constitutional school" he may have been, could have thought one could become an alcoholic without drinking alcohol! And if we say, "Well, what we mean by a specific genetic etiology for developing a disorder on the basis of intake of a certain substance is, *of course*, an inherited disposition to react to the substance differently from the great majority of persons who can imbibe it with impunity," then it is not clear why that implication for alcoholism can be called false. There *is* a hereditary component in alcoholism, so far as we can judge on the present evidence; and it may even be "specific" as our authors are willing to use this adjective in their own polygenic view of schizophrenia.

I am bothered, although of course one knows why they say it, to read that a "dominant gene theory would predict that [the risk in second-degree relatives] would be one-half [that in first-degree relatives]." Since no one can possibly hold a dominant gene theory for schizophrenia itself on present evidence, I assume the only live object of their remark here would be a souped-up dominant gene theory with numerous modifiers and potentiators and *in which the phenotype is schizoid disease* (Heston) *or schizotypy* (Rado, Meehl) *rather than schizophrenia.* I have not seen anyone really work out the mathematics of such a souped-up dominant gene theory which (a) involves multiple polygenic potentiators, (b) takes account of the differential fecundity reduction between schizotypes of the two sexes, and (c) includes (as I would still do despite some negative evidence presented by the authors and others) an additional factor of *environmental potentiation* upon a schizotype who has a schizotypal mother (whether she is diagnosed schizophrenic or not) and that is stronger than schizotypy in father as an environmental potentiator. Absent such a mathematical analysis of what I think is certain to be a more complicated situation than the usual kinds of dominant gene, straightforward penetrance-correction setups with which we are familiar, I don't believe one can say quite what a dominant gene theory would "predict" *numerically* about the ratio of diagnosed schizophrenia in first and second-degree relatives. Of course anybody who propounds such a souped-up theory has an obligation to work out the mathematics, a task that is hardly the responsibility of the authors who hold a polygenic theory.

In their discussion of the MZ and DZ rates in Chapter 2, where they discuss measles and the fact that very similar rates are predictable on environmental grounds, an additional comment is in order. Since almost all biological and social dependencies become decelerated functions (and many are quasi-ogival), if the parametric situation is such that over a sizeable range of high (or low, or both!) values of an independent variable we deal with an output variable near asymptote (or near zero), any research method that involves a combination of measurement error, sampling error, and what might be called sheer "crudity of measures" (low construct validity with respect to the focus of our interest) may bring two rates together without telling us much etiologically. It is not clear to me, without knowing what one can postulate about the genetics of the potential for catching measles and having a sufficiently severe case to be diagnosed, what can be inferred etiologically from the fact that both MZ and DZ twins show high concordance. No doubt the authors can explain this, but I wish they had. Suppose, for instance, that what we might call the "typical" measles susceptibility in a genetic population is sufficient so that a child exposed to the virus will get the disease and will have a case severe enough to be diagnosed as such. And suppose that the measles-susceptibility and severity-susceptibility are polygenic. The dizygotic twins average half of these measles-relevant genes in common with the measles-afflicted proband, and on the parametric situation I am imagining this half will be plenty to yield a very high measles rate. Under such a circumstance MZ-DZ comparisons are simply not illuminating with respect to the contribution of genes to measles susceptibility, are they? It might easily be the case, on a set of rather plausible biometrical assumptions, that a person who carried "counter-measles-genes" at each of, e.g., eight loci would have almost complete resistance to the disease, which would certainly entitle us to say that there were important genetic factors involved; and yet that fact would not be expected to emerge from the usual twin study, would it?

In mentioning my views they appose "schizotypes" to "schizophrenic equivalents," which bothers me because it perpetuates a semantics I consider unfortunate. The received doctrine is understandably (but I argue, wrongly) that the "main concept" is florid schizophrenia and that one should view compensated schizotypes as somehow sort of watered down or subclinical cases of schizophrenia—rather in the sense of the old French psychiatrist's meta-concept *forme fruste*. Now this is a perfectly legitimate way of looking at "latent" or "mild" or "incipient" cases, and I have no wish to impose my theory upon others via semantic stipulations. But the point is that Gottesman and Shields are at this locus referring to Meehl's views, and the language of "schizophrenic equivalents" gives a wrong emphasis to my conception. I do *not* think of the compensated schizotype as, in any usual sense of the word, a "schizophrenic," equivalent or otherwise. I do not consider him a "mild" or "disease-resistant" schizophrenic. He has a certain personality organization developed by social learning on the basis of a subtle neurological (not "psychological!") defect, and so far as I am concerned, *none of the clinical symptoms that enabled descriptive psychiatry to get a foot in the door so as to identify this entity in the first place* are a part of that organization or of that neurological substratum. This seems to be the hardest part of my theory to communicate, so I italicize it—cheerfully admitting that the emphasized sentence is not empirically corroborated (although, I insist, *is* in principle corroborable). If I may use what will seem to some to be a far-fetched analogy (deliberately, so as to drive the point home) it would be rather like saying that an American child who is capable of making a certain kind of nasal sound in his speech more readily than are most other children (on a genetic basis) is a "Gallic equivalent," i.e., he could come under suitable circumstances, to acquire the behavior of speaking French well. For emphasis one might start with schizophrenia and then water it down conceptually. I want to start with schizotaxia, and imagine only a minority of schizotaxic persons having other genes and a life history that (conjointly!) lead them to develop a clinically recognizable psychiatric condition, to wit, the schizophrenic psychosis. Of course epistemologically we start with schizophrenia because that is how we find out that there is something here worth looking into. In the same way, in studying the entity "gout," we start with bouts of arthritis of the big toe, attributable to the deposition of certain salts. However, in the gout-free but genetically gout-prone person, we do not expect to find some kind of attenuated inflammatory tetrad of turgor, rubor, dolor, and calor! I am not poking fun with this example. It really makes the point I want to make. We started our gout research with an inflamed joint; we then got back to the local tissue pathology of deposited urate crystals which are the defining property *locally* of a "gouty" condition; we then begin talking about gout-prone individuals (the great majority of whom never get clinical gout) in terms of their elevated uric acid titer; and even this endophenotypic biochemical dimension is presumably subject to modifiers and environmental factors like consumption of high-purine foods (the "high living" of the folklore) so as to be an imperfect correlate of the gout-disposing genotype. I may be unduly sensitive to the point, but I think not. It is a common complaint, for instance, by European-trained psychiatrists that American clinicians over-diagnose schizophrenia, and that we pay insufficient attention to the "classical (textbook) signs," including those that the great Bleuler made quasi-definitive of the disease. I happen to agree with this criticism, especially when I find some American psychiatrists of "psychodynamic establishment" persuasion calling everybody a schizophrenic who seems to suffer from the adult outcome of a battle-ax double-bind mother! I find that those who favor a more restrictive use of the term "schizophrenia" (I mean, of course, Bleuler's—not Kraepelin's dementia praecox) are surprised to hear me criticize this overuse of "schizophrenia" and then see me clinically identifying "schizotypes" with great clinical abandon. But they should not be surprised. This is a semantic misunderstanding, easy to clear up. The very reason for using words like "schizoid" or "schizotypal" is to permit a distinction between the compensated and the decompensated individual without the necessity for stretching the semantics of the pathology-laden word "schizophrenia" unduly. I don't even feel comfortable labelling a patient with the Hoch-Polatin rubric "pseudoneurotic schizophrenia" if I elicit only the pan-anxiety and anhedonia but not, say, a single micropsychotic episode, or subtle (but still reasonably clear) signs of cognitive slippage, or body image aberrations. How you research something of this sort, and how you assess the evidential weight of family studies, depends intimately upon what it is that you conjecture is inherited. I do not believe that one can, strictly speaking, inherit even a "watered down" or "weak" or "subclinical" schizophrenia. I believe that one can literally inherit Huntington's Disease, and that it makes sense to speak of an early or subclinical case of it, perhaps detectable only psychometrically or by instrumented-and-quantified "soft neurology." I want to put that kind of situation in a different category from what is inherited in the case of the schizophrenia-prone individual. I don't see how one could even "inherit," *stricto sensu,* the elements that enter into the picture of the compensated schizotype as conceived by Rado and myself, inasmuch as these mechanisms, contents and traits are themselves end products of an extraordinarily rich and complicated process of social learning. For this reason I prefer to avoid the conventional term "modifier" for the various polygenic systems that must surely be involved in altering the probability of clinical decompensation, since they do not "modify" the expression of a phenotypic trait at all in my picture of the situation—not even, necessarily, of an endophenotypic trait such as a parametric peculiarity of the synapse. The polygenic systems that alter the probability that I will decompensate if I inherit the dominant schizogene (on the Slater-Heston-Meehl view) need not have any direct effect upon the early links in the causal chain from gene through biochemical endophenotype to neurophysiological endophenotype (e.g., synaptic slippage) to behavior dispositions to the ultimate adult *learned* behavior. They probably enter in a different way as, for example, polygenically determined primary social introversion or an excessive susceptibility to the anxiety experience or a defective hedonic capacity get into the causal chain (on my theory at least) only when we consider the history of social learning in the schizotype's development and the susceptibility of the adult schizotype to clinical decompensation under current (adult) environmental stressors. On this theory, these nonspecific polygenic factors do not *in the least* "modify" the schizogene's *endo*phenotypic expression as schizotaxia, a CNS parametric aberration. They may not even influence the "neurological" indicators of the *exo*phenotype appreciably. Their causal role is later in the chain, and at a "psychological" (molar) level. It does not seem to me that the usual term "modifier" does justice to this kind of picture, and I notice that when I talk to geneticists they become nervous too. It is for that reason that I prefer my term "potentiator," which has the meaning of "raising, *via the social learning process*, the probability of clinical decompensation into diagnosable schizophrenic disease." A well-compensated schizotype is not, on this interpretation, to be thought of as some sort of "schizophrenic equivalent." He may be a person with negligible probability of developing clinical schizophrenia once he has survived intact to a certain age, i.e., the schizogene has missed its chance to have a psychotic effect, by virtue of the particular development his psychodynamics have taken. I am not criticizing the authors for writing their book instead of a book I might have written, a temptation to book reviewers that we properly deplore. But I do wish to highlight the difference between two ways of looking upon the non-psychotic and even non-neurotic individual who carries the "specific schizogene(s)."

One could wish that the authors had done more with the comparison of concordance rates for like sex versus opposite sex DZ pairs. Ever since an early article in which Rosanoff made methodological hay with this finding, I have found it an impressive counterargument against those environmentalists who try to explain the MZ-DZ concordance difference on the (somewhat *ad hoc* but admittedly plausible) counter theory that the physical resemblance of MZ twins and the recognition of them as "identicals" leads to (a) social expectations of similar behavior and (b) more similar treatment than would be true if they were DZ and easily distinguishable. Part of this environmentalist counter-hypothesis has been rendered quite implausible by the important research of Sandra Scarr showing that a number of behavioral characteristics of twin children show intrapair similarities that tend to follow the objective zygosity rather than the mother's beliefs about zygosity when the two are in disagreement. And of course any geneticist can tell you that while other people outside the family often mix them up, the mothers of MZ twins are almost infallible in telling them apart. I do not doubt, however, that determined environmentalists will possess sufficient ingenuity to *ad hoc* Professor Scarr's data (so soon as they hear about them, which many of them seem not yet to have done). But consider what an environmentalist explanation of the MZ-DZ concordance difference must face in the light of the small like-sex/opposite-sex DZ difference. The environmentalist must argue that, in a highly sex-typed culture like ours, and one with such very different kinds, times and settings of social and biological stressors as the culture imposes on the two sexes, differential treatment (within the psychodynamically crucial family, remember!) of two DZ boys born at the same time and reared in the same household is so much greater if they share half their genes than it is when they are genetically identical (even though the family members, and especially mothers, can tell them apart) as to reduce the schizophrenia expectancy in an MZ twin of a schizophrenic proband by a factor of three or four to one, that is a drop from, say, 50-60% to, say, no greater than 15%. But despite this dramatic drop attributable to a marked increase in the differential treatment of two male sibs born at the same time and with around 50% overlap in genes, the *increase in differential treatment* when we move from two boy siblings born at the same time to a boy and a girl born at the same time produces a further percentage decline in concordance that is negligible in size. In order to *ad hoc* this finding, it seems to me one would have to be an environmentalist fanatic. Although perhaps not an absolute hammer-blow destruction, these data together with their summary of the (regrettably small) number of known instances of MZ twins reared apart must surely be taken as greatly weakening two of the favorite counter-genetic hypotheses (identity-confusion and similar treatment). I record here, however, my prediction that the same environmentalist arguments that we have heard repeated regularly since Kallmann's findings slowly forced themselves upon the American psychiatric culture will still be invoked twenty years from now, despite these findings.

I cannot resist the impulse, before concluding my comments on this chapter, to say a word about the current popularity of the "spectrum" terminology. I tend to agree with McKusick that, despite its rapid taking hold in the language of behavior geneticists and their environmentalist opposition, it is not a felicitous semantic contribution. If genetic heterogeneity truly obtains among patients labelled "schizophrenic" (I set aside sheer diagnostic unreliability problems for the moment, although goodness knows that presents terrible difficulties for the researcher, which our authors have manfully striven to surmount—and largely succeeded *for the clinical entity*), then we need two nosological terms, do we not? Until we answer that question about heterogeneity, which presupposes answering the question of specific etiology, especially (see *supra*) how "specificity" is to be construed for a polygenic theory, we do not know whether two labels are needed, let alone to which patients they should be applied. If there is, say, a single nosological entity, disposition toward which was determined by one "big gene," and we then define the concept: *diagnostic error* as it would be defined in neurology or internal medicine, i.e., by reference to a specific etiological agent; then, as McKusick points out, there would be little usefulness either clinically or in theoretical understanding, in talking about a "spectrum." Because if you have the specific gene you would be a schizophrenic, although it might be in a *forme fruste, forme tarde,* or otherwise "atypical clinical picture," familiar notions which are taken for granted in branches of medicine where one is able to present a quasi-explicit *definition* of the disease entity in terms of its etiology and pathology. A person who has early paresis (paretic gold curve, positive spinal Wassermann, "soft" paretic neurology, early parenchymal changes in the brain associated with *Treponema pallidum*) is not said to belong to a "paretic spectrum," he is simply a paretic who isn't very sick yet. *Point:* We either have a specific etiology or we don't. If we do, there is no need to talk about a "spectrum," it is simply the usual question in medicine of how one treats the variety of clinical pictures and complications that "one and the same disease" may present to the clinician. (Bleuler wrote some very clear and sophisticated things about this in 1911.) Of course one realizes that the "spectrum" terminology caught on precisely because of the present etiological and clinical ambiguities, and perhaps no harm is done by it. But one has a tendency to the illusion that the introduction of this language conveyed something substantively or methodologically interesting and incisive; and I fail to see that it has done so. If memory serves, the term "caught on" at or following one of the first big schizophrenia conferences, when many were astonished at findings reported there to the effect that the families of schizophrenic probands showed some tendency to a heightened incidence of nonschizophrenic psychiatric aberration. I understand from colleagues who attended or participated in those conferences that participants expressed misgivings, fearing that we might have to revive the old-fashioned and long discarded generic "neuropathic diathesis" of continental constitutional psychiatry. I am at a loss to understand this reaction, on *either* a monogenic or polygenic view. With apologies for sounding like "I told you so," this is just what I would expect, and what I have long predicted would happen if the studies were done correctly. Suppose, for instance, that there is a specific dominant schizogene that is completely penetrant for a subtle neurological deficit, upon which is superimposed by social learning a collection of *statistically clustered* personality traits that constitute the loose syndrome "schizotype," and that a minority of such schizotypal individuals decompensate into the schizophrenic syndrome proper. I cannot imagine that any psychologist, whatever his major theoretical commitments as regards schizophrenia, would suppose that how much anxiety, rage, polymorph-perverse sexuality, energy level, social dominance, garden-variety social introversion, or mesomorphic toughness a person had would be literally irrelevant to the probability of his decompensating into schizophrenia. Granting this, one is also fairly safe in viewing these variables as partly genetic, by extrapolation from animal data, from the theoretical armchair, and through some human data. But these two high-probability assumptions jointly suggest that if I am clinically schizophrenic, the chances are good that in addition to the specific schizogene (or the scars from my schizophrenogenic mother) I carry a batch of polygenic modifiers of these other kinds mentioned, most of which are (in our culture) *themselves* predisposing to social maladroitness, economic marginality "underachievement," mental distress, neurosis, psychosis, or delinquency. I therefore cannot understand why we should be surprised by the occurrence of "spectrum disorders" in the relatives of schizophrenics, unless we started with an inordinately simpleminded view of what an integrated theory of schizophrenia would look like. If the term "spectrum" encourages conceptual analysis (e.g., what is it that we would require of the potentiators on a theory postulating a specific schizogene?) then it is a healthy bit of semantics; but if it merely makes it easier for us to assimilate what should not have been a surprising familial finding in the first place—without pushing us to ask just *why* we were so surprised—then its introduction will have done us a disservice methodologically.

I shall not comment on the case histories, which make interesting reading but, I believe, contribute less than is often assumed. Despite my being a psychotherapist with (I hope and believe) what is at least the normal fascination with the idiographic question, "How did this unique life develop?," I have rarely been able to persuade myself that case histories are illuminating in considering questions of the present kind, *absent a causal theory-sketch sufficiently rich to generate some sort of quantitative predictions, however crude.* Thus, for instance, the tremendous amount of effort that went into detailed clinical study of the famous Genain quadruplets, about which most behavior geneticists show considerable enthusiasm, leaves me somewhat cold. I really do not know what this kind of material should convince us of, except to bolster something we already know without twin studies, i.e., one and the same patient can present rather strikingly different clinical pictures at different points in time during the "natural history of the disease"; or, perhaps more helpfully, that persons with identical genotypes can differ in varying (and sometimes remarkably great) respects, and the like. For my part, the most important thing about case histories other than their heuristic value (even this, in my jaundiced view, is usually exaggerated), and that's the main value the authors have emphasized, is the fact that a case history, when reliable, can at least show "something is possible." As Bertrand Russell once jokingly said, "A single occurrence of an event establishes its possibility." How many things were not known to be possible but now are known to be possible from reading these case histories I cannot say, but I incline to be skeptical.

Chapter 5 is one of the most impressive chapters of the book, and one hopes that the obscurantist tradition in American psychiatry, which plays down formal nosology without presenting any respectable evidence for a greater power on the part of *competing* concepts, and which largely ignores the genetic data (themselves among the most powerful indicators that the diagnostic system *has* a nonarbitrary, nonadministrative meaning, that it "carves Nature at its joints") will be influenced by these results. It is tiresome to hear daily repeated the cliché (a cliché is bad enough when it is true, and unforgivable when it is false!) that "diagnostic categories in psychiatry are completely arbitrary, and besides are utterly unreliable," when this statement is based upon selective attention to studies so badly designed that rely upon diagnoses by such unskilled (or antidiagnostic!) personnel as to be incapable of shedding appreciable light on the question. It should be obvious that unless we had such a huge population of studies of diagnosis that we must worry about "selecting out the accidentally good ones" (as is the case in item analysis of a large pool of potential test items, or the picking of a small number of variables from a very large set in a multiple regression system, not, of course, the case in question), the value of nosology is better judged by the "good" studies than by the "unfavorable" ones. The favorable studies show that major nosological entities can be intersubjectively judged, and that they possess strong implications for prognosis and for treatment choice (e.g., ECT, phenothiazines). It is pointless to emphasize a study with a small sample, relying on the judgments of American psychiatric residents trained to ignore nosology and to doubt its importance (and not even taught the classic textbook differential diagnostic signs, as anyone acquainted with American psychiatry knows is sometimes the case in training settings). Patients are thus seen superficially and without an opportunity for inpatient study over a sufficiently long time; to then analyze the data with respect to subdivisions of the nomenclature that prejudge whether the subtypes of something like schizophrenia represent different diseases (itself counterevidenced by data in the present book) is, again, nonsensical.

Nor is it illuminating to show that quite a few schizophrenics were formerly labeled differently (the diagnostic shift is more often in that direction) when "does schizophrenia exist?" is the controverted issue. Bleuler made that clinical observation 60 years ago, and he also explained why it works that way. If you want to know something about the accuracy of psychiatric diagnoses, you look to see what the results are from the *best* studies, provided their sample size is large enough to have some faith in the figures as not being merely upward direction sampling error. I challenge psychiatrists and psychologists with an antinosological bias to refute the following claim: The interjudge reliability of a diagnosis of *schizophrenia*—as against other psychiatric diagnoses of functional disorders—is at least as high as typical diagnostic agreements in other branches of medicine; it is higher than the reliability of two clinicians' psychodynamic constructions from interview material or life history data; and it is markedly higher than the reliability of the most popular projective tests relied on by clinical psychologists. (Anyone want to take bets?) In that connection, it is reassuring to note that clinicians Meehl and Mosher, despite the marked difference in their ideological and etiological commitments, showed a reliability as high as that of a well-administered individual intelligence test ($r = .88$), using the seven-point scale of global psychopathology. I trust it will not seem unduly narcissistic for me to suggest that if you have competent and motivated raters making judgments *from decent data*, their judgments will tend to be reliable, and that will usually be the case despite theoretical divergences.

One of the important things about this chapter is that the authors ingeniously "sliced the pie" in different ways. Some psychologists are so methodologically puristic that they consider it sinful to examine whether one way of classifying the material is better than another, despite the scientific principle that one of the main ways to *find out* whether you are "carving Nature at its joints" is whether a certain way of grouping things increases the total orderliness (lawfulness) of the subject-matter system. Any physical scientist knows this, and psychologists—even Skinnerians, who don't like diagnostic entities much—should reread Skinner's classic discussion about the orderliness of cumulative records in his 1938 book! The problem is again similar to that of item analysis, because you realize that if you permit yourself unlimited cut-and-try procedures, especially "blind, empirical," nontheory-motivated gerrymandering, squeezing all of the blood out of what is a pretty anemic statistical turnip, you are in grave danger of seeing order when there is really nothing but chaos. That is a complicated question which involves a more thorough quantitative metatheory concerning *ad hoc* hypotheses than anyone in philosophy of science or statistics has thus far produced. It is therefore, regrettably, a matter of subjective judgment to a considerable extent. I assume most readers will agree with me that the various ways of classifying the diagnostic categories and dimensions employed by Gottesman and Shields each had an excellent prior probability of being good ones (from the armchair and on the basis of previous research) and that they have not overdone the strategy by riffling through all sorts of oddball nonrationally motivated ways of slicing the pie.

We face here a general problem, which if space permitted and it were appropriate, I would develop at some length. It is naive to object, as one frequently hears done, to selection of one of several data-classifications on the empirical basis that it yields greater genetic orderliness. Psychologists and psychiatrists who make this criticism usually display an undergraduate comprehension of philosophy of science. There is nothing viciously "circular" about an argument that begins with an observational-statistical finding, to wit, that one way of classifying events or entities leads to markedly greater empirical lawfulness than another way, and infers therefrom that the first classification is "better." *Per contra*, the most general characterization of scientific inference—cutting across all sciences and certainly not peculiar to problems in behavior genetics—is that we seek to maximize the order in phenomena. The bat "goes with" the whale despite the superficial and more common-sense grouping of whale with pickerel, and bat with bird. Consider psychodynamics: There is nothing phenomenologically obvious about a relation between a patient's inability to discard bent paper clips and his overdocility (with subtle, longer-term refractoriness) in responding to a psychotherapist's interpretations, but psychoanalytically we can conceive that these "go together" by virtue of their common (inferential!) factor of *anality*. Or, to take a nonclinical tough-minded example from another species, if one has doubts as to whether a tentative response-class, e.g., one defined by certain effect on a manipulandum (characteristic of the great majority of response classes studied in experimental psychology these days) is dynamically homogeneous, conjecturing that it actually consists of two response classes that are not only topographically identifiable (a pie-slicing job that could be done "blindly" in a number of arbitrary ways), what he does is to study the orderliness (grain of curve, dependence upon certain contingencies, manipulability via state-variables or discriminative stimuli) of the putative subclasses.

Were it not for certain statistical difficulties with the heritability coefficient—arising partly, if I may comment as a nongeneticist, from inadequate attention to the relation between the employed statistic and the postulated causal model that legitimates it and "makes sense" of it—one might, for example, argue that the best way to choose phenotypic indicators of a postulated genotypic variable, whether polygenic or otherwise, and the way to assign weights to such indicators in an indicator-set, would be the basis of *that choice and weight assignment that maximized heritability*. That is to say, if you are interested in a genotypic dimension and you have some combination of armchair theory, background scientific knowledge, common sense, and previous unintegrated factual research which at least permits you to assign certain indicators to the members of a potential indicator-set on a more or less rational basis, then the further screening of those tentative indicators in or out of the set, and the assignment of weights within the retained set, *ought* in consistency to be made on the basis of the genotypic postulate involved. This process may look "circular" to somebody who does not really understand what circularity is, or how science grows by cleverly "saving the appearances." As one philosopher of science said many years ago, *"There is nothing wrong with arguing in a circle, if it is a big enough circle."* But I repeat, the kind of quasi-circularity involved in this sort of process is not even real circularity in the *vicious* circle sense of the logician, i.e., it is not the case, when the inferential structure is analyzed thoroughly, that one uses a premise in a syllogism which premise is itself the desired conclusion. Finally, while we are on this important topic, it is worth mentioning that the choice of a metric, or the choice of one or another transformation on a phenotypic variable, ought itself to be selected on the basis of some mixture of a causal theory—even a sketchy one—*and the maximizing of order*. It seems to me rather difficult to justify the common practice of picking a more or less arbitrary although conventionally employed nonlinear transformation, such as a log or square root or arc-cosine transformation of phenotypic data, when the latter themselves have a crude and arbitrary metric (e.g., number of pellets hoarded by a rat), and then looking to see whether the offspring of two homozygous strains falls "in the middle." Perhaps I am nitpicking about the mathematics; but it does seem strange that such investigations often attain good results, absent any plausible theory relating the metric of pellet-hoarding via a square-root transformation to a presumed number of genes at a set of loci constituting the polygenic determiners of hoarding behavior.

Similar questions can be raised about the interpretation of MMPI score differences in the present study. The authors' sophistication about diagnosis and about the related problems of "open concepts" and the theory-fact interaction (about which most psychologists are incredibly naive and uninformed) leads one to wish that they had expanded further on this critical point in behavior genetic research on loose syndromes like schizophrenia. There is admittedly a methodological tension between (a) the necessity to avoid a progressive, counterproductive "conceptual drift" via psychometrics or reclassification of patients in the light of selected empirical relationships of increasing orderliness, and (b) the virtues of altering one's initial phenotypic weights on the basis of the empirical relationships observed. There is, for example, nothing wrong with the *concept* involved in saying, "This patient was perhaps not a schizophrenic as we initially thought, as he seems to have made a *complete* recovery." (Bleuler said that schizophrenics often made clinical or social recoveries but insisted that he had never seen so much as one single case in which there was a complete *restitutio ad integrum*; that there would always be some "residual psychic scarring.") Nor is there anything intrinsically wrong in raising the question, "Was co-twin Y a schizotype who remained compensated (hence, "discordant") throughout the risk period?" Those who think there is something illegitimate or "metaphysical" about those questions have, I repeat, not passed beyond an imperfect grasp of scientific method. But of course when a purportedly genetic syndrome is sufficiently loose, or when even the exophenotype held by particular theory to be "definitive" is *not* the full-blown clinical entity, there is a distressingly large element of personal judgment and subjectivity in assigning weights to members of the phenotypic indicator-set. Theoretically this question should, one supposes, have an optimal statistical solution; but none of universal applicability exists at present, to my knowledge. This is, in my view, one of the most important methodological problems, conceivably *the* most important one, in schizophrenia research. Its "solution" by the development of satisfactory techniques for the necessary bootstrapping of fallible phenotypic patterns should have general applicability in other areas of behavior genetics. Thus, as an example, the subclinical manic-depressive ("compensated cyclothyme"), may (I predict will) one of these days come in for as much attention as we are now devoting to the compensated and semicompensated schizotype. Whenever the clinical manifestations of pathology sufficient to bring the patient to psychiatric attention and to attach a certain nosological label are far enough removed *even within the molar behavior domain* from whatever Omniscient Jones would view as "definitive," some combination of internal clustering of phenotypic indicators with maximizing the relation of the entity (thus defined) to the postulated genotypic situation is what is required. It is therefore almost certain that with increasing knowledge, there will occur some degree of shifting in the relative importance given to various phenotypic indicators, sometimes even an introduction of a new one that is weighted heavily, or a dropping of one that is part of the received clinical tradition but that turns out to have too-low construct validity. In the terminology of the epidemiologist, we want phenotypic indicators of high sensitivity and high specificity; and we do not at present know what those indicators are. Although I, as mentioned above, consider the conjunction "thought disorder in the presence of unclouded sensorium" a quasi-pathognomonic inclusion test for clinical schizophrenia, it would be a mistake for me to infer that I must look upon thought disorder as the most powerful indicator (at the molar level) that *exists*. I consider it not only possible but rather likely—and recent unpublished research from Israel encourages me in this belief—that a sufficiently souped-up "soft neurology" may be a more powerful indicator *of the schizoid genotype* than thought disorder, whether the latter is assessed clinically or psychometrically. But associated with all such thinking is the lurking danger of a conceptual drift (associated, as psychologists go about their work, with psychometric drift) that would mean that we had slowly begun (unwittingly) to study something other than schizophrenia and its psychological or biological substratum. I do not suggest, however, that the authors have fallen victim to this possibility.

The remarkable results achieved by Dr. Essen-Möller might well be assigned reading for all first-year clinical psychology students, as it should almost suffice by itself to counteract the usual undergraduate brainwashing that psychiatric diagnosis is meaningless and that genes are irrelevant to mental illness. This great Swedish clinician evidently knows precisely what he is looking for, and whatever it is, he proves its importance. It seems almost "too good to be true," given the usual correlational baselines we have learned to live with in behavior science, that when an index case was diagnosed as schizophrenic by Dr. Essen-Möller's "solid-gold, hard-core, dementia-praecox-like" nosological standards, *every single one of the MZ co-twins* was also (blindly) diagnosed as schizophrenic or "schizotypal" by him—and this without an untoward inflation of schizophrenia-rate among the DZs. A truly virtuoso performance!

My scientific joy at this result is not unmixed, since the beautiful MZ concordance (and associated MZ/DZ ratio's potent evidence of heritability) is here attained by Essen-Möller's magisterial application of a (deliberately) "narrow" conception of the nosological entity. The findings put a further burden of proof upon those of us who entertain a broader conception of schizophrenia (*not*, I must reiterate, the same thing as our postulating the entity called "schizotypy"). I would like to work through the statistical implications of Dr. Essen-Möller's amazing display of clinical acumen—implications which I confess are presently unclear to me. On a souped-up dominant gene theory, suppose Essen-Möller's "hard-core," close-to-Kraepelin category were composed entirely of schizotypes who carry, in addition to the schizogene, heavy *multiple* loadings of the (several) nonspecific polygenic potentiators that theory requires. Would a biometrical geneticist predict this subgroup to show a higher MZ/DZ concordance ratio than a more inclusive group who have the dominant schizogene but are clinically more "peripheral" because they carry (on the average) fewer of the malignant polygenic loadings? This theoretical question must be answered, bear in mind, under the constraint that strong environmental influences are holding down the MZ concordance for the broader group by a sizeable amount. Writing under a publisher's deadline, I find that a rigorous causal and mathematical consideration of this important query must, alas, be put aside for now. But I will say at least something about it, at the risk of seeming to "explain away" evidence against my complicated variant of dominant-gene theory. I admit that this evidence does speak, *prima facie*, against that theory, since the MZ/DZ ratio is one crude index of "heritability," so that a clinical diagnostic emphasis which yields a high MZ/DZ index is to that extent an "optimizer of genetic orderliness," following the methodological pie-slicing rule defended *supra*. But I feel justified in qualifying my cheerful admission of counter-evidence by indicating where I think some quantitative ambiguities lie, urging that they deserve thorough biometric scrutiny before we conclude flatly that the continental, quasi-Kraepelinian entity should be the "genetically pure" object of future research efforts.

Adopting *arguendo* my conjecture in the preceding paragraph that the "hard-core" cases *are* hard-core (and, hence, identified by Essen-Möller with confidence and accuracy) *because* they carry a heavy genetic loading on *each of several* polygenic potentiators, his perfect MZ-concordance presents no great puzzle. The more malignant polygenes a schizotype has, the less unfavorable his "environment" (early nurturing regime plus adult stressors) has to be in order for him to decompensate (cf. "essential epilepsy" contrasted with convulsive disorders having a brain-damage etiology, and remembering that *all* of us have a seizure threshold that can be exceeded to produce a *grand mal* fit given a severe enough environmental insult, e.g., ECT, metrazol). Presumably the causal situation underlying severity-concordance data is an attenuated form of what yielded Essen-Möller's 100% MZ value.

The stomachache comes with the low DZ-value (since he could have achieved a "spurious" MZ rate by the strategy of calling everybody schizophrenic—but trivially reducing to nothing the conventional index of heritability). What might we expect, speaking quasi-quantitatively in the absence of a powerful statistical model for the complexities of a "souped-up" dominant-gene theory? I (timidly) commend the following rough conjectures to the attention of those who possess competencies in statistical genetics that I lack as yet: The "soft" or "peripheral" schizotypes are those who carry the dominant schizogene but relatively fewer of the polygenic potentiators. Further, on this complex theory, the word "fewer" refers not merely to an overall *count* of "favorable/adverse" alleles at all relevant loci. On my "psychologically"-oriented views, an equally important quantitative question is the *distribution* of these "bad genes" over *different polygenic systems*. These systems are "different" in that they control higher-order molar-behavior dispositions that may have nothing "physiologically" in common, but nevertheless converge causally—in the patient's *psychological* development and *social* history—to potentiate schizophrenia [= to raise the probability of a schizotype's decompensating]. I trust that my thus denying the fungibility of multiple polygenic potentiators does not seem contrived or fanciful. While it is not, I gather, a common mode of thinking among *non*behavior geneticists, it seems to me quite plausible in the behavior domain. Suppose, to put some flesh on the bones, we assume that both the rage-parameter and the anxiety-parameter are important polygenic potentiators tending, *ceteris paribus*, to decompensation in one who carries the dominant schizogene. A schizotype inheriting a +2 sigma hostility-parameter and a −1 sigma anxiety-parameter may remain compensated (albeit a somewhat cranky, litigious pain in the neck to employers and land-lords!). Another, with these two temperamental parameters reversed, resigns himself to a menial job and a passive, withdrawn social existence, remaining "adjusted" at that marginal level. But a third schizotype has the bad genetic luck to inherit +1 sigma parameters for both rage and fear, and this one is in for trouble. He is less able than average to fight, but also less able to "give in." He will resent his incapacity to relate and achieve, but he will be anxiously inhibited from doing anything effective about it. The (rage → fear) sequence of traditional psychodynamics will be a chronic problem for him ("I am afraid that I will hurt somebody"), as will the less-recognized but equally important (fear → rage) sequence emphasized by Rado ("I am almost always fearful, except when I get angry,"—cf. schizotype Hitler's seemingly deliberate use of this autotherapeutic technique, working himself into a rage in order to counteract the massive inhibition of thought and speech that often threatened to impair his functioning when conferring with a visiting diplomat.)

The index case of a twin study, *being* a schizophrenic, is more likely to resemble our third example than either of the other two, despite the fact that our third man's "mean sigma score" (treating the polygenes wrongly as fungible) is the same as theirs (= +1 sigma). Where did our index case get his two +1 sigma polygenic potentiators? More probably than not, either (a) both were largely inherited from the same parent (*not* the parent from whom he got the schizo-gene, because that parent was very likely a compensated schizotype, as the proband → parent figures and the fecundity statistics show); or (b) the genes for the two affective parameters came from both parents, but not "randomly" distributed across systems, e.g., he may have received the big dose of rage-param-eter polygenes from his schizotypal mother—the battle-ax compensated type—and the big dose of fear-parameter polygenes from his non-schizotypal (but somewhat neurotic) father. As I understand what the geneticists tell me, such a "piling-up" of relevant polygenes on one side (rather than a more symmetrical "mix" in parental origin) will tend to yield higher average values in the sibship than would be true for the fungible situation. Hence, on the average, this theory leads to a "boosted" DZ concordance, moving the DZ value up closer to the MZ for those clinical judges who diagnose "schizophrenia" more freely than con-tinental practitioners, there being no systematic influence tending to raise their MZ figure correspondingly. (Side comment for geneticists: Do these causal conjectures fit the traditional conception of a "modifier" closely enough to justify using that word?)

Taking a different tack, what would happen if a super-Radovian clinician *could* spot schizotypy perfectly (i.e., regardless of decompensation): On a dominant gene model, the MZ concordance would be 100%, and the DZ 50%. But such a clinician might nevertheless "do worse" than Essen-Möller, since they both hit 100% on the MZs whereas Essen-Möller would—*even if the dominant gene theory of schizotypy were correct*—run considerably lower than 50% for the DZ cases, not labeling those (among the schizotypal 50%) with "good polygenic luck" as schizophrenic by his strict clinical criteria. So an MZ/DZ or (MZ −DZ) index would favor Essen-Möller even if the compensated schizotype were a real entity and were accurately diagnosable.

In that connection, there is a serious problem about the index to be used. In comparing the six clinical judges, Gottesman and Shields employ the simple *ratio* MZ:DZ of concordances. I do not suggest that this is "wrong;" but I should like to have seen more discussion of their basis for choosing it (over, say, the MZ −DZ difference). Not only can the "distance" between two judges be different as between these indexes, but even the *order* of judges can rather easily be re-arranged by one's index choice. In Table 5.3, consider the concordance for (a) "definite schizophrenia in both twins." To get the six-judge order determined by index (MZ −DZ) into the authors' MZ:DZ index order requires two permutings (and, I note egocentrically, shifts Judge PM from fifth place to the middle position). Employing concordance (b), "inclusive of ? Sc in either twin," the MZ:DZ and (MZ −DZ) orders again differ by two permutings. We note further that the indexes differ from each other by the same permutation number as MZ:DZ does from DZ-rate itself in cases (a), and in cases (b) the (low) DZ-rate order is exactly the same as that of MZ:DZ but for a paired ranking (1.5 for judges KA and LM). The denominator DZ exerts heavy influence on a ratio index, of course; we recall the theorem of elementary calculus that the derivative of a quotient has its denominator *squared* in the denominator. These problems of choosing the "optimum" index relating two concordances are as difficult as they are important, and we must try to avoid indexes that have a mathematically built-in feature capable of loading the dice. My suspicion is that a ratio index inflates the significance of DZ-concordance interpretively, but I primarily want to raise the question for thorough analysis. Its importance is enhanced by Gottesman and Shields' suggesting (from Table 5.3 orderings) that either a "narrow" (Judge JB) or "wide" (Judge PM) diagnostic usage leads to a poor heritability result.

Of course the ratio index is a linear function of the difference index *for fixed DZ base*; but it is both nonlinear and *nonmonotonic* when we allow both MZ and DZ to move around considerably. To give the reader some notion of the complexity of this "index" issue, suppose that x = MZ concordance; y = DZ concordance; $u = \frac{x}{y}$ i.e., Gottesman-Shields' choice of index, MZ/DZ; and $v = x - y$ simple (MZ −DZ) difference. How does the ratio index change with changes in the difference index? The derivative of the former with respect to the latter turns out to be

$$\frac{du}{dv} = \frac{1}{y} \frac{1 - (x/y)(dy/dx)}{1 - (dy/dx)}$$

which is a can of worms, especially when we remind ourselves that dy/dx will surely depend in no *simple* fashion upon the value of x (the DZ concordance). To put in my psychologist's two cents worth again (since I don't belong to their trade union, the geneticists can ignore me with a clear conscience), it seems to me that adoption of indexes like MZ/DZ or MZ −DZ (or, for that matter, the ubiquitous and sometimes strangely behaved heritability coefficients—and why *two*?) ought ideally to be itself based on theory. And by "theory" here I mean something more substantive than a statistician's discussion of how components of variance can be algebraically separated and pooled. The choice of a numerical index is rather like the choice of a scientific instrument for a certain purpose; and we understand how that purpose is rationally served by the chosen instrument in terms of a substantive theory concerning the causal structure of the domain. It may be that I am poorly instructed as to the "causal-theoretical" rationale underlying an index like MZ:DZ, and that the genetics literature somewhere explains it all quite clearly. If so, reviewers of this book will no doubt enlighten me, and in the process others (some of whom seem to me almost equally unclear but not as distressed about it) will have their education furthered along with mine.

A final comment on Essen-Möller's results is to reassure ourselves that bootstrapping from clinical judgment (when the clinicians "believe in diagnosis" and have bothered learning how to do it) is not a foolish way to spend research time and money.

My only criticism of this chapter is a minor literature disagreement which, if anything, favors the authors' own position. They mention the 1956 work by Schmidt and Fonda as one of the studies taken to suggest that psychiatric diagnosis is unreliable, and while no doubt some muddleheads have cited it for that purpose, that is *not* the burden of Schmidt and Fonda's own conclusion, as the reader will find if he goes back to the original paper. As indicated by Schmidt and Fonda's title, "The reliability of psychiatric diagnosis: A new look," and the introduction of the paper which points out that they themselves, having previously asserted without documentation that psychiatric diagnosis is unreliable, thought better of this statement and decided to investigate it more thoroughly, their interpretation is distinctly pro-nosology. The last sentence of their article reads: "It is concluded that satisfactory reliability has been demonstrated for some of the psychiatric diagnoses, but that this carries no implications regarding their semantic validity or usefulness." Quite apart from what Schmidt and Fonda thought they had shown, when schizophrenia is the subject matter, it would be inappropriate to cite their study on the "unreliable" side of the argument, since a tetrachoric r they computed on the dichotomy "schizophrenia/nonschizophrenia" came to .95. This is surely better than many diagnoses in internal medicine, and vastly better than the kinds of personological judgments and psychodynamic inferences (not to mention predictions about psychotherapy!) which are routinely relied on in clinical psychology and psychiatry in the United States. The Schmidt and Fonda study, either in terms of its data or in terms of those authors' conclusions, should not be cited with studies indicating unreliability.

In Chapter 6, I understand why the authors say, "It would seem appropriate to use more than one criterion for severity in any one study so long as the construct remains incompletely defined." But I am troubled by that language, which seems to suggest that severity is some sort of "construct," comparable methodologically to the postulated schizogene(s) or a postulated schizophrenogenic mother or a postulated virus for multiple sclerosis. There isn't any good reason I can think of, or that the authors present, for supposing "severity" to be that kind of construct at all. It is not that kind of construct in nonpsychiatric medicine. "Severity," which can range from a patient who is accidentally discovered in the course of an insurance examination or a mass tuberculosis or cancer screening, to have an unsuspected pathology even though he had no complaints (or, more dramatically, the man who is killed by a truck and who, when autopsied, is found to have one kidney's lumen completely occluded by a gigantic renal calculus filling it, the famous "staghorn kidney," without so much as a history of complaints of mild backache!) to an individual who is moribund and receiving the last rites. We need not assume that "severity," which is basically an administrative or "clinical handling" type of concept to begin with, designates any theoretically homogeneous entity. For this reason, the authors' language in speaking of using more than one "criterion *for* severity" strikes me as possibly misleading. It sounds as though the methodological problem presented by the severity concept is on all fours with the methodological problem of assigning differential indicator weights for the presence of the schizogene, or of deciding whether the colloidal gold curve should be thrown into the clinical hopper along with the blood and spinal Wassermann and washed-out nasolabial folds and Argyll-Robertson pupils as different indicators of the presence of general paresis. In fact, I think the word "criterion" is undesirable in this kind of situation, whether we have a hypothetical construct (such as a disease entity or germ or gene) in mind or not. "Criterion" means that which defines or which suffices definitively to include or exclude. (In psychometrics, it also has the weaker meaning of "used to validate.") That is not the situation with loose genetic syndromes or, for that matter, with any kind of syndrome in which the underlying pathology is either unknown or unreliably judgeable, and consequently is not useable "operationally" as an explicit definition of the entity. Any time we deal with open concepts, whether in behavior genetics or elsewhere, we ought to use some more modest word like "indicator" rather than the strong word "criterion."

I do not mean to suggest that the study of correlations in severity between twins or other family members is unprofitable—far from it. Such correlations are illuminating as to the proper theoretical model; and any kind of theory—whether genetic or nongenetic—explaining concordance and discordance values in the clinical entity as diagnosed must do justice to more detailed and refined assessments of severity. (Even this remark requires a caveat, however: In other branches of medical genetics, and in medicine generally, "how sick" somebody is with Disease X is not the focus of much attention when the disease as yet is hardly understood. And even diseases whose etiology and pathology are very well understood, such as those due to a known invading microorganism, exhibit wide variation in severity, most of which attracts rather little interest on the part of medical researchers.) All I am objecting to is the language labeling severity a "construct" and implying that the various administrative, social and psychological variables that have been employed in studying it are fruitfully viewed as multiple criteria for severity. The point is that there is no strong reason to believe that severity is a "theoretical" dimension at all. Rather than being a dimension for which we desire good indicators, *it* is itself in the role of a (global, best-available) "indicator" of the known and unknown theoretical entities underlying it. As such, of course, it is presently a worthwhile thing to study. But it will not, I presume, be found occurring in a full-fledged nomological network explaining schizophrenia, having been replaced by the genetic and environmental variables it now does duty for in our state of incomplete knowledge. The problem is analogous to the vexed problem of psychotherapy outcome criteria, which some psychologists persist in treating as though it were a problem of assigning some kind of optimal weights in an outcome "index." Absent a theoretical or empirical showing that some such variable as "outcome" has homogeneity, that there is in fact some sort of entity (whether taxonomic or continuous) that is being "estimated" or "approximated" by the various outcome measures, it is not clear why anybody wants to make a composite "outcome" measure. If the effect of psychotherapy is to make Jones feel less anxious because he is more openly aggressive to his father, as a result of which an MMPI anxiety measure will show him to be improved and a family's rating on "pleasantness" will show him to have become worse, this does not present any kind of problem in "weighting" these two measures, nor any paradox in theoretically understanding them. They are simply not indicators of some generic "good" or "bad" outcome; one of them is a measure of anxiety and the other is a measure of overt hostility. Whether or not this is an improvement depends upon the therapeutic contract. It is not subject to empirical study of the sort we would do when trying to find out whether spilling gravy on your vest or having grandiose delusions are as powerful indicators of the disease entity "G.P.I." as was stated in the old neurology textbooks prior to the development of accurate serological methods for identifying the paretic patient with high validity. In the theory of schizophrenia, for instance, I conceive of several polygenic modifiers or potentiators that may contribute to a patient's "holding together" until his thirties or forties (as in the paranoid subtype) and hence contribute to a rating of "lesser severity" by *one* set of clinical criteria (e.g., later age of onset, less psychometric evidence of severe overinclusiveness, or other types of gross hebephreniclike thought), such as high intelligence plus mesomorphic toughness plus high aggressiveness plus high social dominance plus relatively low social fear. Yet these factors can all be responsible for the tendency to act out aggressively by taking a shot at the President of the United States—presumably another indicator of "severity." This reasoning suggests that rather than studying concordance of severity, or treating various indicators as somehow "competing" or "alternative" *measures* of a putative construct "severity," what we really want would be a rather extensive set of clinical and psychometric measures of avowedly different theoretical entities. No doubt such a list would show that some cases presently considered discordant as to "severity" were in fact concordant psychologically, others not; and *most* cases (whether concordant or discordant using present indexes of global "severity") would be concordant for some variables and discordant for others. (I pass by the obvious point that on that kind of multivariate profile basis, "discordance" would be a quite arbitrary cut on each dimension measured, as it would also on any profile-similarity index adopted.) For example, suppose a patient manages to stay out of a mental hospital for ten years because he employs extrapunitive, socially dominant, counteraggressive defensive techniques (along with just a dash of subclinical paranoid ideation) and is therefore able to manage his anxiety and prevent himself from a gross, socially observed psychotic breakdown. Would we call him concordant or discordant for "severity" with a twin who spent those intervening ten years in a psychiatric hospital? I think this kind of example shows us why the severity concept does not deserve to be treated as a *theoretical* construct with respect to which we must find optimal indicators for genetic research purposes.

In this chapter the authors say that "It is not merely a tendency to be schizoid that is inherited and is in some way causally related to schizophrenia, but something more besides." I confess I do not fully understand this statement, or the sentences that follow it. The tendency to be schizoid is a dispositional concept and I don't know just what it would mean to speak of "something else" as being an important contributory factor, when that something else is also said to be "under considerable genetic control." Is there perhaps some sort of conceptual confusion here? If being schizoid is having a personality makeup predisposed to schizophrenia, and what is inherited is a disposition [= "tendency," they call it] to that state or condition, I find it hard to grasp what a something else, also under genetic control, and also *disposing to* being schizoid or to being schizophrenic (either one!) means conceptually. But this may be partly because I cannot conceive, as mentioned *supra*, that even being schizoid is inherited. Being schizoid is having certain kinds of behavior and mental life, which include form (e.g., type of defense mechanism, parameters of various sorts) and content (e.g., social fear, apprehension about being disliked, deficient basic oral trust, low feelings of personal worth); none of this could possibly be "in the genes." One of the reasons we need an additional terminology for designating individuals who are neither clinically schizophrenic (even "pseudoneurotic") or phenotypically schizoid—I set aside here entirely the reliability and validity of those assessments—is precisely that if such a person is the MZ twin of a clearly schizophrenic individual, we need a term to refer to whatever he is dispositionally that is not the same as being either schizophrenic *or having a schizoid personality in terms of the usual nomenclature.* What term should be used for this purpose unfortunately hinges upon something not entirely conventional or stipulative, to wit, a tentative decision as to whether MZ co-twins who seem not to be schizoid would still be judged non-schizoid given a sufficiently thorough and high construct-valid assessment. Of course nobody knows, and the authors do not claim to know, the answer to this question.

With regard to their interesting suggestion that not all relevant genes get "switched on" until certain psychosomatic states are reached, I have no expertise to evaluate that within the received causal frame of physiological genetics; but I don't quite see how it follows from their empirical findings under discussion at that place in the text. I think the hooker comes from the previous sentence, where they speak of ". . . something more besides." Why does it have to be *besides*? Wouldn't it be better to say something else instead? If (as I believe Professor Gottesman has agreed in conversations with me) delusions and hallucinations and catatonic posturizing cannot possibly be transmitted by the genes, and if the Radovian "scarcity economy of pleasure" and exaggerated social fear of the schizothyme cannot, strictly speaking, be inherited either; then what is inherited is not "something else besides" a disposition to schizophrenia or schizoidness. Rather the genes determine something else instead, to wit, the underlying neurological or biochemical aberration (or whatever else it is) that is the biological substructure corresponding endophenotypically to a *higher-order disposition to acquire the disposition: schizoid personality makeup.* We can hardly suppose that the switching on or off of the schizogene(s) can switch on or off the acculturated, developed, organized adult schizoid personality, can we?

One understands why, toward the end of this chapter, they agree with Slater in excluding the alcoholic hallucinosis of one MZ co-twin, and I do not quarrel with that decision. On the other hand, one can raise here a question similar (perhaps identical?) with the question I raised *supra* in discussing such factors as influenza or brain damage in the precipitation of a schizophrenic break, from which the patient may subsequently fail to recover after he has recovered from the acute infectious disease. Not wishing to beg any genetic question, it is not clear what should be done with alcoholic hallucinosis, which is—even by clinicians who employ the term, and many do not recognize it despite its listing in DSM-II—seen as importantly different from delirium tremens or alcoholic Korsakoff syndrome. We are not today in a position to say that prolonged excessive use of alcohol *cannot* be a stressor leading to a schizophrenic psychosis in a genetically predisposed individual, who might or might not have remained compensated without overindulgence, and had he decompensated, might instead have done so *with a different clinical picture*. There simply isn't any good evidential basis for us to decide whether or not the alcoholic hallucinosis syndrome is an atypical schizophrenia whose atypicality is due to the particular chemical mode of its precipitation. For all we know, everybody with alcoholic hallucinosis, if the syndrome were optimally defined as in the eyes of Omniscient Jones, would be an "alcoholically precipitated atypical paranoid schizophrenia." I don't assert this, I simply say that nobody knows whether it is so or not. If a college student develops an irreversible schizophrenia following use of LSD, he is nonetheless a schizophrenic. Some people become almost psychotically paranoid under the influence of the amphetamines, and my impression has been that many of such are rather obvious schizotypes, although not all. The causal chain might very well run: Schizotypy → alcoholism → brain damage → "alcoholic hallucinosis," this last event being converged upon by two causal arrows, one from the (penultimate) brain damage and the other from the schizotypy that initiated the whole chain. It is a nice example where neither fact nor theory nor an optimal bootstrapping procedure for cleaning up a loose clinical syndrome of unknown multiple etiology suffice to tell us how a case ought to be classified. Which illustrates the Feyerabendian point that our theories play a critical role in how we classify our "facts."

For myself, Chapter 7 on psychometric contributions to genetic analysis elicited the most ambivalent affective responses. I will not dilate upon the obvious advantages of employing psychometrics in genetic research, since it is perfectly clear from the authors' MMPI profiles how useful it was. One hopes that this chapter will lead psychologists and psychiatrists who study the genetics of mental disorder in the future to employ an objective instrument of respectable validity like the MMPI, instead of relying upon clinically popular devices of low objectivity whose validity is not corroborated (in some instances one should rather say, whose *in*validity has been strongly evidenced by the research literature). One must consider the fact that the MMPI is an objectively scorable and to considerable degree even objectively interpretable psychometric device; that it requires negligible time on the part of the investigator and only an hour or so by the typical examinee; that its construct validity is supported by a vast empirical literature; and that it is (to an unprejudiced reader of the record) at least as valid as any existing personality assessment device—considerably more valid than most of them. I here permit myself a dogmatic remark: There is little excuse for anybody subsequently to enter upon a costly and difficult study in human behavior genetics and fail to include the MMPI in his battery. I should opine that although that was already fairly obvious 15 years ago, Gottesman and Shields' work in Chapter 7 makes it blindingly clear.

With respect to the problem of circularity, the chapter provides us with a beautiful example of mutual support between claims of (a) psychometric validity, (b) the empirical meaningfulness or legitimacy of an open concept, and (c) the verisimilitude of a substantive theory. Many psychologists still evince a rather low level of sophistication about this extremely important methodological matter, despite the (unanswered) arguments by Cronbach and myself of 17 years ago that one cannot, except quite artificially, disentangle the logical and evidential problems of validating an instrument and corroborating a substantive theory about a construct variable, which the instrument purports to measure. One still hears it argued that you either have to validate the measuring instrument "independently" (whatever that means!) before you can plug it into a theoretical network, or you have to (again, "independently") corroborate the theory and then go about "validating" the measuring device. If this can be done, it is of course a delightful state of affairs; although one must point out that the alleged "independence" will not, when closely scrutinized, ever turn out to be "theory independent," at most it may be a relative independence of the particular *portion* of theory that we have our eye on for the present purpose. But a realistic and sophisticated methodological appraisal of the research situation, especially in the case of open concepts and loose syndromes in behavior genetics, shows it usually to be otherwise. What we have is an interconnected set of constructs in which validity claims for an instrument and verisimilitude claims for a substantive theory concerning a variable, which the instrument supposedly measures (and which, note well, the theory "contextually" defines!) are all one ball of wax. It is thoroughly misleading to ask what appears to be the sophisticated question, "Do these impressive MMPI findings on schizophrenic co-twins tend to support the construct validity of the MMPI Sc scale, or do they tend to support the nosological claim that such an entity as schizophrenia exists, or do they tend to support the conjecture that genes have something to do with schizophrenia?" There is no either-or about it. The point is that the findings support all three of these things at once. Neither the logic nor the statistics of this should cause any intellectual distress, although I gather they often do. Again, there is nothing "viciously circular" about the factual claims, nor is there anything "circular" or "tautologous" about the definition—objections we still meet within this area, despite repeated efforts by numerous writers to clarify the matter.

As a result of the analyses carried out by Gottesman and Shields in this chapter on psychometrics, we now have better reasons than we had before for believing that there is such an entity as schizophrenia, that the MMPI has respectable construct validity for detecting it, and that genes have something to do with it. I would not even be prepared to say whether one of these three assertions receives more incremental support from the chapter's data than one of the others. Of course, one must play fair about it. By the same token, the negative psychometric results on thought disorder in nondiagnosed relatives tend to discorroborate a conjunction of assertions, to wit, that thought disorder is a "core" aspect of the schizotypal makeup (absent clinical decompensation), that schizotypal thought disorder is genetic, and that the Object Sorting Test has moderate to strong construct validity for schizotypal thought disorder. At least one of these antecedently plausible assertions would seem, given the authors' discouraging data, to be false. I leave the choice to the reader. One notices an asymmetry here (which Sir Karl Popper would find displeasing, but I cannot help that, until his fellow logicians clean it up); the positive findings on the MMPI are taken to corroborate all three assertions, whereas the negative findings on OST clearly refute only the conjunction of three. This being a consequence of the logic of conjunction, negation, and *modus tollens*, I am afraid I cannot be held responsible for it! There is a "heads-I-win-tails-you-lose" aspect here which is, I suggest, built into the very nature of the research situation. What are we to do with a psychometric device that characteristically indicates a relatively lesser amount of "thought disorder" in schizophrenic patients who have paranoid ideation, since for a high IQ person to think so "crazily" as to have delusions (and in some instances to "support" them by hallucinations) can hardly be considered, from a *clinician's* standpoint, to be other than a "disorder of thought?" Until that still disputed matter is cleaned up one does not know just what to think about the authors' findings on the Object Sorting Test. I believe that we do not have at present any psychometric device for detecting mild amounts of thought disorder in the schizotype, just as we do not have any accurate means of measuring schizoid anhedonia. Until we do, we won't know what to make of this kind of negative result. But lest I seem to *ad hoc* the unexpected negative findings, I must record the judgment that these results do argue somewhat against my inclusion of "cognitive slippage" as a fundamental feature of the schizotypal makeup. To that extent the chapter must be received in evidence as against some of my theoretical opinions.

The apparent insensitivity of the MMPI to well compensated schizotypy was disappointing. One does not, of course, know exactly what to predict psychometrically about the DZ co-twins, even on a (modifier-and-potentiator) dominant gene theory, unless he relies upon the *non*psychometric diagnosis of schizotypy, which would rather defeat the purpose of including psychometrics as a separate, and putatively more sensitive, schizotypal indicator. Setting aside that possibility (which, however, the authors quite properly include as one way of looking at their data), let me spell out somewhat the source of the DZ unpredictability. There were 26 DZ co-twins on whom MMPI profiles were available. Let us neglect any influence of selective attrition, keeping in mind that this influence has a reasonably good chance of pulling down the DZ scores because of noncooperation of schizotypal co-twins, but also pushing in the opposite direction because of the greater likelihood of a decompensated DZ co-twin being available by virtue of his clinical status (e.g., in hospital). We are neglecting clinical status. Proportions are, alas, among the most unstable statistical estimators, although not as bad as correlation coefficients and beta weights. If there were no selective attrition, and ignoring clinical status of the 26 DZ co-twins, one has a theoretical expectancy of 13 schizotypes on a dominant gene model. If we compute the standard error of this expected number np by the usual formula $\sqrt{npq} = 2.6$, we see that a single standard error random sampling fluctuation of $(k - np)$ ranges from 10.4 to 15.6 schizotypes among the DZ co-twins. There is one chance in three that we are even outside these numerical limits in the schizotype rate. Now let us suppose averages of a T-score approximately 70 (higher than that suggested by previous research or by the MZ data) for schizotypes, and $T = 50$ for "normals." The expected mean T-score for all DZ co-twins would then run around 60, but the above computation in sampling fluctuation on a dominant gene hypothesis could easily take us down to the limits of one standard error fluctuation, and that would be only around $T = 56$. Of course if one sets up a more traditional "confidence interval" for the purpose of refuting the dominant gene hypothesis psychometrically, and were on that basis to expand his range of tolerance around the theoretical expectancy of 60 to a 2-sigma fluctuation (95% confidence), the mean value of the DZ's could go down almost to the Minnesota norm $T = 50$, without postulating an outlandish sampling fluctuation. Let me emphasize that I am not in any way trying to make out the psychometrics to be better than they are. They are quite good enough to make the authors' point without any statistical finagling or fudging on my part. But one could be tempted to argue strongly against a dominant gene model on the basis of these results, unless he thought through the actual numerical situation that obtains when working with samples of this size. I am not greatly distressed about my own monogenic theory in looking at the DZ values, since extrapolating (very roughly!) from previous research it is not to be firmly expected that the Sc scale would yield a mean any higher than 1 standard deviation above the Minnesota normals.

What is discouraging, to anyone (like myself) contemplating the use of Minnesota's huge mass of MMPI file data ($N = 10,000$ patients) in an intermediate-stage reliance on MMPI to "bootstraps" other more direct indicators of the schizotaxic nervous system, is the fact that the MZ co-twins who were clinically normal, or diagnosed something other than schizophrenia, only managed to eke out a T-score on Scale 8 of 63 (Table 6.7). That suggests rather strongly that the extant MMPI Sc scale is insufficiently responsive to schizotypy as such, i.e., that the discriminations are achieved by virtue of "psychopathology" items.

Of course, however disappointed one may be about this he is not entitled to be too surprised. The Sc scale was derived using patients diagnosed in the late 1930's on the basis of formal diagnoses by a fairly conservative psychiatric staff, prior to the publication of such influential papers as Hoch and Polatin or Eisenstein, or the work by Rado on the schizotypal organization. Inspection of the item content tells us how important some rather florid psychotic items are in contributing to the Sc score. But let us not overdo that argument. The scale does contain quite a number of "subclinical" items that seem to reflect the phenomenology of the schizotype and that one would expect to show at least an average trend even in compensated cases. But these items are definitely in the minority, so we are requiring that a rather small tail wag the psychometric dog. *Suggestion*: Break Scale 8 up for genetic research. Whether the MMPI pool is qualitatively rich enough to permit construction of a "schizotypal key" that will reflect the personality organization and the subclinical phenomenology without requiring presence of any of the classical textbook diagnostic signs of diagnosable schizophrenia is doubtful. And, pushing the argument further, it is also by no means assured that any item pool composed of verbal responses is in principle constructable that would have sufficiently high discriminative validity for the compensated schizotypal condition to be satisfactory in behavior genetics research. But I intend to try it.

Chapter 8 on the environment will doubtless elicit complaints against the authors because of its relative brevity, but it does not appear from what they *did* do by way of unscrambling environmental stressors or nurturing factors that spinning it out further would have been profitable. No doubt the qualitative impressions from the case studies helped make this a short chapter, as would the overall implications of the research they review (e.g., foster child data, MZ twins separated, father versus mother, and the like). I am gratified by the appearance of social dominance as a predictor variable, it being one of the important polygenic potentiators on my conjectural list of factors that determine whether a schizotype remains clinically compensated. I here record a high-confidence prophecy that the same will be found when such variables as anxiety-proneness, polymorph perverse sexuality, "garden variety social introversion," rage-readiness, sex drive, energy level, mesomorphy, and hedonic potential are adequately researched. We already know that heterosexual aggressiveness in one of its aspects, to wit, marriage behavior, is the biggest single element on premorbid adjustment scales—almost enough by itself to be used as a scale-surrogate.

I think it important that we regularly distinguish between (a) those environmental factors that are part of the early nurturing environment and therefore play a role in forming the schizotypal organization that emerges from the patient's childhood (these being the factors customarily stressed by antigenetic psychodynamicists); and (b) those stressors that are exerted in adolescence and adult life upon the schizotypal organization resulting from the sequence of child development as influenced by factors (a). One must be careful not to infer anything about one of these "environmental domains" on the basis of research findings concerning the other. Here again, what sort of causal theory you have in mind for the relations between schizophrenia, schizoid disposition, and the genes will properly influence how you draw your causal arrow diagram.

I think the establishment psychodynamicist might justly complain that the authors do not sufficiently allow for the richness and subtlety of the psychodynamic theory in their discussion. When we criticize the theory, we have to allow it its own inner coherency and not require it to predict facts that it really doesn't claim to predict. One may be skeptical about whether the subtle intrafamily psychodynamisms play the critical role that establishment psychodynamics gives them in the causation of mental disorder, but if they did, the kind of information one can get from formal diagnosis plus psychometrics plus even a respectable clinical case history would not have any guarantee to reveal it.

We do have a fascinating paradox that presents itself to us, given the total body of evidence of all kinds available today: How is it that almost half of the

MZ twins of schizophrenics will manage to live out their lives without developing schizophrenia but the "obvious causal candidates" among neither socio- and psychodynamic factors in a nurturing environment, nor situational stressors imposed upon the adult schizotype appear, whether looked at statistically or idiographically, to be clearly identifiable as making the difference between the schizophrenic proband and his compensated MZ co-twin of identical genotype? *Point*: The concordance statistics themselves prove a major influence upon clinical status to be *non*genetic; but when we look in a variety of ways for conceptually treatable aspects of the "environment," we draw practically a blank. This is surely a puzzler for the theoretician. The first possibility—and here, as I indicated *supra*, I tend to line up somewhat with the psychodynamicist "against" our authors—is that nothing but a very detailed knowledge of the life history in its unique idiographic sequence of events (including *inner* events!) would provide the evidentiary material for a causal understanding of the differ- ence If this were so, and I am willing to assume that it is at least sometimes so for the discordant pairs, one cannot infer that a more thorough going study by intensive depth interview methods of the psychotic and nonpsychotic adult pairs would even give us the information. My own determinist beliefs lead to an unjustified optimism about reconstructing the past when "history" (whether of a life, a nation, or a religious or political movement) is the subject matter. It is not obscurantist, nor incompatible with a faith in determinism, to have doubts about the feasibility of such reconstruction. There may be critical events, in systems involving what Langmuir called "divergent causality," that are literally unavailable to the investigator from a study of the patients, or any kind of documents, twenty years after the critical event occurred. If a geologist finds a somewhat unusual kind of formation occurring in a given type of rock, he may or may not be able to spin out some plausible causal hypotheses within his overall theoretical framework; but even if he can do so, the most detailed knowledge of physics and chemistry plus the overall generalizations about the history of the earth, about how igneous, metamorphic, and sedimentary rocks are formed, and the like, will not enable him to infer with confidence just how this one strange chunk of rock got the way it is.

I do not think this kind of analogy is far-fetched, given a theory of schizo- phrenic decompensation as complicated as I believe we are going to need even to deal with "the general case." A major difficulty is that we have to conceptualize "the environment," which is literally "everything causally efficacious except the genome proper," largely by *parameters* or by *classes* of events. Whereas a filled-out psychodynamic comprehension of why Jones underwent an acute schizophrenic break at age 23 but his genetically identical twin did not, would be to know the full psychological details of a particular event in the decom- pensated twin's life history, *and that we may never know, even on the basis of a thousand hours of intensive psychotherapy*. Thus, for example, one interested in the role of sexual conflict in producing schizophrenic decompensation might do a statistical study of "familial puritanism" in relation to diagnostic concordance, and find nothing. Frustrated in this nomothetic parametric effort, he might shift to a more intensive study of families e.g., whether a seductive compensated schizotype mother tended to put the one twin in more frequent or intense oedipal binds than she did the other. But even this might not yield anything, whether treated statistically over sets of families, or strictly idiographically. After spending thousands of hours of his brain power and perhaps thousands of dollars of taxpayer money, our eager-beaver investigator reluctantly concludes that "There seems to be nothing about the sexual environment that helps to explain the difference." Suppose that similar investigations are conducted by many competent persons into other aspects of both the nurturing and adult stressing environments, covering such plausible sectors as academic achievement, social acceptability, neighborhood atmosphere, financial success, and intercurrent disease, with similar negative results. If such should be the long-term outcome of numerous similar investigations, one might be tempted to conclude that there is "nothing environmental" that bears upon the question of compensation versus decompensation, an obviously preposterous conclusion given the high discordance rate for MZ twins! This dreadful fantasy of negative results "no matter what we look at in the environment" is not likely to materialize, but it hardly seems excludable on the basis of the evidence thus far gathered by our authors and others.

Consequently we should set our switches in advance not to be completely baffled by a paradox: Failure to find positive environmental correlates of decompensation, despite the clearly established power of environmental influences collectively as shown by the genetic statistics themselves. I don't think it will be too hard for a genetically oriented psychotherapist to make sense of these facts, scientifically frustrating though they would be. Neither a rough measure of familial puritanism, nor a measure of a schizotypal mother's differential seductiveness as between two MZ twin boys, covers the possibility of a critical event such as Twin *A* receiving a completely unexpected low grade in his physical education class (due—let's really run it into the ground—to a clerical error!) several months following his first heterosexual experience. It doesn't take much for a schizotypal mind to connect up these two happenings in some sort of crazy, hypochondriacal, and guilt-ridden fashion. *Without anything else* being "systematically" different between him and his MZ co-twin, his aberrated CNS may take it from there and snowball it into a psychosis that appears, say, a year later. I do not myself find this kind of "happenstance" at all implausible, and therefore I was pleased to see our authors' explicit emphasis upon "chance" factors in their discussion of the environment. The causal model for an integrated theory of schizophrenia would surely involve social feedback loops, autocatalytic processes, and powerful critical episodes initiating chains of divergent causality, perhaps the most important of this third kind of causal relation

being idiographic content features that alter the subsequent psychological meaning of interpersonal events *that may have the same sort of "average value," parametrically speaking, for both members of a discordant MZ pair*. Thus, for example, 6 months after our hypothetical Twin *A* in the preceding example has schizotypically concluded that "sexual drainage" caused him to fail the gym course, both twins are present at a dinner table conversation. The father says, innocently and with no notion of his critical schizophrenogenic role, "If there is anything I cannot stand, it is a boy who is a sissy." Now the schizotypal snowball really gets going for Twin *A*, whereas father's casual remark has negligible psychological significance for co-twin *B* (who has neither failed in gym nor visited a prostitute). I find it hard to think of any kind of statistical analysis of case history material or neighborhood characteristics that would tease out this kind of thing. And if one combines these "chance" factors with the possibility (some would say likelihood) that the schizogene(s) can "switch on and off" as a function of intercurrent biochemical states, quite possibly including states induced by momentary stressors, the elements of "psychological-social coincidence" can loom very large indeed. Every psychotherapist who has treated schizophrenics knows that the patients themselves sometimes connect a momentary resurgence of anxiety or confusion with what would to a normal mind be a very minor happenstance; and while I certainly do not wish to rely heavily on these anecdotal connections (see *supra*), neither would I be willing to dismiss them as of no evidential weight. Following close upon father's "sissy" remark, random episode E_1 (a waitress momentarily ignores him in favor of a customer that she knows well) ticks off in our Twin *A* a 2-hour increase in the blood level of norepinephrine. Due to his oddball dietary obsessions, which in turn went back to his reading a pamphlet (which didn't *happen* to fall into co-twin *B*'s hands) at age 14, he also is running an unusually high level of organic acid X at the time. These concurrent alterations in the intracellular milieu of the schizogene, and note that they are neither physiologically nor psychologically related, nor attributable to any *systematic* characteristics of the environment, "switch on" the cerebral schizogenes and as a result the patient undergoes an increase in his pan-anxiety, his anhedonia, and his tendency to cognitive slippage. The last straw: While he is in this state, which is a deviation from his usual schizotypal norm, his girlfriend breaks a date with him, speaking rather roughly on the telephone (because she is embarrassed, and in order not to feel defensive she becomes aggressive). *Result:* snowballing in the aversive direction, dangerously consolidating the mixed-up schizoid complex: "I am bad and weak because being sexually drained I have become a sissy as my father said, which is why waitresses prefer others to me, as does my girlfriend, and hence all women. I'm a hopeless nothing." Twin *A* is now well on the way to clinical decompensation.

I must emphasize that envisioning these kinds of "idiographic unpredictabilities" and even after-the-fact "unexplainabilities"—not quite the same thing, as Scriven and others have convincingly shown—does not require that we postulate any sort of radical biological or psychological indeterminism. We just have to recognize unblinkingly the rather obvious fact that even a so-called "thorough in-depth" life history is extremely deficient as regards details, and that we have nothing to guarantee that this deficiency could be made up by *any* amount of ingenuity or expenditure on the part of an investigator concerned to reconstruct the past. The kinds of "influences," "variables" and "factors" that social scientists are usually able to assess are almost always, when carefully scrutinized, classes of variables or kinds of episodes. In terms of learning theory, we deal in most psychological and sociological research with parameters of the elaborate "social Skinner box" that constitute a person's nurturing environment and adult life. The numbers we can correlate with outcome (e.g., psychosis) are in the nature of average values, they do not point to what may be critical events but rather to stochastic features of discriminative and eliciting stimuli or schedules of reinforcement contingencies.

If I were to present an engineer or physicist with a "molar" problem involving a chair made out of a specific kind of metal, riveted or welded in a specified way, telling him I was going to catapult it from the roof of the Physics Building at 12:32 P.M. in a direction so it would land on the stone steps of the Administration Building, and require him to predict whether and how it would fracture, he would probably decline the invitation. *A fortiori*, he would decline if the object to be catapulted were a century old oak-case grandfather clock! But that, brethren, is about the situation when we are trying to understand why Twin *A* became schizophrenic and MZ Twin *B* did not.

I may conclude this discussion of "chance" factors determining which twin decompensates by quoting from my paper read at the 1971 MIT Conference on Prospects for Schizophrenia Research, where I said

> But that there are *some* schizophrenics who act more like *some* anxiety-neurotics than they do like *other* schizophrenics is an important fact to think about and to explain. It is, however, very weak evidence against anybody's theory of schizophrenia, genetic or otherwise. I understand from the mathematicians that there is a clear negative answer to this question; and it is found, for example, in the theory of certain stochastic processes such as random walks. Whether a particle pursuing a random walk ends up (after a specific finite time) in a so-called "absorbing state" [=a state which it cannot leave] is a "dichotomous end-result" comparable to whether a schizotype ends up in a state hospital, enters and leaves, or wins the Pulitzer Prize for poetry; but the mathematics of such situations teaches us that it would be a mistake to presume a dichotomous etiological basis for this important outcome difference. In that connection, I mention in passing that we should not assume, even in the case of MZ twins, that a clinical discordance must always have some "big" life-history factor discernible. That Twin *A* becomes schizophrenic and Twin *B* remains "healthy" may be a random walk problem—I myself would bet that *many* of them are.

I am not, in the light of these reasonings, entirely happy with the authors' statement "MZ 22*A* spoke convincingly of how ordinary contact with a young man at work reawakened sexual conflicts which then escalated into overt thought disorder and a relapse. We can hardly call a happening as subjectively apperceived and idiosyncratically elaborated as this an objective etiological stress. It would not be feasible to design environments so as to insulate people against these kinds of stresses." It seems to me that this conflates a technological problem in social engineering (admittedly important to be clear about) with what is primarily a theoretical issue. I do not see any reason why the girl's contact with the young man should not be considered an "objective" etiological stress, if it is the objective condition for the actualization of (an also objective) disposition. The disposition is admittedly aberrated, because the individual is schizotypal; but its activation is nonetheless "objective," is it not? I think that the behavior geneticist should take these kinds of situations as the tort lawyer takes them: It is no defense, in an action for damages produced by a light blow on the head, for the defendant to show that plaintiff had a rare disease called "eggshell-skull." The tort lawyer's maxim is, "You must take the plaintiff as you find him."

As a final comment on this chapter, it may be that the customary distinction between "necessary" and "sufficient" factors requires more detailed methodological analysis before it can be powerfully applied in behavior genetics, especially to the examination of a polygenic theory. I venture the opinion that this is one of the rare instances in which the technical contributions of philosophers of science may actually be of some positive value to the scientific theorist or investigator. Both in law and in medicine, examples are currently being subjected to intensive philosophical analysis, and the outcome of these investigations is not presently foreseeable. But for reasons similar to those that led me to needle the authors gently on ambiguities I discern in what "specific genetic etiology" means for the polygenic case, I would say that there is a slight conceptual murkiness in the distinction between necessary and sufficient causal factors in their chapter on environment. To detail that argument would go too far afield so I shall content myself with noting that, in a fascinating transatlantic exchange with Professors Gottesman and Shields, I found myself delineating some half-dozen possible mathematical meanings of the term "specific etiology" (not, by the way, confined to the genetic case), any one of which is semantically defensible and which might have a methodologically fruitful application, extending from the crystal-clear case of a dichotomous factor (e.g., the mutated gene at the Huntington's chorea locus) to an attenuated meaning (formulable only in terms of mixed partial derivatives) that is *so* weak one hesitates to apply the term "specific etiology" at all. I hope shortly to publish the results of this analysis.

My only comments on Chapter 9 concern the question of "other diagnoses." Had they been classified as part of a continuum of schizophrenic psychopathology, the authors would have been severely criticized for "loading the dice" in favor of a genetic emphasis, and I certainly do not want to suggest that *for their purposes* these other diagnoses should be thus subsumed. But lacking any substantive specification of what it is that is schizo-specific in their polygenic model, we cannot be sure that such cases would be excluded by Omniscient Jones. On a big-gene model with numerous modifiers and potentiators, one may be in doubt as to whether this or that individual case is "atypical schizophrenia" or any of the other official and unofficial labels that some of us are fond of employing, perhaps carelessly at times (e.g., "pseudoneurotic schizophrenia," "pseudopsychopathic schizophrenia"); but one does have a fairly clear *meaning* for the monogenic concept, in the sense that whatever the ambiguities of the clinical phenotype, even including a beautifully compensated schizotype detectable by no presently available methods, the individual belongs to the group if he has the schizogene, otherwise not. I want to urge, not that a polygenic model *could* not provide such a basis (would it be genuinely taxonomic?), but that when the psychological or biochemical *nature* of what is allegedly schizo-specific but nevertheless polygenic has not been even speculatively indicated, one has no rational basis for deciding whether a severely depressed or chronically delinquent or episodically alcoholic person does or does not "belong," does one?

I have less to say about the final chapter on genetic theorizing, because most of what occurs to me upon reading it would be repetition of methodological or substantive issues that have arisen in my comments on the previous chapters. For reasons given *supra*, I think the authors concede too much by saying "On the face of it, such observations [of a disproportionally higher incidence and prevalence of schizophrenia in the lowest social classes] provide strong support for the role of social stressors as causes of schizophrenia." While they go on to criticize that interpretation, I cannot agree that this correlation provides even "on the face of it" strong support for much of anything. It is as it stands an ambiguous fact, so far as etiology is concerned. I trust I have given sufficient reasons for that opinion above.

In their discussion of our colleague Dr. Leonard Heston's view that about 50% of the first-degree relatives of schizophrenics have a mental abnormality and that these are cases of "schizoid disease," the authors are, of course, correct in mentioning the difficulty that arises from our having no reliable way to identify a *case* of such "schizoid disease" without reference to his relatedness to a schizophrenic. But they go on to say, "If the concept is defined broadly enough to encompass abnormalities in 50% of schizophrenics' parents, sibs, and children, and then generalized, the population base rate will be exaggerated and include many false positives." While Dr. Heston would of course agree that his interpretation is at present speculative—betting on our monogenic horse!—in examining the merits of that substantive position, are we really entitled to assert confidently that the population base rate will be "exaggerated" and include many false positives? That depends upon the gene frequency, and on a view such as

Heston's or mine, the "clinical penetrance" will run very low. Although there would be a vicious circularity in *arguing* that two actual cases of apparently "anxiety neurosis" syndrome are nevertheless different because one is manifested by the MZ co-twin of a known schizophrenic and the other not, *speculating* that this is in reality always the situation (although we do not presently know how to corroborate or discorroborate it) does not commit us (substantively) to an *impossibly* high base rate, does it? Apart from genetic investigations, we already know (e.g., from Piotrowski and Lewis' follow-up studies of rediagnosis, or from Peterson's work on the MMPI profiles of misdiagnosed VA patients) that some schizotypes do appear clinically indistinguishable from, say, "anxiety neurosis" or "depression" when first seen by a psychiatrist. The point is that one does not know how many "false positives" would exist among the patients that Heston considers "schizoid disease," let alone those that Rado or I would label "compensated schizotypes" (some of these not even showing the pseudoneurotic schizophrenia syndrome) unless he already has somehow assigned a numerical value to the hypothetical dominant gene population frequency.

In discussing the psychotic relatives in Ødegaard's study the authors say: "If transmission were monogenic, we would usually expect to find (a) an excess of schizophrenia and no excess of any other abnormality and (b) an unambiguous bimodal distribution of affected and unaffected in the relatives of probands." As a literal statement, if "monogenic" is taken to mean one "big gene" whose "clinical penetrance" is being held down *by modifiers that have no other behavioral relevance*, this statement is unexceptionable. But nobody holds such a theory. We already realize that you can't make a dominant gene theory fit the facts without finageling with penetrance (or, saying it a different way, but one that comes in my view to the same thing, my not making my theoretical distinction between schizophrenia and schizoid disease, schizoidia, or schizotypy); and that is what the pushers of a monogenic theory, such as Slater, Heston and myself have regularly done. So we begin by arguing that the facts require modifiers (in my terminology, schizophrenic potentiators). Now is there any plausible basis to suppose that these modifiers and potentiators are *psychologically irrelevant* when found in nonschizotypes? I cannot conceive that this would be the case. It seems to me that the theoretical and clinical grounds for denying it are, in the aggregate, so massive and interlocking that it would take a fair amount of research of above-average caliber to convince me of the contrary. As to their role in the schizotype, surely no one, whatever his etiological view of schizophrenia, believes that the major human emotions—even in their "nonpathological" form and intensity—have *nothing* to do with the phenomena of schizophrenia. All you have to do is go through a state hospital or treat a schizophrenic patient to know that rage, fear, sex, dependency, shyness, pride and so forth play a role in the character and intensity of his symptoms. Of course this recognition does not contradict an etiological theory making the schizogene(s) a necessary condition for developing the disorder. It just recognizes the rather trivial and unexciting fact that schizophrenics, being people, tend by and large to have pretty much the same things on their minds that the rest of us do! (It takes a psychiatrist or psychologist to make a theoretical mountain out of this particular molehill.)

Gottesman and Shields' own data, taken together with the other studies they review, indicate at least one psychological variable—not identifiable as such with schizophrenia or schizoidia—to wit, social dominance, that plays an important role in concordance, course and severity. But we have independent evidence, both in humans and in animals, that social dominance is to some considerable degree inherited, presumably on a polygenic basis. Hence I conclude, until further notice, that one of the potentiators of schizophrenia (given schizotypy) is very probably social dominance. And I cannot imagine that anxiety proneness, rage readiness, social introversion, and the whole list of variables which we already know something about, have some ability to measure psychometrically, and have some evidence (on humans and animals) to consider partly hereditary, are utterly beside the point in potentiating a schizophrenia. (It seems to me that this might be considered even more plausible on a polygenic theory, although I do not wish to press the point). Now, unless the authors believe that these temperamental dispositions have nothing to do with mental disorder *outside* "the schizophrenic spectrum," why would they want to say what they say in the sentence quoted? I would expect the contrary, either on a polygenic view or on a main-gene view with multiple potentiators, namely: The collaterals of schizophrenic individuals ought to be expected, given sufficiently accurate assessment, to show a heightened incidence of practically every kind of behavior aberration except possibly any that might result from abnormally low contributors to "social fear," e.g., the true, Cleckley-type hard-core psychopathic deviate.

As to the second half of their sentence, that one ought to find an unambiguous bimodal distribution of affected and unaffected in the relatives of probands, I do not believe this has been shown by the authors, or by anybody else. How "unambiguous" a bimodal distribution looks depends upon the degree of latent curve separation, and very importantly on the unreliability of the continuous variable employed as an indicator. It takes a separation of around three standard deviations to produce a clear "valley" in a joint distribution having equal base rates, and with a separation of two standard deviations, inspection will just barely detect the bimodality with infallible measures. With asymmetry of base-rates it becomes still harder to discern. So far as I am aware, detailed working out of the relationship between observed bimodality and a latent taxonomic situation has yet to be rigorously done; and the occurrence of bimodality has been shown by Murphy to be neither a necessary nor a sufficient condition for a single (qualitative, dichotomous) etiology. I don't mean to push my dominant gene view *a priori*; I merely want to assure that no factual implications are improperly drawn from a dominant gene theory that is sufficiently complex to remain in the running given the present data. I will go a little farther than this, hoping thereby to shape the direction of future research on the competing genetic models. The differential impact of schizophrenia on fecundity between the sexes, taken together with clinical experience working with compensated and pseudoneurotic schizotypes of the two sexes, leads me to the conjecture that on a dominant gene theory, there ought to be quite a few more schizophrenics who got the schizogene from mother than from father. Speaking of the schizogene and not clinical schizophrenia, I am going to stick by that conjecture awhile yet, despite some evidence to the contrary. As I suggested in my 1962 APA presidential address, a plausible family pattern (which also explains much of the intrafamily dynamics commonly assigned etiological status) is that of a compensated schizotypal mother—especially one with a heavy dose of dominance and aggressiveness, *which is partly what keeps her compensated*—and a neurotic father, who provides high anxiety and low dominance polygenes (along with poor identification for a male child). Be that as it may, one does not need such speculations to draw some tentative inferences from the well-corroborated sex difference in schizophrenia's effect on fecundity. I have enough of the psychodynamicist in me to find it incredible that a schizotypal mother is, on the average, no more psychologically malignant a feature of the nurturing environment than is a schizotypal father. I say this despite the authors' summary of evidence, with the exception of Reed *et al.*, suggesting the contrary. (I must once again remind the reader that schizotypy is not the same as clinical schizophrenia.) I have been unable to find in the literature—my local geneticist colleagues, including Professor Gottesman, tell me that the reason I cannot find it is that it doesn't exist—a mathematical analysis of the expected situation when we conceive a dominant gene to be potentiated by polygenic systems that tend more often to come from the *other* parent rather than from the one from whom the dominant gene is inherited. Because if the schizotypal parent also had many of the potentiating polygenes, he would tend to have been schizophrenic, and consequently would tend not to have *been* a parent at all! We then postulate the psychodynamic hypothesis of a greater environmental potentiating factor if mother is a schizotype than if father is. I predict that nothing less complicated than this model will do justice to the facts when we have (a) sufficiently accurate taxonomic statistics, and (b) applied them to respectably valid indicators of compensated schizotypy. The sort of causal model I have in mind is just too complicated to permit the kinds of traditional geneticist implications of monogenicity that the authors express in the quoted sentence.

I seem to have a blind spot for an argument that my geneticist friends keep giving me, which relies on the low gene frequency for nonpsychiatric disorders with single gene inheritance. Gottesman and Shields repeat this argument using the example of cystic fibrosis, pointing out that it is 20 times rarer than schizophrenia. Since everybody makes this argument, I feel considerable pressure to accept it (especially from experts in a field where I have no real expertise), despite my persisting inability to fully understand its logical structure. As mentioned *supra*, I think we ought to be prepared to find that schizophrenia is kind of "funny" from a geneticist's standpoint, i.e., that it violates some of his ordinary numerical expectations. And nobody cites to me any actual laws of either population genetics or physiological genetics that will assure me that a gene for subtle neurological integrative defect could not occur 20 times as frequently as the gene for cystic fibrosis. I suspect that the whole set of analogies to other genetic conditions may be misleading us, because their application to schizophrenia is not based upon a sufficient recognition of the differences that are quite likely to arise between behavior genetics and other branches of genetics by virtue of the complex social learning involved—that what is inherited is not merely the first step in a three, four or five link metabolic chain (of the kind with which we are familiar in the physiological genetics of Mendelizing mental deficiencies), but a "something" that affects certain functional parameters of the CNS (and in a rather subtle way, initially) *so that what we call the "pathology" of the clinical disease is learned social pathology*, related only derivatively to the fourth or fifth order dispositions that are, strictly speaking, "heritable." I have such a hard time conveying that point of view to geneticists and psychologists that I am almost ready to conclude that my thinking is haywire. But for the record, I rashly record here the prediction that it will be necessary to "think differently" about not only schizophrenia but manic-depression, unipolar depression, compulsion neurosis and all other learned behavior disorders whose syndromes are defined by certain *psychological* contents, themes, object-cathexes, self-concepts, and preferred mental mechanisms. The causal and statistical model needed is, I predict, just not the same *sort* of model as is involved when we deal with the diabetic's defect in carbohydrate metabolism or the PKU child's inability to handle a normal phenylalanine dietary intake.

   In this chapter, which to one with my concerns is one of the most fascinating and stimulating in the book, I wish the authors had permitted themselves to speculate a bit on what kinds of psychological dispositions they believe to be polygenically inherited and yet schizospecific, especially since elsewhere in the book—including the remarks in this chapter mentioned immediately above—they seem to exclude from the list of polygenic loci those that bear on psychological dimensions of the kind that we know are important in other kinds of behavior disorder. If whatever is the polygenic system that makes specifically for schizophrenia does not overlap with whatever makes people anxious or angry or depressed or sexually perverse or shy or submissive or whatever, what *is* it behaviorally? On my view that is part of the difficulty in any formulation that makes the inherited trait(s) *behavioral at all*, which I think they are not. They *get to be behavioral* only as a result of several further processes. Basically, they are "neurological," in a sufficiently sophisticated sense of that term. I daresay that Gottesman and Shields would reply to this criticism that we have only a little evidence to make such speculations fruitful at the present time. I cannot quarrel with that response, since I am not one to tell others that they have to adopt the same philosophy of science as I do. But with so beautiful a set of data, and such a mastery of the empirical research literature, I think they could have permitted themselves somewhat more free-wheeling speculation in their final chapter. And—trusting that this is not hitting below the belt—I entertain the dark suspicion that one reason they did not speculate about what the specifically schizoid disposition is (in its inheritable component) is that had they done so, they would have been in somewhat of a conceptual bind because of the combination of their polygenic view, their schizospecific view, and their exclusion from the list of what would otherwise be the most plausible candidates. As to their discussion of the "trajectories" of two hypothetical twins, all I need say is "Good!—and matters must be at least as complicated as this, in the eyes of Omniscient Jones."

Well, this afterword has almost become a book review. I have devoted space very largely to caveats, queries, obscurities, challenges, and (very few) real "complaints." I trust this approach does not convey a negative, carping impression. The great merits of this book, reflecting the talent and dedication of its authors, are sufficiently obvious, so I stated at the beginning that only a brief summary of them would be appropriate. The highest praise of a scientific work is to take it seriously; and that means, especially for us neo-Popperians, to *criticize* it, in Sir Karl's honorific sense of that word. The history of science is a history of errors, since all theories are lies. The point is to abandon the black lies as quickly as possible, and to change the gray lies in the direction of lighter grays—to improve a decent theory's "verisimilitude." If, as I believe, a genetic theory of schizophrenia has (by this book) been definitively corroborated, the task now put before us is to determine what is inherited and how. Empirical research on these difficult questions will, I think, be wasteful unless we explore the *conceptual* possibilities imaginatively, and prepare ourselves for some "new problems" for which the received models and methods of genetics may not be adequate, because of the socio-psychological learning processes that enter the picture. My belief in the importance of that methodological thesis is what led me (given Professor Gottesman's highly "permissive" invitation) to indulge so freely in unresearched substantive conjectures.

To repeat what I said initially, this book, definitive for its purpose, perhaps represents the final step in a purely "behavioral" study of the genetics of schizophrenia. I do not myself believe that further twin or other family studies will add materially to what is surely a sufficient body of evidence for a genetic interpretation of this disease. And I cannot convince myself that further studies

will add significantly to our ability to choose between competing genetic
models, unless and until more powerful phenotypic indicators of the relevant
gene(s) become available. The "validation" of such indicators will involve complicated methodological issues, including the Cronbach-Meehl "bootstraps
effect," about which many of my brethren are still regrettably unclear. At the
strictly "psychological" level, there may be some merit in pursuing the construction of personality inventories of MMPI format whose *content* is beamed more
specifically at the schizotypal deviations in thought, in soft neurology, and
body-image aberrations. I hazard the guess that measures of hedonic deficit, to
which hardly any psychometric effort has thus far been devoted (because of
psychologists' inexcusable ignoring of Rado's seminal contributions) should
be assiduously pursued. I have myself constructed a tentative list of inventory
items based partly upon the literature but mainly upon my clinical experience
with schizotypes. I would not advise anyone to invest much research time trying
to construct devices of a psychological sort beamed at the detection of the
"interpersonal aversiveness" which is such a dramatic feature of disintegrated
schizotypy (but almost equally striking in the decompensated case called
pseudoneurotic), because while it has its qualitatively unique flavor (different
from, for example, sociologically based feelings of inferior status, or "garden-
variety polygenic social introversion"), I think it unlikely that these subtle
phenomenological differences can be satisfactorily tapped by question and
answer item material. Difficulty in relating to other people is a ubiquituous
feature among mentally aberrated persons. Unless we can tap into the very
special quality of the schizotype's deep, pervasive, recalcitrant, and preverbal
lack of basic "oral trust," which I doubt we can do with a personality test item
(or, for that matter, a Rorschach or TAT), we are going to get too many false
positives among nonschizotypal mental patients. I have considerably more faith,
on either my own theory or anybody else's, in the improvement of measurement
techniques for phenotypic indicators less "psychological," closer to the gene(s).
Even in the behavior domain, I should anticipate that psychophysiological
measures will do better than personological ones, although I hasten to add that
the widespread interest in measures of anxiety seems to me wholly misplaced
and, in fact, somewhat naive. Surely *nobody*, on any theory of schizophrenia,
supposes that anxiety-proneness is in any way schizospecific? I am puzzled by
the direction of research efforts in this field, since there are so many variables we
might be looking at that one would expect us to begin by looking at whatever
appears, on available clinical and experimental evidence, to have better prior
probabilities of ringing the bell. I would think it better strategy to study
malocclusion or the tendency to "swallow the wrong way" both clinical correlates of schizotypy by some reports (I agree) than anxiety-proneness. The first
two "far-out" guesses have at least a fighting chance of being interesting (as
pleiotropic); but the third does not. We should research variables that we have

reason for thinking may be more or less specific to whatever makes schizo-phrenia behaviorally, prognostically, and *genetically* different from unipolar or bipolar affective psychoses, psychoneuroses, sociopathy, the psychophysio-logical disorders, and the like. Concentrating on anxiety, whether measured psychometrically or physiologically, when one is interested in the genetics of schizophrenias strikes me as rather like concentrating on fever when one is interested in the genetics of, say, the rheumatoid diseases. But perhaps there is something important going on here that I obtusely fail to understand.

I would look to a refinement and objectification of the "soft neurology" as perhaps the most plausible single area to focus on while remaining in the domain of the exophenotype. Granted the perennial ambiguity of "soft neurology" in the decompensated case, the longitudinal research of Barbara Fish and the ongoing study of Israeli kibbutz children of schizophrenic mothers lend, I think, sufficient quantitative support to the folklore of clinicians (from Kraepelin through Bleuler to Schilder) to justify pursuing this lead. Being a biochemical ignoramus, I shall say nothing here about research possibilities for the endo-phenotype, except to suggest that on my speculative theory of schizotaxia, single-neuron studies of synaptic parameters might be very profitable, being only one or two steps removed from the postulated "biochemical lesion." Even here, however, the psychogeneticist may still play a critical role, if he can come up with molar behavior indicators having sufficiently high validity for compensated schizotypy. I predict that this will only be possible if "nonpathological" behavior dimensions are treated conjointly via sophisticated "bootstrapped" taxonomic statistics.

AUTHOR INDEX

Numbers in italics refer to the pages on which the complete references are listed.

A

Abe, K., 26, 211, 232, 234, *349*
Alanen, Y. O., 46, 47, 48, 49, 225, 257, *349*
Albee, G. W., 46, 294, *356*
Allen, G., 25, 305, 340, 342, *349*
Allen, M. G., 34, *349. 359*
Alpers, M., 19, *353*
Anderson, V. E., 22, 323, 329, *349, 359*
Angst, J., 325, *349*
Ash, P., 209, *349*
Astrup, C., 256, *349*
Atkinson, G. F., 34, 61, *355*

B

Bagley, C. R., 309, 325, *351*
Barrai, I., 324, *351*
Barton, W. E., 1, *355*
Beard, A. W., 325, *362*
Bell, R. Q., 47, *349*
Benaim, S., 130, 300, *350*
Birley, J.L.T., 209, 211, 298, *350, 364*
Bleuler, E., 1, 11, 15, 16, 251, 284, *350*
Bleuler, M., 11, 47, 338, *350*
Block, J., 262, *350*
Böök, J. A., 4, 17, *350*
Bradford, N. H., 259, *350*
Bratfos, O., 50, *350*
Briggs, P. F., 64, 272, *350*

Broen, W. E., 286, *350*
Brooke, E. M., 11, 208, 242, *361*
Brown, G. W., 11, 16, 46, 211, 298, *349, 350*
Bruetsch, W. L., 12, *350*
Brush, S., 240, *359*
Bruun, K., 36, *358*
Burks, B. S., 36, *359*
Butcher, J. N., 259, *350*

C

Cabot, R. C., 209, *350*
Cameron, N., 284, *350*
Campbell, M. A., 323, *352*
Cancro, R., 207, *350*
Carter, C. O., 8, 9, 321, 322, *350, 351*
Carter, M., 323, *351*
Cassar, J., 256, *359*
Cederlöf, R., 61, 62, 308, *351*
Cerasi, E., 256, *351*
Clancy, M., 256, 258, 285, 289, *357*
Clark, J., 57, 256, *351*
Clausen, J. A., 4, 46, *351*
Clayton, P. J., 325, *364*
Cohen, S., 34, *349*
Cole, J. O., 1, *355*
Collins, R. L., 317, *352*
Conn, J. W., 2, *358*
Connell, P. H., 325, *351*

417

SUBJECT INDEX[1]

[1] For Chapter 4 (Case Histories) only the *Comment* sections have been indexed.

423

TWIN PAIR INDEX